FIELD PRACTICES FOR WASTEWATER USE IN AGRICULTURE

Future Trends and Use of Biological Systems

Innovations in Agricultural and Biological Engineering

FIELD PRACTICES FOR WASTEWATER USE IN AGRICULTURE

Future Trends and Use of Biological Systems

Edited by

Vinod Kumar Tripathi, PhD
Megh R. Goyal, PhD, PE

APPLE
ACADEMIC
PRESS

First edition published 2021

Apple Academic Press Inc.
1265 Goldenrod Circle, NE,
Palm Bay, FL 32905 USA

4164 Lakeshore Road, Burlington,
ON, L7L 1A4 Canada

CRC Press
6000 Broken Sound Parkway NW,
Suite 300, Boca Raton, FL 33487-2742 USA

2 Park Square, Milton Park,
Abingdon, Oxon, OX14 4RN UK

First issued in paperback 2021

Library and Archives Canada Cataloguing in Publication

Title: Field practices for wastewater use in agriculture : future trends and use of biological systems / edited by Vinod Kumar Tripathi, PhD, Megh R. Goyal, PhD, PE.

Names: Tripathi, Vinod Kumar, editor. | Goyal, Megh Raj, editor.

Series: Innovations in agricultural and biological engineering.

Description: Series statement: Innovations in agricultural and biological engineering | Includes bibliographical references and index.

Identifiers: Canadiana (print) 20200312693 | Canadiana (ebook) 20200312820 | ISBN 9781771889087 (hardcover) | ISBN 9781003034506 (ebook)

Subjects: LCSH: Sewage irrigation. | LCSH: Sewage—Purification. | LCSH: Sustainable agriculture. | LCSH: Biotechnology.

Classification: LCC TD760 .F54 2021 | DDC 628.3/623—dc23

Library of Congress Cataloging-in-Publication Data

Names: Tripathi, Vinod Kumar, editor. | Goyal, Megh Raj, editor.

Title: Field practices for wastewater use in agriculture : future trends and use of biological systems / edited by Vinod Kumar Tripathi, Megh R. Goyal.

Other Titles: Innovations in agricultural and biological engineering.

Description: First edition. | Palm Bay, FL, USA : Apple Academic Press, 2021. | Series: Innovations in agricultural and biological engineering | Includes bibliographical references and index. | Summary: "Field Practices for Wastewater Use in Agriculture: Future Trends and Use of Biological Systems discusses the growing importance of wastewater application in the field of agriculture. Addressing the tremendous need for the irrigation sector to reduce the demand for freshwater in agriculture, this volume looks at wastewater as a source for agricultural irrigation. The volume is divided into four sections: current and emerging issues in wastewater use in agriculture, wastewater management with biological systems, effective field practices for wastewater use, and case studies that provide information on scientific analytical studies on the environment under the influence of wastewater quality from different pollution sources. This book sheds light on the vast potential of wastewater use in agricultural irrigation while also considering safety of the agricultural products for human consumption. Much emphasis has also been given to technological aspects for the treatment of wastewater to protect our environment for better public health protection. Key features: Provides information to planners and policymakers through case studies Offers solutions for the treatment of wastewater generated from different sources Considers sustainability issues and the protection of the environment Examines the issue of water quality from certain sources and with specific techniques"-- Provided by publisher.

Identifiers: LCCN 2020036803 (print) | LCCN 2020036804 (ebook) | ISBN 9781771889087 (hardcover) | ISBN 9781003034506 (ebook)

Subjects: LCSH: Sewage irrigation. | Sewage--Purification. | Sustainable agriculture, | Biotechnology.

Classification: LCC TD760 .F54 2021 (print) | LCC TD760 (ebook) | DDC 631.5/87--dc23

LC record available at https://lccn.loc.gov/2020036803

LC ebook record available at https://lccn.loc.gov/2020036804

ISBN: 978-1-77188-908-7 (hbk)
ISBN: 978-1-77463-768-5 (pbk)
ISBN: 978-1-00303-450-6 (ebk)

ABOUT LEAD EDITOR

Vinod Kumar Tripathi, PhD
Assistant Professor, Department of Farm Engineering at Banaras Hindu University, Varanasi, Uttar Pradesh, India

Vinod Kumar Tripathi, PhD, is an Assistant Professor in the Department of Farm Engineering at Banaras Hindu University, Varanasi, Uttar Pradesh, India. He has worked at the ICAR–Indian Institute of Water Management, Bhubaneswar, and the Centre for Water Engineering and Management at the Central University of Jharkhand, Ranchi, India. Dr. Tripathi developed a methodology to improve the quality of produce with drip irrigation by utilizing poor quality water. His areas of interest are geo-informatics, hydraulics in micro irrigation, and development of suitable water management technologies for higher crop and water productivity. Dr. Tripathi acts as organizing secretary for one national and two international conference. He has published over 35 peer-reviewed research papers in national and international journals with high impact factors, eight books, over 25 book chapters, and several bulletins. He received the Rekha Nandi and Bhupesh Nandi Prize from The Institution of Engineers (India), Kolkata; Jain Irrigation Award; and Distinguish Service Certificate Award from the Indian Society of Agricultural Engineers, New Delhi, India. He obtained his MTech (Irrigation and Drainage Eng.) from G.B. Pant University of Agriculture and Technology, Pantnagar, India, and his PhD from the Indian Agricultural Research Institute, New Delhi.

ABOUT SENIOR-EDITOR-IN-CHIEF

Megh R. Goyal, PhD, PE
*Retired Professor in Agricultural and Biomedical
Engineering, University of Puerto Rico,
Mayaguez Campus; Senior Acquisitions Editor,
Biomedical Engineering and Agricultural Science,
Apple Academic Press, Inc.*

Megh R. Goyal, PhD, PE, is a Retired Professor in
Agricultural and Biomedical Engineering from the
General Engineering Department in the College of
Engineering at University of Puerto Rico-Mayaguez Campus; and Senior
Acquisitions Editor and Senior Technical Editor-in-Chief in Agricultural and
Biomedical Engineering for Apple Academic Press Inc.

Since 1971, he has worked as Soil Conservation Inspector (1971);
Research Assistant at Haryana Agricultural University (1972–75) and the
Ohio State University (1975–79); Research Agricultural Engineer/Professor
at Department of Agricultural Engineering of UPRM (1979–1997); and
Professor in Agricultural and Biomedical Engineering at General Engineering
Department of UPRM (1997–2012). He spent a one-year sabbatical leave in
2002–2003 at the Biomedical Engineering Department, Florida International
University, Miami, USA.

He was the first agricultural engineer to receive a professional license
in Agricultural Engineering in 1986 from the College of Engineers and
Surveyors of Puerto Rico. On September 16, 2005, he was proclaimed as the
"Father of Irrigation Engineering in Puerto Rico for the twentieth century" by
the ASABE, Puerto Rico Section, for his pioneer work on micro-irrigation,
evapotranspiration (ET), agroclimatology, and soil and water engineering.
During his professional career of 50 years, he has received awards such as
Scientist of the Year, Blue Ribbon Extension Award, Research Paper Award,
Nolan Mitchell Young Extension Worker Award, Agricultural Engineer
of the Year, Citations by Mayors of Juana Diaz and Ponce, Membership
Grand Prize for ASAE Campaign, Felix Castro Rodriguez Academic
Excellence, Rashtriya Ratan Award and Bharat Excellence Award and Gold
Medal, Domingo Marrero Navarro Prize, Adopted Son of Moca, Irrigation

Protagonist of UPRM, Man of Drip Irrigation by Mayor of Municipalities of Mayaguez/Caguas/Ponce and Senate/Secretary of Agriculture of ELA, Puerto Rico. The Water Technology Centre of Tamil Nadu Agricultural University in Coimbatore, India recognized Dr. Goyal as one of the experts "who rendered meritorious service for the development of micro-irrigation sector in India" by bestowing *Award of Outstanding Contribution in Micro Irrigation.*" This award was presented to Dr. Goyal during the inaugural session of the National Congress on "New Challenges and Advances in Sustainable Micro Irrigation" on March 1, 2017, held at Tamil Nadu Agricultural University. On August 1 of 2018, ASABE bestowed on him Netafim Award for Advancements in Micro-irrigation.

He has authored more than 200 journal articles and more than 80 books, including *Elements of Agroclimatology* (Spanish) by UNISARC, Colombia; two *Bibliographies on Drip Irrigation.*

He received his BSc degree in Engineering in 1971 from Punjab Agricultural University, Ludhiana, India; his MSc degree in 1977 and the PhD degree in 1979 from the Ohio State University, Columbus; his Master of Divinity degree in 2001 from Puerto Rico Evangelical Seminary, Hato Rey, Puerto Rico, USA. A prolific author and editor, he has written more than 200 journal articles and textbooks and has edited over 80 books. He is the editor of three book series published by Apple Academic Press: Innovations in Agricultural & Biological Engineering, Innovations and Challenges in Micro Irrigation, and Research Advances in Sustainable Micro Irrigation. He is also instrumental in the development of the new book series Innovations in Plant Science for Better Health: From Soil to Fork.

ABOUT THE BOOK SERIES: INNOVATIONS IN AGRICULTURAL AND BIOLOGICAL ENGINEERING

Under this book series, Apple Academic Press Inc. is publishing book volumes over a span of 8–10 years in the specialty areas defined by the American Society of Agricultural and Biological Engineers (www.asabe.org). Apple Academic Press Inc. aims to be a principal source of books in agricultural and biological engineering. We welcome book proposals from readers in areas of their expertise.

The mission of this series is to provide knowledge and techniques for agricultural and biological engineers (ABEs). The book series offers high-quality reference and academic content on agricultural and biological engineering (ABE) that is accessible to academicians, researchers, scientists, university faculty and university-level students, and professionals around the world.

Agricultural and biological engineers ensure that the world has the necessities of life, including safe and plentiful food, clean air and water, renewable fuel and energy, safe working conditions, and a healthy environment by employing knowledge and expertise of the sciences, both pure and applied, and engineering principles. Biological engineering applies engineering practices to problems and opportunities presented by living things and the natural environment in agriculture.

ABE embraces a variety of the following specialty areas (www.asabe.org): aquaculture engineering, biological engineering, energy, farm machinery and power engineering, food, and process engineering, forest engineering, information, and electrical technologies, soil, and water conservation engineering, natural resources engineering, nursery, and greenhouse engineering, safety, and health, and structures and environment.

For this book series, we welcome chapters on the following specialty areas (but not limited to):

1. Academia to industry to end-user loop in agricultural engineering.
2. Agricultural mechanization.
3. Aquaculture engineering.
4. Biological engineering in agriculture.

5. Biotechnology applications in agricultural engineering.
6. Energy source engineering.
7. Farm to fork technologies in agriculture.
8. Food and bioprocess engineering.
9. Forest engineering.
10. GPS and remote sensing potential in agricultural engineering.
11. Hill land agriculture.
12. Human factors in engineering.
13. Impact of global warming and climatic change on agriculture economy.
14. Information and electrical technologies.
15. Irrigation and drainage engineering.
16. Micro-irrigation engineering.
17. Milk Engineering.
18. Nanotechnology applications in agricultural engineering.
19. Natural resources engineering.
20. Nursery and greenhouse engineering.
21. Potential of phytochemicals from agricultural and wild plants for human health.
22. Power systems and machinery design.
23. Robot engineering and drones in agriculture.
24. Rural electrification.
25. Sanitary engineering.
26. Simulation and computer modeling.
27. Smart engineering applications in agriculture.
28. Soil and water engineering.
29. Structures and environment engineering.
30. Waste management and recycling.
31. Any other focus areas.

For more information on this series, readers may contact:

Megh R. Goyal, PhD, PE
Book Series Senior Editor-in-Chief
Innovations in Agricultural and Biological Engineering
E-mail: goyalmegh@gmail.com

OTHER BOOKS ON AGRICULTURAL AND BIOLOGICAL ENGINEERING BY APPLE ACADEMIC PRESS, INC.

Management of Drip/Trickle or Micro Irrigation
Megh R. Goyal, PhD, PE, Senior Editor-in-Chief

Evapotranspiration: Principles and Applications for Water Management
Megh R. Goyal, PhD, PE, and Eric W. Harmsen, Editors

Book Series: Research Advances in Sustainable Micro Irrigation
Senior Editor-in-Chief: Megh R. Goyal, PhD, PE

Volume 1: Sustainable Micro Irrigation: Principles and Practices
Volume 2: Sustainable Practices in Surface and Subsurface Micro Irrigation
Volume 3: Sustainable Micro Irrigation Management for Trees and Vines
Volume 4: Management, Performance, and Applications of Micro Irrigation Systems
Volume 5: Applications of Furrow and Micro Irrigation in Arid and Semi-Arid Regions
Volume 6: Best Management Practices for Drip Irrigated Crops
Volume 7: Closed Circuit Micro Irrigation Design: Theory and Applications
Volume 8: Wastewater Management for Irrigation: Principles and Practices
Volume 9: Water and Fertigation Management in Micro Irrigation
Volume 10: Innovation in Micro Irrigation Technology

Book Series: Innovations and Challenges in Micro Irrigation
Senior Editor-in-Chief: Megh R. Goyal, PhD, PE

Volume 1: Management of Drip/Trickle or Micro Irrigation
Volume 2: Sustainable Micro Irrigation Design Systems for Agricultural Crops
Volume 3: Principles and Management of Clogging in Micro Irrigation
Volume 4: Performance Evaluation of Micro Irrigation Management
Volume 5: Potential Use of Solar Energy and Emerging Technologies in Micro Irrigation

Volume 6: Micro Irrigation Management: Technological Advances and Their Applications
Volume 7: Micro Irrigation Engineering for Horticultural Crops
Volume 8: Micro Irrigation Scheduling and Practices
Volume 9: Engineering Interventions in Sustainable Trickle Irrigation
Volume 10: Management Strategies for Water Use Efficiency and Micro Irrigated Crops
Volume 11: Fertigation Technologies In Micro Irrigation

Book Series: Innovations in Agricultural & Biological Engineering
Senior Editor-in-Chief: Megh R. Goyal, PhD, PE
- Dairy Engineering: Advanced Technologies and Their Applications
- Developing Technologies in Food Science: Status, Applications, and Challenges
- Emerging Technologies in Agricultural Engineering
- Engineering Interventions in Agricultural Processing
- Engineering Interventions in Foods and Plants
- Engineering Practices for Agricultural Production and Water Conservation: An Interdisciplinary Approach
- Engineering Practices for Management of Soil Salinity: Agricultural, Physiological, and Adaptive Approaches
- Engineering Practices for Milk Products: Dairyceuticals, Novel Technologies, and Quality
- Evapotranspiration
- Field Practices for Wastewater Use in Agriculture
- Flood Assessment: Modeling and Parameterization
- Food Engineering: Emerging Issues, Modeling, and Applications
- Food Process Engineering: Emerging Trends in Research and Their Applications
- Food Processing and Preservation Technology: Advances, Methods, and Applications
- Food Technology: Applied Research and Production Techniques
- Handbook of Research on Food Processing and Preservation Technologies: Volume 1: Nonthermal and Innovative Food Processing Methods
- Handbook of Research on Food Processing and Preservation Technologies: Volume 2: Nonthermal Food Preservation and Novel Processing Strategies
- Modeling Methods and Practices in Soil and Water Engineering

- Nanotechnology and Nanomaterial Applications in Food, Health and Biomedical Sciences
- Nanotechnology Applications in Dairy Science: Packaging, Processing, and Preservation
- Novel Dairy Processing Technologies: Techniques, Management, and Energy Conservation
- Novel Strategies to Improve Shelf-Life and Quality of Foods
- Processing of Fruits and Vegetables: From Farm to Fork
- Processing Technologies for Milk and Milk Products: Methods, Applications, and Energy Usage
- Scientific and Technical Terms in Bioengineering and Biological Engineering
- Soil and Water Engineering: Principles and Applications of Modeling
- Soil Salinity Management in Agriculture: Technological Advances and Applications
- State-of-the-Art Technologies in Food Science: Human Health, Emerging Issues and Specialty Topics
- Sustainable Biological Systems for Agriculture: Emerging Issues in Nanotechnology, Biofertilizers, Wastewater, and Farm Machines
- Technological Interventions in Dairy Science: Innovative Approaches in Processing, Preservation, and Analysis of Milk Products
- Technological Interventions in Management of Irrigated Agriculture
- Technological Interventions in the Processing of Fruits and Vegetables
- Technological Processes for Marine Foods, From Water to Fork: Bioactive Compounds, Industrial Applications, and Genomics

CONTENTS

CONTRIBUTORS

Ajay Bharti
Associate Professor, Department of Civil Engineering, North Eastern Regional Institute of Science and Technology (NERIST), Nirjuli, Itanagar– 791109, Arunachal Pradesh, India, Mobile: +91-0360-2257919, E-mail: abt@nerist.ac.in

Mayuri Chabukdhara
Assistant Professor, Environment Pollution and Management Laboratory, Department of Environmental Biology and Wildlife Sciences, Cotton University, Pan Bazaar, Guwahati – 781001, Assam, India, Mobile: +91-9436962824, E-mails: mayuri.chabukdhara@gmail.com; mayuri_chabukdhara@yahoo.co.in

Pebbeti Chandana
PhD Research Scholar, Department of Agronomy, Tamil Nadu Agricultural University, Coimbatore – 641003, Tamil Nadu, India, Mobile: +91-8790698658, E-mail: chandananareddy660@gmail.com

Djadouni Fatima
University Professor and Researcher, Biology Department, Faculty of Natural Sciences and Life, Mascara University, Mascara – 29000, Algeria, Mobile: +0021-354-0039761, E-mail: Sdjadouni@gmail.com

N. O. Gopal
Professor and Head, Department of Agricultural Microbiology, Tamil Nadu Agricultural University, Agricultural College and Research Institute, Madurai – 625104, Tamil Nadu, India, Mobile: +91-9442014093, E-mail: nogopal1964@gmail.com

Megh R. Goyal
Distinguished Professor (Retired) in Agricultural and Biomedical Engineering from College of Engineering at University of Puerto Rico-Mayaguez Campus; and Senior Technical Editor-in-Chief in Agricultural and Biomedical Engineering for Apple Academic Press Inc., P.O Box 86, Rincon-PR – 006770086, USA, E-mail: goyalmegh@gmail.com

Sanjay K. Gupta
Technical Superintendent, Environmental Engineering Laboratory, Department of Civil Engineering, Indian Institute of Technology, Hauz Khas, New Delhi – 110016, India, Mobile: +91-11-26596443, E-mail: a26685@civil.iitd.ac.in

Garima Jhariya
PhD Research Scholar, Department of Farm Engineering, Institute of Agriculture Sciences, Banaras Hindu University Varanasi, Uttar Pradesh – 221005, India, Mobile: +91-9453486849, E-mail: garima2304@gmail.com

Nivedita Khawas
M.Tech. Research Scholar, Department of Water Engineering and Management, Central University of Jharkhand, Brambe, Ranchi – 835205, Jharkhand, India, Mobile: +91-7677758701, E-mail: nivedita.iwem@gmail.com

Shrikaant Kulkarni
Assistant Professor, Vishwakarma Institute of Technology, Department of Chemical Engineering, 666, Upper Indira Nagar, Bibwewadi, Pune – 411037, India, Mobile: +91-9970663353, E-mail: shrikaant.kulkarni@vit.edu

Anil Kumar
Assistant Professor, Center for Life Sciences, Central University of Jharkhand, Brambe, Ranchi – 835205, Jharkhand, India, Mobile: +91-9955273226, E-mail: anil.kumar@cuj.ac.in

Ashish Kumar
PhD Research Scholar, Department of Farm Engineering, Institute of Agricultural Sciences, Banaras Hindu University, Varanasi – 221005, Uttar Pradesh, India, Mobile: +91-8874482123, E-mail: ashish48179@gmail.com

Krishnamurthy Kumar
Professor, Department of Agricultural Microbiology, Tamil Nadu Agricultural University, Coimbatore – 641003, Tamil Nadu, India, Mobile: +91-9443572956, E-mail: azollakumar@rediffmail.com

Pravendra Kumar
Professor, Department of Soil and Water Conservation Engineering, College of Technology, G. B. Pant University of Agriculture and Technology, Pantnagar, Uttarakhand, India, E-mail: pravendrak_05@yahoo.co.in

Y. Lavanya
PhD Research Scholar, Department of Agronomy, Tamil Nadu Agricultural University, Coimbatore – 641003, India, Mobile: +91-8500144406, E-mail: yenabalalavanya@gmail.com

Sushobhan Majumdar
Former PhD Research Fellow, Department of Geography, Jadavpur University, Kolkata – 700032, West Bengal, India, E-mail: sushobhan91@gmail.com

Devendra Mohan
Professor, Department of Civil Engineering, Indian Institute of Technology, Banaras Hindu University Varanasi, Uttar Pradesh – 221005, India, Mobile: +91-9451173756, E-mail: devmohan9@gmail.com

Arvind K. Nema
Professor, Environmental Engineering Laboratory, Department of Civil Engineering, Indian Institute of Technology – Delhi, Hauz Khas, New Delhi – 110016, India, Mobile: +91-11-26596423, E-mails: aknema@gmail.com; aknema@civil.iitd.ac.in

Pankaj K. Pandey
Assistant Professor, Department of Agricultural Engineering, North Eastern Regional Institute of Science and Technology (NERIST), Nirjuli, Itanagar– 791109, Arunachal Pradesh, India, Mobile: +91-9612615313, E-mail: pkpnerist@gmail.com

Bright Jeberlin Prabina
Assistant Professor, Department of Agricultural Microbiology, Tamil Nadu Agricultural University, Agricultural College and Research Institute, TNAU, Madurai – 625104, Tamil Nadu, India, Mobile: +91-9442054951, E-mail: jebajegesh@yahoo.co.in

A. S. Raghubanshi
Professor, Institute of Environment and Sustainable Development, Banaras Hindu University, Varanasi – 221005, Uttar Pradesh, India, E-mail: raghubansh@gmail.com

Amitava Rakshit
Assistant Professor, Department of Soil Science and Agricultural Chemistry,
Institute of Agricultural Science, Banaras Hindu University, Varanasi – 221005, Uttar Pradesh, India,
Mobile: +91-9450346890, E-mail: amitavabhu@gmail.com

Kishan S. Rawat
Scientist-E, Center for Remote Sensing and Geo-Informatics, Satyabama University, Chennai – 600119,
Tamil Nadu, India, Mobile: +91-8681882352, E-mail: ksr.kishan@gmail.com

K. Kiran Kumar Reddy
PhD Research Scholar, Professor Jayashankar Telangana State Agricultural University,
College of Agriculture, Rajendranagar, Hyderabad, India, Mobile: +91-7981644675,
E-mail: kiranreddy.msc94@gmail.com

Sushil K. Shukla
Assistant Professor, Department of Environmental Sciences,
Central University of Jharkhand, Brambe, Ranchi – 835205, Jharkhand, India,
Mobile: +91-8969271817; 9454061220, E-mail: shuklask2000@gmail.com

Ram Mandir Singh
Professor, Department of Farm Engineering, Institute of Agriculture Sciences,
Banaras Hindu University Varanasi, Uttar Pradesh – 221005, India,
Mobile: +91-9452562607, E-mail: mandirsingh@rediffmail.com

Sudhir K. Singh
Assistant Professor, K. Banerjee Center of Atmospheric and Ocean Studies, IIDS,
University of Allahabad, Allahabad – 211002, Uttar Pradesh, India,
Mobile: +91-9793414696, E-mail: sudhirinjnu@gmail.com

S. Sivaranjani
PhD Research Scholar, Soil Science Discipline, Forest Ecology and Climate Change Division,
Forest Research Institute, Dehradun, New Forest – 248006, Uttarakhand, India,
Mobile: +91-9486401829, E-mail: ranjani.agri@gmail.com

Vinod Kumar Tripathi
Assistant Professor, Department of Farm Engineering, Institute of Agricultural Sciences,
Banaras Hindu University, Varanasi – 221005, Uttar Pradesh, India,
Mobile: +91-8987661439, E-mails: vktripathi@bhu.ac.in; tripathiwtcer@gmail.com

Shweta Upadhyay
PhD Research Scholar, Institute of Environment and Sustainable Development,
Banaras Hindu University, Varanasi – 221005, Uttar Pradesh, India,
Mobile: +91-7839001700, E-mail: iesdbhushwetaupadhyay@gmail.com

ABBREVIATIONS

%Na	percent sodium
AAS	atomic absorption spectrophotometer
AFM	atomic force microscopy
AHD	Aswan high dam
Al	aluminum
ANFIS	adaptive neuro-fuzzy inference system
ANN	artificial neural network
ARGs	antibiotic resistance genes
ARPs	antibiotic-resistant pathogens
As	arsenic
B	boron
Be	beryllium
BOD	biological oxygen demand
BWFs	blue water footprints
C/N ratio	carbon/nitrogen ratio
CC	correlation coefficient
CCME	Canadian Environmental Quality Guideline
CCUBGA	Centre for Conservation and Utilization of Blue Green Algae
Cd	cadmium
CE	coefficient of efficiency
CGWB	Central Ground Water Board
CH_4	methane
CIAE	Central Institute of Agricultural Engineering
CO	carbon monoxide
Co	cobalt
CO_2	carbon dioxide
CO-4	Coimbatore 4 (TNAU variety)
COD	chemical oxygen demand
Cr	chromium
CRP	conservation reserve program
CTAB	cetyl trimethyl ammonium bromide
CU	coefficient of uniformity
Cu	copper
CWU	crop water use

DAP	di-ammonium phosphate
DI	daily intake
DO	dissolved oxygen
DOC	dissolved organic carbon
DRV	discharge ratio variation
DWs	dry weight
E. coli	*Escherichia coli*
EC	electrical conductivity
EPS	exo-polysaccharide
EPS	extracellular polymeric substance
ET	evapotranspiration
EU	European Union
F	fluoride
FAO	Food and Agricultural Organization
FCE	ferric chloride-induced crude extract
Fe	iron
Fe^{2+}	ferrous ion
FIS	fuzzy inference system
FL	fuzzy logic
GHG	greenhouse gases
GIS	geographical information system
GOI	government of India
GPS	global positioning system
GT	gamma test
GWFs	green water footprints
GWWB	groundwater board of West Bengal
GYWFs	grey water footprints
Hg	mercury
HI	hazard index
HQ	hazard quotient
HRI	health risk index
IE	irrigation efficiency
IPCC	intergovernmental panel on climate change
IR	increasing rate
IWQI	irrigation water quality index
K	potassium
kDa	kilo Dalton
KMC	Kolkata Municipal Corporation
KVIC	Khadi and Village Industries Commission

kWh	kilo watt-hour
LCA	life cycle assessment
Lh^{-1}	liter per hour
Li	lithium
lph	liter per hour
LS	lignin sulfonates
LSNT	late spring nitrate test
MAR	magnesium absorption ratio
MF	membership function
MH	magnesium hazard
Mha	million hectares
MLD	million liters per day
MLP	multilayer perceptron
Mn	manganese
Mo	molybdenum
MRTN	maximum return to N
MSL	mean sea level
MSW	municipal solid waste
MT	metric tons
MT	million ton
N	nitrogen
N_2O	nitrous oxide
$NaCO_3$	sodium carbonate
$NaHCO_3$	sodium bicarbonate
NaOH	sodium hydroxide
NCS	nano-calcium silicate
NDDI	normalized difference dispersal index
NGWB	National Groundwater Board
NH_3	ammonia
Ni	nickel
NO_3-N	nitrate-nitrogen
NPCMR	National Policy for Management of Crop Residues
–OH	hydroxyl group
OPEC	organic pollutants of emerging concern
P	phosphorus
PARE	pooled average relative error
Pb	lead
PCR	polymerase chain reaction
pH	potential of hydrogen

PLFAs	phospholipids fatty acids
ppm	parts per million
PPNT	pre-plant nitrate test
PSNT	pre-side dress nitrate test
RA	residual alkalinity
RMSE	root mean square error
rpm	revolution per minute
RSC	residual sodium carbonate
SAR	sodium absorption ratio
SCC	sudden climate change
SDI	subsurface drip irrigation
Se	selenium
SEPA	State Environmental Protection Administration
SMS	straw management system
SO_2	sulfur dioxide
SSC	suspended sediment concentration
STMS	secondary treated municipal sewage
STPs	sewage treatment plants
STRR	short tandemly repeated repetitive
TDS	total dissolved solids
Tg	tera grams
THQ	target hazard quotient
TSE	treated sewage effluent
TSK	Tagaki-Sugeno-Kang
TSS	total suspended solids
TWh	tera watt-hours
TWW	treated municipal wastewater
V	vanadium
VOA	volatile organic acids
WF	water footprint
WFN	water footprint network
WHO	World Health Organization
WUE	water use efficiency
WW	wastewater
Zn	zinc

Microorganism Symbols

| *A* | *Acinetobacter* |
| *B* | *Bacillus* |

C	*Candida*
Chr	*Chromobacterium*
E	*Escherichia coli*
N	*Neisseria*
Pr	*Propionibacterium*
Ps	*Pseudomonas*
Sh	*Shigella*
Sl	*Salmonella*
T	*Taenia*
Tr	*Trichoderma*
V	*Vibrio*

PREFACE

Wastewater (WW) use can have many applications, such as irrigation of agricultural land, aquaculture, landscape irrigation, urban, and industrial uses, recreational and environmental uses, and artificial groundwater recharge. Generally, WW can be used for all purposes for which freshwater is used, but it requires suitable treatments. With few exceptions worldwide, WW use applications are restricted to non-potable uses.

As the demand for fresh water is increasing in other priority sectors, freshwater availability for irrigation is becoming scarce. Therefore, the use of poor quality water for irrigation becomes an attractive option. Approximately 20 Mha(million hectares) of land in the world is irrigated with treated, untreated, or partially treated WW. Especially in arid and semi-arid countries, WW irrigation is expanding rapidly.

WW use in agriculture is an old tradition. Besides increasing water stress, drivers for the expansion include: increasing urbanization, growing urban WW flows due to the expansion of water supply and sewerage services, and more urban households are engaged in agricultural activities that could be intensified with additional sources of irrigation water. Freshwater is either unavailable or too expensive, and WW treatment is not keeping up with urban growth, and urban farmers often have no alternative but to use highly polluted water. Many of them belong to the urban poor, who depend on agricultural activities as a source of income and employment generation as well as food security,

In most cases, the irrigated lands are on the outskirts of the city or near urban areas, where the WW is easily available without any additional cost. WW also provides an opportunity to apply a reduced rate of fertilizers to crops. Protection of the environment from wetland disposal is an added advantage. There is ample scope of improving soil fertility and increasing productivity by application of WW and adopting appropriate nutrient management strategies. The use of WW and nutrient management technology would increase the productivity of rain-fed crops without affecting the soil health and demand for freshwater. Therefore, there is an urgent need for a new green revolution in rain-fed agro-ecosystems to feed the teeming millions.

In this book volume, major emphasis is on: (i) WW use in agriculture: emerging issues and current/future trends, (ii) WW management with biological systems, (iii) field practices for WW use, and (iv) two case studies to provide information on the scientific use of WW under special industrial effluent utilization to protect the environment. Additional topics on the estimation of runoff and sediments through soft computing tools like ANN, etc., with the measures to control soil erosion and conserve water, have been discussed. The target of this book project is to provide information on the safe use of WW for irrigation. The health protection of agricultural producers and consumers has been the primary focus through this book volume.

We understand that the present book provides an opportunity for readers for enhancing their knowledge with successful experiences. We sincerely hope that the community engaged in addressing the conservation of freshwater for agriculture will focus on issues related to conservation agriculture and come out with appropriate implementable-action plans. This book is a valuable reference to academicians, researchers, water managers, planners, and policymakers of the irrigation sector to take decisions related to agricultural water management of a watershed.

At 51ˢᵗ Annual Convention of the Indian Society of Agricultural Engineers (ISAE) during February 16–18, 2017 at CCS Haryana Agricultural University, Hisar (Haryana), the seed for this book was planted in an informal meeting among both of us. In February of 2018, Megh R. Goyal (Senior Technical Editor-in-Chief for Apple Academic Press Inc.,) visited Fauna International in Tuticorin (Tamil Nadu-India: <https://www.faunaintl.com/>), who are renowned suppliers of dried flowers/plant plants/seeds, etc. (See image next page: Courtesy Bipin Jain) that are byproducts from solid agriculture waste. This visit helped to shape our vision for this book volume.

The contributions of all cooperating authors to this book volume have been most valuable in the compilation. Their names are mentioned in each chapter and in the list of contributors. We appreciate you all for having patience with our editorial skills. This book would not have been written without the valuable cooperation of these investigators, many of them are renowned scientists.

The goal of this book volume is to guide the world science community.

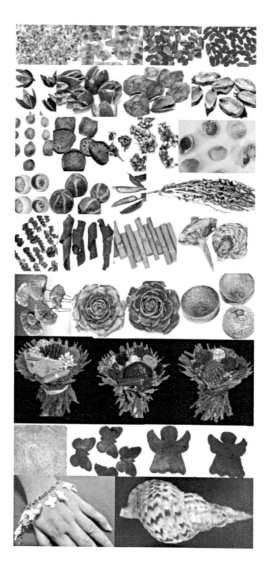

We will like to thank editorial staff at Apple Academic Press, Inc., for making every effort to publish this book when all are concerned with issues related to the use of WW.

Also, we would like to thank our families who have taught us the importance of working hard, having clear goals, and standing for what we believe is right. It is a lesson that guides us in everything we do. Last but not least, we wish to thank our wives for their understanding and patience throughout this book project.

As educators, there is a piece of advice to one and all in the world: *"Permit that our Almighty God, our Creator, Provider of all and excellent Teacher, guide us to the intelligent use of wastewater in agriculture...; and Get married to your profession...."*

—Vinod Kumar Tripathi, PhD
Megh R. Goyal, PhD, PE

Part I:
Prospects of Wastewater Use in Agriculture

WATER FOOTPRINTS OF SUSTAINABLE AGRICULTURE: AN APPROPRIATION OF FRESH AND WASTEWATER

SHWETA UPADHYAY and A. S. RAGHUBANSHI

ABSTRACT

Freshwater scarcity and overexploitation are irrefutable issues. Practically in southern and eastern Asian regions, withdrawal of water for the agriculture sector accounts for 95%, a much higher fraction than the typical global agricultural water use (70%). In India, water withdrawal of 761 km³ (19%) is the highest water withdrawal in the world. It is expected from the Indian agriculture sector to avoid the cultivation of water-intensive crops and to improve the inefficient cultivation and irrigation systems. Therefore, it is necessary to quantify the freshwater appropriation in the agrarian system for sustainable water consumption and production. The Water Footprint (WF) is an appropriate indicator to track changes in consumption and production patterns of freshwater resources. WF comprises of three different components such as Green (G), Blue (B), and Grey (Gy) that makes the water assessment complete in agreement with the water footprint network (WFN) in accordance with the latest ISO 14046. The green WF deals with the consumed quantity/volume of rainwater (total rainwater evapotranspiration (ET) + incorporation) during the agricultural production process. Blue WF is an indicator of fresh surface/groundwater (consumptive water use), while grey WF is a volume of freshwater required assimilating pollutant load based on natural background concentration and existing ambient water quality standards (water pollution). WF reduction approaches (such as infield rainwater harvesting, mulching, and conservation tillage) can increase green water. Drip, sprinkler, and deficit irrigation can save water without compromising crop yield. Grey WF can be optimized by less use of fertilizer and organic farming. The objective is to promote various sustainable agricultural practices that can reduce/minimize the WFs in agricultural production. Therefore, WF may act as a potential indicator in the holistic endorsement of sustainable agriculture.

1.1 INTRODUCTION

Global freshwater is a rare natural resource that is facing a huge pressure due to increasing water use consumption and water pollution. Freshwater is critically essential, but it is being affected by human-driven activities and anthropogenic climate change. Moreover, human-driven activities consume and pollute a significant quantity of freshwater. Freshwater scarcity has been evident in various parts of the globe [25, 36, 59, 62, 85, 86]. Global maximum water use is by agricultural production, but the industrial and domestic sectors also consume and pollute a substantial quantity of freshwater [35]. The demand for freshwater has increased due to increasing population, urbanization, rapid industrialization, economic development, agriculture expansion, a shift in cropping and land use patterns, and change in irrigation and drainage systems during the recent years. Agricultural activities are enormous water consuming and a source of chemical pollution. Therefore, both the land and the amount of freshwater get effected [30].

The water footprint (WF) deals with the quantity of water consumption per unit cropped area as a quantifiable indicator of water volume and its quality affected by water pollution [31, 35]. Therefore, issues related to water depletion and degradation can be effectively resolved by efficient water management. The WF has been proposed as a metric to indicate the human appropriation of indirect and direct usage of freshwater. The freshwater consumption is measured in terms of water being evaporated/incorporated into the product or freshwater pollution per unit time. Globally, the agriculture sector accounts for 99% of consumptive WF as the largest freshwater user [34]. Therefore, to raise the agricultural water productivity (which means "more crop per drop"), the WF may serve as a potential way to minimize the pressure on the global freshwater resources [61, 66]. The WF concept was introduced and incorporated by Hoekstra in 2002 [37]. Hoekstra et al. [31] also understood the significance of the global characteristic of freshwater by unseen water use behind each product for its consumption and trade.

The assessment of WF assists in improved management of global freshwater resources. The multidimensional quality of WF as an indicator deals with the empirical calculation of water being used "when and where" in the entire sequence of a final product. WFs deal with water appropriation and act as a geographically precise indicator in the context of land and amount of water utilization for crop production. According to the *Water Footprint Assessment Manual* [87], WF has basically three components: green, blue, and grey WF. These are discussed in this section.

1.1.1   GREEN WATER FOOTPRINTS (GWFs)

Green water is the freshwater that is primarily accumulated within the soil and root zone, and is readily available for plant growth and development. Moreover, GWF is the quantity of rainwater that is evapotranspired (by plant or field area) in addition the water assimilated in a crop that is harvested. Briefly, it is the volume of rainwater utilized/incorporated in agricultural crops (Figure 1.1).

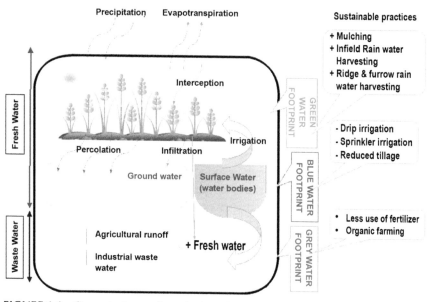

FIGURE 1.1   Conceptual water footprint in agriculture.

1.1.2   BLUE WATER FOOTPRINTS (BWFs)

Bluewater footprint (BWF) comprises of consumptive freshwater usage (evaporated/incorporated), its quantity of surface and groundwater for agricultural production. It is emphasized that the water abstracted from surface/groundwater in one catchment, which is returned to another catchment/sea (water, which is not reverted to the catchment from where it was withdrawn).

1.1.3 GREYWATER FOOTPRINTS (GYWFs)

Greywater footprint (GYWFs) is the most critical component in the assessment of WF because it considers the wastewater (WW) or the water pollution and its dilution to meet the standards of its background water quality. GYWF indicates the freshwater pollution in the full supply chain in the production of a product. While calculating GYWF, the amount of freshwater is necessary to dilute pollutants and assimilate a load of pollutants as per existing ambient water quality standards.

Therefore, based on these concepts for various constituents of WF, water productivity (ton/m³) for crop production can be estimated. The water productivity of the agricultural system is an inverse relation with the green-BWFs (m³/ton) of crop production. Based on the literature review on water productivity, it has four basic categories:

1. **Field Studies:** Wherein field measurements are done to determine/estimate seasonal water use and crop yield [60, 64, 69, 72, 73, 93, 94]. Water productivity studies done by field measurements have a relatively small number of arenas. Therefore, all field measurements are restricted to specific local conditions, such as soil characteristics, climate, and different water management practices.

2. **Modeling Studies:** It comprises of crop growth modeling and soil water equilibrium used to estimate/determine different constituents and nexus of seasonal crop water balance [5, 6].

3. **Remote Sensing Studies:** These are established on satellite data usages to estimate spatial variations in water productivity [7, 10, 98–100]. Satellite-based data estimate water productivity over large areas and it provides a collective measured data-set with the simulation models [26, 65, 68].

4. **Combination of Studies:** It comprises of field measurements and data collection from modeling/satellite for crop water productivity. Recently, these studies are receiving tremendous attention as well as consideration, in contrast with minimizing water pollution (the grey WF) per unit crop yield. GYWF per unit crop yield exclusively

varies and depends on different local specific agricultural practices [14, 53, 55].

Previous research studies on agriculture WF considered only blue and green components of WF neglecting the GYWF. Most of the studies were conducted at the coarse spatial resolution and had treated the world, continents [43, 74] and countries [70, 75] similarly. Recent studies considered a high spatial resolution for the global water consumption in different agricultural crops. It eventually segregated the water consumption for crop production into two basic components: green and blue water use [17, 28, 45, 48, 67, 76, 77].

Mekonnen [52] has evaluated and quantified the green, blue, and GYWFs in a spatially-explicit way for global crop production during 1996–2005 as 7,404 billion m^3 per year (Green: 78%, Blue: 12%, and Grey: 10%). The WFs (Gm^3 per year) for different agricultural crops were sequenced as (wheat = 1087) > (rice = 992) > (maize = 770), respectively. Globally, WFs of rain-fed agriculture crops are 5173 Gm^3 per year (91% green and 9% grey), compared to 2230 Gm^3 per year (48% green, 40% blue, and 12% grey) for WF of irrigated agriculture crops. Total WFs (Gm^3 per year) in different countries were in the order: (India = 1047) > (China = 967) > (USA = 826), respectively. Later, Mekonnen et al. [57] gave the basic concept of global WF that were taken as benchmark standards for 124 different crops based on earlier study targeting to reduce crop water consumption and pollution per unit crop production [39, 40].

Xinchun et al. [90] emphasized the importance to reduce the WF of grain product (WFGP, m^3/t). The crop-specific WFs have been estimated by different researchers for, namely: tomatoes [12], wheat [46, 55, 77], rice [13], cut flowers [56] and coffee/tea [11]. Researchers in different regions had also paid attention to the grain crop WF, such as maize [78], rice [29], wheat, and maize [25], and main grain irrigated crops [79]. These studies showed that WFs of agricultural crops vary massively within and across the regions [8, 17, 23, 32, 53, 54, 77]. Therefore, detailed studies incorporating several technologies and various agriculture practices can give a clear sketch of water conservation and its production in agricultural systems. For instance, green water productivity can be enhanced by utilizing sustainable rainwater harvesting and supplemental irrigation by decreasing soil water evaporation. While, declining the BWF by different irrigation techniques (such as drip and subsurface irrigation), along with effective control of different weeds. GYWF can be reduced by less use of fertilizer and promoting organic farming. These approaches can solve the issue of water management in the agricultural systems.

This book chapter suggests a combination of various sustainable agricultural practices that can reduce or minimize the WFs in agricultural production.

1.2   IMPORTANCE OF WATER FOOTPRINT (WF)

The freshwater is a highly scarce natural resource and its demand has already exceeded its rate of supply. Today, freshwater consumption has transgressed the safe operating space of humanity. The WF of humanity has already transgressed its sustainable limits at various locations; and its unequal distribution is a serious concern. The agriculture sector alone contributes to 92% of water consumption and pollution, out of which 89% is exclusively contributed by the crop production component. Therefore, the agricultural WF for crop production can resolve the puzzle of fresh and WW sustainability. The influence of climate change can be felt severely via freshwater, because less water will be captured or preserved in snow and ice, with frequent extreme events causing floods and droughts. An extensive study of WF will give us a clear idea of freshwater resources for sustainable and equitable distribution across the globe, thus leading to the knowledge of WF of every single product either being small or larger. And it will raise awareness for both the consumers as well as the businesses (who are producer, processor, trader, or seller of those products) at different stages of the supply chain. Globally, freshwater resource is getting depleted due to excessive consumption and pollution. The WF can also do "water profiling," whereby, displaying a nexus exists between the daily consumption of goods and the problems (related with the water decline and pollution) from the production to the consumers. WF assessment has its major focus to analyze water use and pollution by production, consumption, and trade-off several water-consuming goods and services as a novel aspect of research and development. The WF is a priority-based tool to make a transparent impact. WF assessment consists of following diverse phases [35]:

1. The first phase is to define the purpose and scope of assessment.
2. The second phase deals with accounting. It quantifies the volume of water being used, its timing, and location.
3. The third phase deals with sustainability at a local and global level for water management by accessing the social, economic, and environmental impacts of WFs.

4. The fourth phase deals with the WF recommendations, assessment regarding possibilities to reduce different components of WFs. The sustainability assessment is also an integral component of the WF assessment.

The WFs (water productivities) of crop production need a better understanding of geographic and temporal variations for assessment of problems and solutions related to WF. WFs also comprise of estimation/evaluation of different agricultural practices, policies, and governance contributing to the sustainable, efficient, and reasonable allocation of WF across different societies.

1.3   CALCULATION OF GREEN, BLUE, AND GREY WATER FOOTPRINTS (GYWF) FOR GROWING OF CROPS

The agricultural sector is a maximum water consuming sector, and these products have relatively higher WFs. Therefore, agricultural products need a precise evaluation and assessment of the process starting from growing of a crop. The WF is the amount of water being used to produce a unit quantity of the product (m^3/t) or the amount of water per year ($m^3/year$) of a given demarcated area (e.g., catchment or nation), by the specific individual or a community. The methodology till today comprises of two different basis: volumetric approach given by WF network (WFN); and life cycle assessment (LCA) approach that is recently being developed by Vanham and Bidoglio [83, 84].

-   **Volumetric Approach:** It is given by WFN developed by Hoekstra and Hung [33]; it includes quantification of the effective water quantity by: (i) its goals and the scope of estimation; (ii) accounting; (iii) its sustainability assessment; and (iv) the response formulation.
-   **LCA Approach:** It is described in the recent ISO-14046 [42]. It comprises of: (i) its goals, scope, and definition; (ii) different inventory assessment; (iii) the impact assessment; and (iv) the interpretation.

Vanham [83] has reported several differences among the two different methodologies, such as WFN has a water management-focused approach, while LCA has a specific product-focused approach. However, all these different methods have been compared by Manzardo et al. [50]. In this chapter, Figure 1.2 gives a complete illustration of the method involved for "calculation of WF in case of crop production."

1.3.1 THE TOTAL WATER FOOTPRINT (WF) FOR CROP PRODUCTION

The WF of a process in agricultural crop production is the summation of its different components, that is, green, blue, and GYWF. The WF is the ratio of crop water use (CWU) per unit crop yield. The green component in the WF of growing a crop (*WF proc, green*, m³/ton) is a ratio of the green component in the CWU (*green*, m³/ha) divided by the crop yield (*Y*, ton/ha). The blue component (*WF proc, blue*, m³/ton) is a ratio of blue components in CWU (*blue,* m³/ha) divided by the crop yield (*Y*, ton/ha).

For convenience, the WFs in agriculture in water volume per unit mass are expressed as m³/ton or liter/kg. Yields can be determined from the yield statistics for annual crops, i.e., FAO (Food and Agricultural Organization). While for the perennial crops, we must consider its annual yield during its complete lifespan for the crop production. It is assumed that the first year of planting and the end of life span the values are considered as zero or very low. While, yields are considered as maximum or highest after few years of this crop growth. Therefore, the average annual CWU over its complete life span is accounted for different crop water usage.

$$\text{WF}_{\text{proc, Total}} = \text{WF}_{\text{proc, Green}} + \text{WF}_{\text{proc, Blue}} + \text{WF}_{\text{proc, Grey}}$$

$$= \text{I} + \text{II} + \text{III} \tag{1.1}$$

The third and critical component is the GYWF in agricultural production for growing crops (*WF proc, grey*, m³/ton). It is computed as per hectare of chemical application within the field area (*AR*, kg/ha), multiplied by leaching and its run-off fraction (α), which is divided by the acceptable maximum concentration (*cmax*, kg/m³), subtracting the natural concentration of different pollutants (*cnat*, kg/m³), which is finally divided by crop yield. The water pollutants generally comprise of various types of fertilizers, pesticides, and insecticides. Though considering the 'waste flow' to the freshwater bodies, it is generally a smaller portion of the total application of fertilizers, chemicals, or pesticides to the agricultural fields. While in these calculations, only the most critical pollutants with highest water volume are taken into consideration.

The complete CWU (m³/ha) of green and blue component are estimated by collection of evapotranspiration (*ET*, mm/day) data throughout the whole growing period (Figure 1.2). The *ET green* denotes the green water ET and *ET blue* denotes blue water ET. In the given computation, the factor of 10 is considered to convert the water depths in mm into the water volumes per unit land area (m³/ha). This is estimated for the period from the day of planting (day 01) to the day of harvesting (length of growing period (*lgp*) in days). The estimated water use varies significantly for different crop types and varieties because of differences in the length of growing period.

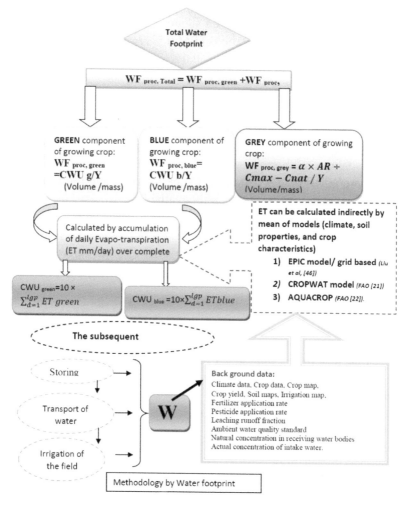

FIGURE 1.2 Step wise procedure for calculation of water footprint in crop production (WF $_{proc, Total}$) in an agricultural system. (Legend: CWU = crop water use; ET = evapotranspiration; Lgp = long growing period (natural concentration for the pollutant); WF$_{proc, blue}$ = water footprint blue; WF$_{proc, green}$ = water footprint green; WF$_{proc, grey}$ = water footprint grey; WFA = water footprint assessment; Y = yield).

The ET from the agrarian field can be either measured/estimated using a model based on different empirical methods. It is usually expensive and cost-effective. Therefore, the approximations for ET are done based on modeling that uses data of climatic conditions, its soil properties, and different crop characteristics. Several, new alternative ways are used to estimate *ET* different researchers, such as, most frequently used EPIC model [88, 89] and grid-based model [45].

The CROPWAT model has been suggested by the Food and Agriculture Organization of the United Nations (FAO) [21], which is exclusively based on the method explained by Allen et al. [1]. In addition, AQUACROP model was developed for estimating the crop growth along with *ET* under water-deficit situations by FAO [22].

The approximation of green, blue, and GYWF of a growing agricultural crop requires massive amount of data-sets and data sources. Local data from the crop field location are always preferred. However, the location-specific data is usually too laborious and needs effort, time, and money. Therefore, generally rough estimates pertaining to nearby location/region/national averages are considered as these are easily available.

TABLE 1.1 Suggested Options for Reducing the Agricultural Water Footprint in Crop Production

---

1. Reducing Green Water Footprint of Crop Production:

Soil mulching to reduce evapotranspiration from the topsoil surface.

Promoting rain-fed agriculture (for improving agricultural practices) to increase land productivity (yield/ton/ha). Water productivity can only be improved by green water footprint reduction as rainwater availability is constant and for a limited duration of time.

2. Reducing Blue Water Footprint of Crop Production:

Selection of irrigation technique, which has lower evapotranspiration loss of blue water from the field.

Selection of various crop or crop varieties as per local climate to reduce the water needed for irrigation.

Decreasing the amount of water being used for irrigation, i.e., deficit/supplementary irrigation/no irrigation at all.

Reducing evapotranspiration losses of water stored as surface water, i.e., reservoirs.

Optimizing timing and improving the irrigation scheduling for water application in the field.

3. Reducing Grey Water Footprint of Crop Production:

Promoting organic farming in reduce the application of chemicals (non-organic fertilizers and pesticides) in the field.

Optimizing timing and technique of adding chemicals so that a minimum amount leaches from the agricultural field.

Hence, in the calculations described under this section, authors of this chapter did not distinguish between green and blue water integrated into the harvested crops. The components of WF can simply be known by the water fraction of the harvested crop, e.g., vegetables: 90–95%, fruits: 80–90% on wet basis. The green-blue ratio in the water that is integrated in the agricultural crop can be estimated as equivalent to the ratio of *CWU green*: *CWU blue*.

The BWF in this section delineates only the ET of irrigation water from its cropped field. While, it eliminates the evaporation of water from non-natural surface, water reservoirs constructed for storage of irrigation water and the evaporation from transport channels that fetch the water from the place of its abstraction to the agricultural field. The water storage and transport have their WF, which is very important for growing the agricultural crop in the field. Hence, the evaporation losses mentioned above needs to be ideally incorporated for overall WF estimation, when WFs of the harvested agricultural crop are determined (Table 1.1).

1.4 REDUCTION OF GREEN AND BLUE WATER FOOTPRINTS (BWFS) IN AGRICULTURE: CASE STUDIES

Suggested Options for reducing the agricultural WF in crop production are listed in Table 1.1. Several studies have been conducted for the reduction of green water footprint (GWF) via mulching. Mulching is a technique of covering the topsoil surface (either organic or synthetic) to decrease ET each unit of yield in crop production [51, 58, 95, 96]. It can conserve moisture, downgrade weed growth, reduce soil erosion, and improve soil environment. Mulching (installing mulches) can help to advance crop yield and enhance water use. Sharma et al. [71] evaluated the maize-wheat cropping system in the sub-mountainous North-western Himalayan regions of India, and they reported that mulching is beneficial for enhancing moisture and for nutrient conservation to improve soil conditions and productivity. Feng et al. [19] have reported the benefits of straw mulching to improve plant growth by increasing soil nitrogen availability.

Arora et al. [4] reported the importance of crop residues like straw mulching, which affects the hydro-thermal regime of soils by regulating soil moisture and temperature, along with controlling the weeds population. Totin et al. [82] have reported the benefits of 'rice-straw' mulch in improving irrigation water use efficiency (WUE) in rice fields. Zhang et al. [92] reported the significance of plastic mulching as an essential agricultural practice for improving crop productivity. Therefore, mulching improves not only the soil moisture ambient but also helps in adequate soil temperature and soil nutrient accumulation and distribution, with regulation and control mechanism of root and shoot equilibrium [44, 49, 91].

Studies on blue water management for irrigated crops by Tindula et al. [81] have reported that sprinkler-, drip-, and mini-sprinkler irrigation as the primary method that can be used efficiently by farmers. Thompson [80] has shown potential water saving in agriculture by subsurface drip irrigation (SDI) for low WUE. Al-Said et al. [2] have shown the positive effect of drip versus sprinkler irrigation on vegetable yield. Researchers have also studied deficit irrigation or supplementary irrigation to improve WUE [41, 63]. This indicates that the efforts are to increase the ratio of green to BWF for improving agricultural WF.

Recently, Chukalla et al. [15] published a thorough report of the possible ways to reduce the green and BWFS (together) of growing crops for different management practices under different environment conditions (systematic model-based). The study comprised of different environmental conditions (such as arid along with semi-arid, sub-humid, and humid) at distinctive locations of Israel, Spain, Italy, and the UK for different soil types (such as sand, sandy loam, and silty clay loam). Also, maize, potato, and tomato crops were considered for three different management practices, such as furrow-, sprinkler-, drip-, and SDI along with no mulching, organic/synthetic mulching and four different irrigation techniques/strategies (such as full, deficit, supplementary, and no irrigation). The results were compared with control (furrow irrigation, full irrigation, no mulching). The consumptive WFs were reduced by 8–10% under drip irrigation into account, 13% with organic mulching, and 17–18% with sub-surface drip irrigation in combination with organic mulching compared with 28% for drip, sub-surface drip irrigation in combination with synthetic mulching. A significant reduction in overall green and BWFS were obtained compared with the control.

Zhuo et al. [97] has further elaborated this work by investigating effects of variable agricultural management practices on diverse water-efficient indicators, irrigation efficiency (IE), crop WUE, and green and BWFS. Organic and synthetic mulching enhanced WUE by 4% and 10%, respectively, and compacted the volume of blue WF by 8% and 17%, respectively, while similar yields. Whereas, drip along with sub-surface drip irrigation enriched the IE and WUE, but the drip irrigation had a moderately high BWF. Hence, the advancement in single water efficiency indicator may cause deterioration in another. Therefore, synergies, and trade-off within and among different agricultural WF reduction approaches can be explored for further study.

1.5 REDUCTION OF GREY WATER FOOTPRINT (GYWF) IN FIELD: A CASE STUDY

Mekonnen and Hoekstra [54] reported that global agricultural crop production adds three quarter of the nitrogen related with GYWF in world. Increasing population and elevated living standards are promoting more of anthropogenic-driven nitrogen application in the agriculture sector with the raise in demand for agricultural products (such as food, fiber, biofuel, etc.). While, GYWF per ton of crop production has a limitation in the context of assimilation capacity of freshwater for the appropriation of GYWF. WF assessments of nitrate-nitrogen are contemplated to be major pollutant of freshwater under diverse cropping systems [31, 35]

Deurer et al. [16] and Green et al. [27] conducted WF studies to evaluate the leaching of agro-chemicals for crop production, displayed nitrate-nitrogen (NO_3-N) as the main pollutant, e.g., grey- WF is controlled by nitrate leaching. Whereas, mostly GYWF computations are based on the naive assumption that an average 10% of the nitrogen applied in fertilizer is lost throughout the leaching [14]. It is an approximate estimation, which voluntarily ignores different factors, such as; different agricultural practices, soil types, and interaction between different agrochemicals in the soil, and local soil hydrology. Moreover, considerable improved measurements and modeling are needed to authenticate the GWF.

In general, the crop yield depends on the addition of anthropogenic-nitrogen (such as nitrogen fertilizers), part of which is certainly lost through nitrogen leaching and agricultural runoff, resulting in the ground as well as surface water pollution. Interpretation of freshwater appropriation is done by diluting the water pollutant load entering different water bodies [14, 18, 38]. Hoekstra et al. [35] indicated the concept of greywater appropriation to incorporate a load of pollutants to influence the ambient water quality benchmarks of GYWF. Franke et al. [20] distinguished three tiers to evaluate the GYWF from diffused pollution:

- The *tier-1* is constructed on the expert-based suppositions, on which sections of applied or surplus nitrogen in the soil will leach or run off by background influences. It gives a primarily rough calculation of the nitrogen load deprived of tagging the interface and modification of various chemical constituents in the soil or lengthwise its path flow [9, 53].
- The additional superior *tier-2* approach for approximating grey WFs from diffused pollution is grounded on the nitrogen equilibrium approach, employing a basic model methodology [47, 54].
- The *tier-3* approach contemplates physical and biochemical procedures using an advanced water and nutrient equilibrium model (e.g., APEX model).

1.6 SUMMARY

WF is an important indicator for appropriation of water use along the entire chain of the agricultural products either direct or indirectly. Each agricultural product has its own WF (small or large). Hence, it is important that a site, region, and location specific agriculture practices shall be suggested to reduce WFs in agricultural system. It is suggested to determine all trade-off and synergies that exist within different sustainable agricultural practices of WFs. WF is a beneficial indicator, which represent quality and quantity of water in agricultural systems. Research on WF requires sophisticated technology and equipments; and it is labor and costly. Therefore, whatever available studies are used from temperate countries or developed nations. Despite of having few limitations, it is still used efficiently and equitability in the estimation of agricultural WFs. Till date, the absolute picture is not yet clear in terms of global estimation and standardization, thus indicating the need for additional research in the region especially in the tropical regions of the word.

ACKNOWLEDGMENT

The author would like to thanks, Banaras Hindu University for providing funding for the current research work.

KEYWORDS

- evapotranspiration
- life cycle assessment
- sustainable agriculture
- wastewater
- water footprint
- water production

REFERENCES

1. Allen, R. G., Pereira, L. S., Raes, D., & Smith, M., (1998). *Crop Evapotranspiration: Guidelines for Computing Crop Water Requirements* (p. 56). FAO Irrigation and Drainage, Food and Agriculture Organization (FAO), Rome. (Accessed on 20 June 2020).
2. Al-Said, F. A., Ashfaq, M., Al-Barhi, M., Hanjra, M. A., & Khan, I. A., (2012). Water productivity of vegetables under modern irrigation methods in Oman. *Irrigation and Drainage, 61*, 477–489.
3. Amir, J., & Sinclair, T. R., (1991). Model of water limitation on spring wheat growth and yield. *Field Crops Research, 28*, 59–69.
4. Arora, V. K., Singh, C. B., Sidhu, A. S., & Thind, S. S., (2011). Irrigation, tillage, and mulching effects on soybean yield and water productivity in relation to soil texture. *Agricultural Water Management, 98*, 563–568.
5. Asseng, S., Keating, B. A., Fillery, I. R. P., Gregory, P. J., Bowden, J. W., Turner, N. C., Palta, J. A., & Abrecht, D. G., (1998). Performance of the APSIM wheat model in Western Australia. *Field Crops Residue,57*, 163–179.
6. Asseng, S., Turner, N. C., & Keating, B. A., (2001). Analysis of water-and nitrogen-use efficiency of wheat in a Mediterranean climate. *Plant and Soil, 233*(1), 127–143.
7. Biradar, C. M., Thenkabail, P. S., Platonov, A., Xiao, X., Geerken, R., Noojipady, P., Turral, H., & Vithanage, J., (2008). Water productivity mapping methods using remote sensing. *Journal of Applied Remote Sensing, 2*, 23522–23544.
8. Brauman, K. A., Siebert, S., & Foley, J. A., (2013). Improvements in crop water productivity increase water sustainability and food security-a global analysis. *Environmental Research Letters, 8*(2), 24030–24034.
9. Brueck, H., & Lammel, J., (2016). Impact of fertilizer N application on the grey water footprint of winter wheat in a NW-European temperate climate. *Water, 8*, 356–359.
10. Cai, X., Thenkabail, P. S., Biradar, C. M., & Platonov, A., (2009). Water productivity mapping using remote sensing data of various resolutions to support "more crop per drop." *Journal of Applied Remote Sensing,3*, 33523–33557.
11. Chapagain, A. K., Hoekstra, A. Y., & Savenije, H. H. G., (2006a). Water saving through international trade of agricultural products. *Hydrology and Earth System Sciences, 10*(3), 455–468.

12. Chapagain, A. K., Hoekstra, A. Y., Savenije, H. H. G., & Gautam, R., (2006b). The water footprint of cotton consumption: An assessment of the impact of worldwide consumption of cotton products on the water resources in the cotton producing countries. *Ecological Economics, 60,* 186–203.

13. Chapagain, A. K., & Hoekstra, A. Y., (2011). The blue, green, and grey water footprint of rice from production and consumption perspectives. *Ecological Economics, 70*(4), 749–758.

14. Chapagain, A. K., & Orr, S., (2009). An improved water footprint methodology linking global consumption to local water resources: A case of Spanish tomatoes. *Journal of Environmental Economics and Management, 90*(2), 1219–1228.

15. Chukalla, A. D., Krol, M. S., & Hoekstra, A. Y., (2015). Green and blue water footprint reduction in irrigated agriculture: Effect of irrigation techniques, irrigation strategies, and mulching. *HydrologyandEarth System Sciences, 19,* 4877–4891.

16. Deurer, M., Green, S. R., Clothier, B. E., & Mowat, A., (2011). Can product water footprints indicate the hydrological impact of primary production?: A case study of New Zealand kiwifruit. *Journal of Hydrology,408,* 246–256.

17. Fader, M., Gerten, D., Thammer, M., & Heinke, J., (2011). Internal and external green-blue agricultural water footprints of nations, and related water and land savings through trade. *Hydrology and Earth System Sciences, 15,* 1641–1660.

18. Falkenmark, M., & Lindh, G., (1974). How can we cope with the water resources situation by the year 2015? *Ambio., 1974,* 114–122.

19. Feng, K. S., Siu, Y. L., Guan, D. B., & Hubacek, K., (2012). Assessing regional virtual water flows and water footprints in the yellow river basin, China: A consumption-based approach. *Applied Geography, 32,* 691–701.

20. Franke, N., & Mathews, R., (2013). *C & A's Water Footprint Strategy: Cotton Clothing Supply Chain* (p. 58). Water footprint network/C & A Foundation, Netherlands/Zug, Switzerland.

21. FAO, (2010). *CROPWAT 8.0 Model.* FAO, Rome. http://www.fao.org/land-water/databases-and-software/cropwat/en/ (accessed on 20 June 2020).

22. FAO, (2010). *AQUACROP 3.1; FAO,* Rome. www.fao.org/nr/water/aquacrop.html (accessed on 20 June 2020).

23. Finger, R., (2013). More than the mean-a note on heterogeneity aspects in the assessment of water footprints. *Ecological Indicator, 29,* 145–147.

24. Ge, L. Q., Xie, G. D., Li, S. M., Zhang, C. X., & Chen, L. X., (2010). Study on production water footprint of winter-wheat and maize in the North China plain. *Resource Science, 32*(11), 2066–2071. (In Chinese).

25. Gleick, P. H., (1993). Water and conflict: Fresh water resources and international security. *International Security, 18,* 79–112.

26. Grassini, P., Hall, A. J., & Mercau, J. L., (2009). Benchmarking sunflower water productivity in semiarid environments. *Field Crops Research, 110,* 251–262.

27. Green, S. R., Deurer, M., Clothier, B., & Andrews, S., (2010). Water and nitrogen movements under agricultural and horticultural lands. In: *Farming's Future: Minimizing Footprints and Maximizing Margins.* Occasional Report No. 23. Fertilizer and Lime Research Center, Massey University, Palmerston North, New Zealand; 23rd Annual FLRC Workshop. http://flrc.massey.ac.nz/workshops/10/paperlist10.htm (accessed on 20 June 2020).

28. Hanasaki, N., Inuzuka, T., Kanae, S., & Oki, T., (2010). Estimation of global virtual water flow and sources of water withdrawal for major crops and livestock products using a global hydrological model. *Journal of Hydrology, 384*, 232–244.

29. He, H., Huang, J., Huai, H. J., & Tong, W. J., (2010). The water footprint and its temporal change characteristics of rice in Hunan. *Chinese Agricultural Science Bulletin,26*(14), 294–298. (In Chinese).

30. Herath, I., Green, S., Singh, R., Horne, D., Vander-Zijpp, S., & Clothier, B., (2013). Water foot printing of agricultural products: A hydrological assessment for the water footprint of New Zealand's wines. *Journal of Cleaner Production, 41*, 232–243.

31. Hoekstra, A. Y., & Chapagain, A. K., (2008). *Globalization of Water: Sharing the Planet's Freshwater Resources* (p. 111). Blackwell Publishing, Oxford, UK.

32. Hoekstra, A. Y., & Chapagain, A. K., (2007). Water footprints of nations: Water use by people as a function of their consumption pattern. *Water Resource Management, 21*, 35–48.

33. Hoekstra, A. Y., & Hung, P. Q., (2002). *Virtual Water Trade: A Quantification of Virtual Water Flows Between Nations in Relation to International Crop Trade* (p. 120). Value of water research report series no. 11, UNESCO-IHE, Delft, Netherlands. https://waterfootprint.org/media/downloads/Report11_1.pdf (accessed on 20 June 2020).

34. Hoekstra, A. Y., & Mekonnen, M. M., (2012). The water footprint of humanity. *Proceedings of the National Academy of Sciences,109*, 3232–3237.

35. Hoekstra, A. Y., Chapagain, A. K., Aldaya, M. M., & Mekonnen, M. M., (2011). *The Water Footprint Assessment Manual: Setting the Global Standard* (p. 228). Earth scan, London, UK. https://waterfootprint.org/media/downloads/TheWaterFootprintAssessmentManual_2.pdf (accessed on 20 June 2020).

36. Hoekstra, A. Y., (2013). *The Water Footprint of Modern Consumer Society* (p. 208). London: Routledge.

37. Hoekstra, A. Y., (2003). Virtual water trade. In: *Proceedings of the International Expert Meeting on Virtual Water Trade.* Value of water research report series no. 12, UNESCO-IHE, Delft, Netherlands; www.waterfootprint.org/Reports/Report12.pdf (accessed on 20 June 2020).

38. Hoekstra, A. Y., (2008). *Water Neutral: Reducing and Offsetting the Impacts of Water Footprints.* Value of water research report series no. 28; UNESCO-IHE, Delft, the Netherlands. https://research.utwente.nl/en/publications/water-neutral-reducing-and-ofsetting-water-footprints (accessed on 20 June 2020).

39. Hoekstra, A. Y., (2013). *The Water Footprint of Modern Consumer Society* (p. 224). Routledge, London, UK.

40. Hoekstra, A. Y., (2013). *Wise Freshwater Allocation: Water Footprint Caps by River Basin, Benchmarks by Product and Fair Water Footprint Shares by Community* (p. 68). Value of water research report series no. 64; UNESCO-IHE, Delft, The Netherlands.

41. Hoean, P. I., Ogunewe, W. N., & Dike, R. I., (2008). Effect of tillage and mulching practices on soil properties and growth and yield of cowpea (*Vigna unguiculata*(L), Walp) in southeastern Nigeria. *Agro. Sci., 7*, 118–128.

42. International Organization for Standardization (ISO), (2014). *ISO-14046: Environmental Management: Water Footprint-Principles, Requirements and Guidelines.* International Organization for Standardization: Geneva, Switzerland; https://www.iso.org/standard/43263.html (accessed on 20 June 2020).

43. L'vovich, M. I., White, G. F., & Turner, B. L. I., (1990). *Use and Transformation of Terrestrial Water Systems, in: The Earth as Transformed by Human Action: Global and Regional Changes in the Biosphere Over the Past 300 Years* (pp. 235–252). Cambridge University Press, New York.

44. Liu, G. C., Yang, Q. F., Li, L. X., Fan, T. L., Zhao, X. W., & Zhu, Y. Y., (2008). Study on soil water effects of the techniques of whole plastic-film mulching on double ridges and planting in catchment furrows of dry land corn. *Agricultural Research in the Arid Areas, 6*, 18–28.

45. Liu, J., & Yang, H., (2010). Spatially explicit assessment of global consumptive water uses in cropland: Green and blue water. *Journal of Hydrology, 384*, 187–197.

46. Liu, J., Williams, J. R., Zehnder, A. J. B., & Yang, H., (2007). GEPIC modeling wheat yield and crop water productivity with high resolution on a global scale. *Agricultural Systems,94*(2), 478–493.

47. Liu, J., Zehnder, A. J. B., & Yang, H., (2009). Global consumptive water use for crop production: The importance of green water and virtual water. *Water Resource Research, 45*, 15; E-article W05428. doi: 10.1029/2007WR006051.

48. Liu, C., Kroeze, C., Hoekstra, A. Y., & Gerbens-Leenes, P. W., (2012). Past and future trends in grey water footprints of anthropogenic nitrogen and phosphorus inputs to major world rivers. *Ecological Indicators,18*, 42–49.

49. Malhi, S. S., Nyborg, M., Solberg, E. D., Dyck, M. F., & Purveen, D., (2011). Improving crop yield and N uptake with long-term straw retention in two contrasting soil types. *Field Crops Research, 124*(3), 378–391.

50. Manzardo, A., Mazzi, A., Loss, A., Butler, M., Williamson, A., & Scipioni, A., (2016). Lessons learned from the application of different water footprint approaches to compare different food packaging alternatives. *Journal of Cleaner Production, 112*, 4657–4666.

51. Mao, D., Wang, Z., Luo, L., & Ren, C., (2012). Integrating AVHRR and MODIS data to monitor NDVI changes and their relationships with climatic parameters in Northeast China. *International Journal of Applied Earth Observation and Geoinformation, 18*, 528–536.

52. Mekonnen, M. M., & Hoekstra, A. Y., (2011). *National Water Footprint Accounts: The Green, Blue and Grey Water Footprint of Production and Consumption* (p. 94). Value of water research report 50; UNESCO-IHE Institute for Water Education; Delft, the Netherlands. https://ris.utwente.nl/ws/portalfiles/portal/5146139/Report50-NationalWaterFootprints-Vol2.pdf.pdf (accessed on 20 June 2020).

53. Mekonnen, M. M., & Hoekstra, A. Y., (2011). The green, blue, and grey water footprint of crops and derived crop products. *Hydrology and Earth System Sciences, 15*, 1577–1600.

54. Mekonnen, M. M., & Hoekstra, A. Y., (2015). Global gray water footprint and water pollution levels related to anthropogenic nitrogen loads to fresh water. *Environmental Science and Technology, 49*(21), 12860–12868.

55. Mekonnen, M. M., & Hoekstra, A. Y., (2010). The green, blue, and grey water footprint of crops and derived crop products. *Hydrol. Earth Syst. Sci., 15*, 1577–1600.

56. Mekonnen, M. M., & Hoekstra, A. Y., (2012). A global assessment of the water footprint of farm animal products. *Ecosystems, 15*(3), 401–415.

57. Mekonnen, M. M., & Hoekstra, A. Y., (2014). Water footprint benchmarks for crop production: A first global assessment. *Ecological Indicators, 46*, 214–223.

58. Ogban, P. I., Ogunewe, W. N., Dike, R. I., Ajaelo, A. C., Ikeata, N. I., Achumba, U. E., & Nyong, E. E., (2008). Effect of tillage and mulching practices on soil properties and

growth and yield of cowpea (*Vigna unguiculata* (L), Walp) in Southeastern Nigeria. *Journal of Tropical Agriculture, Food, Environment and Extension, 7*(2), 118–128.

59. Oki, T., & Kanae, S., (2006). Global hydrological cycles and world water resources. *Science, 313*, 1068–1072.

60. Oweis, T., Zhang, H., & Pala, M., (2000). Water use efficiency of rain fed and irrigated bread wheat in a Mediterranean environment. *Agronomy Journal,92*, 231–238.

61. Passioura, J., (2006). Increasing crop productivity when water is scarce-from breeding to field management. *Agricultural Water Management, 80*, 176–196.

62. Postel, S. L., (2000). Entering an era of water scarcity: The challenges ahead. *Ecological Applications, 10*, 941–948.

63. Qin, W., & Chi, B., (2013). Long-term monitoring of rain fed wheat yield and soil water at the loess plateau reveals low water use efficiency. *PLoS One, 8*(11), e78828.

64. Rahman, S. M., Khalil, M. I., & Ahmed, M. F., (1995). Yield-water relations and nitrogen utilization by wheat in salt-affected soils of Bangladesh. *Agricultural Water Management, 28*, 49–56.

65. Robertson, M. J., & Kirkegaard, J. A., (2005). Water-use efficiency of dry land canola in an equi-seasonal rainfall environment. *Australian Journal of Agricultural Research, 56*, 1373–1386.

66. Rockström, J., (2003). Water for food and nature in drought-prone tropics: Vapor shift in rain-fed agriculture. *Philosophical Transactions of the Royal Society of London B: Biological Sciences, 358*(1440), 1997–2009.

67. Rost, S., Gerten, D., Bondeau, A., Lucht, W., Rohwer, J., & Schaphoff, S., (2008). Agricultural green and blue water consumption and its influence on the global water system. *Water Resources Research, 44*(9). Online: https://agupubs.onlinelibrary.wiley.com/doi/full/10.1029/2007WR006331 (accessed on 20 June 2020).

68. Sadras, V., Baldock, J., Roget, D., & Rodriguez, D., (2003). Measuring and modeling yield and water budget components of wheat crops in coarse-textured soils with chemical constraints. *Field Crops Research, 84*, 241–260.

69. Sadras, V. O., Grassini, P., & Steduto, P., (2007). *Status of Water Use Efficiency of Main Crops* (p. 47). SOLAW Background Thematic Report; Food and Agricultural Organization, Rome, Italy. http://www.fao.org/fileadmin/templates/solaw/files/thematic_reports/TR_07_web.pdf (accessed on 20 June 2020).

70. Seckler, D., Amarasinghe, U., Molden, D. J., De Silva, R., & Barker, R., (1998). *World Water Demand and Supply, 1990–2025: Scenarios and Issues.* IWMI Research Report 19, IWMI, Colombo, Sri Lanka. https://cgspace.cgiar.org/handle/10568/39802 (accessed on 20 June 2020).

71. Sharma, B. R., Rao, K. V., Vittal, K. P. R., Ramakrishna, Y. S., & Amarasinghe, U., (2010). Estimating the potential of rain fed agriculture in India: Prospects for water productivity improvements. *Agricultural Water Management, 97*(1), 23–30.

72. Sharma, D. K., Kumar, A., & Singh, K. N., (1990). Effect of irrigation scheduling on growth, yield, and evapotranspiration of wheat in sodic soils. *Agricultural Water Management,18*, 267–276.

73. Sharma, T., Kiran, P. S., Singh, T. P., Trivedi, A. V., & Navalgund, R. R., (2001). Hydrologic response of a watershed to land use changes: A remote sensing and GIS approach. *International Journal of Remote Sensing, 22*(11), 2095–2108.

74. Shiklomanov, I. A., (1993). *World Fresh Water Resources, in: Water in Crisis: A Guide to the World's Fresh Water Resources* (pp. 13–24). Oxford University Press, Oxford, UK.

75. Shiklomanov, I. A., & Rodda, J. C., (2004). *World Water Resources at the Beginning of the Twenty-First Century* (p. 25). IH Series; Cambridge University Press, Oxford, UK.

76. Siebert, S., & Döll, P., (2008). *The Global Crop Water Model (GCWM): Documentation and First Results for Irrigated Crops* (p. 42). Report No. 07; Institute of Physical Geography University of Frankfurt (Main), Frankfurt.

77. Siebert, S., & Döll, P., (2010). Quantifying blue and green virtual water contents in global crop production as well as potential production losses without irrigation. *Journal of Hydrology, 384*(3), 198–217.

78. Sun, S. K., Wu, P. T., Wang, Y. B., & Zhao, X. N., (2013a). Temporal variability of water footprint for maize production: The case of Beijing from 1978 to 2008. *Water Resource Management, 27*(7), 2447–2463.

79. Sun, S. K., Wu, P. T., Wang, Y. B., & Zhao, X. N., (2013b). The impacts of inter annual climate variability and agricultural inputs on water footprint of crop production in an irrigation district of China. *Science of the Total Environment, 444*, 498–507.

80. Thompson, T. L., Pang, H. C., & Li, Y. Y., (2009). The potential contribution of subsurface drip irrigation to water-saving agriculture in the Western USA. *Agricultural Sciences in China, 8*, 850–854.

81. Tindula, G. N., Orang, M. N., & Snyder, R. L., (2013). Survey of irrigation methods in California in 2010. *Journal of Irrigation and Drainage Engineering-ASCE, 139*, 233–238.

82. Totin, E., Stroosnijder, L., & Agbossou, E., (2013). Mulching upland rice for efficient water management: A collaborative approach in Benin. *Agricultural Water Management, 125*, 71–80.

83. Vanham, D., Bouraoui, F., Leip, A., Grizzetti, B., & Bidoglio, G., (2015). Lost water and nitrogen resources due to EU consumer food waste. *Environmental Research Letters, 10*(8), e-article 084008.

84. Vanham, D., & Bidoglio, G., (2013). Review on the indicator water footprint for the EU28. *Ecological Indicators, 26*, 61–75.

85. Vörösmarty, C. J., McIntyre, P. B., & Gessner, M. O., (2010). Global threats to human water security and river biodiversity. *Nature, 467*, 555–561.

86. Wada, Y., Van, B. L. P. H., & Viviroli, D., (2011). Global monthly water stress, Part II: Water demand and severity of water stress. *Water Resources Research, 47*, 17, e-article W07518, doi: 10.1029/2010WR009792.

87. Hoekstra, A. Y., Chapagain, A. K., Aldaya, M. M., & Mekonnen, M. M., (2009). *Water Footprint Manual: State of the Art* (p. 131). Water Foot Print Organization; Enschede, the Netherlands.

88. Williams, J. R., Jones, C. A., Kiniry, J. R., & Spanel, D. A., (1989). The EPIC crop growth model. *Transactions of the American Society of Agricultural Engineers, 32*(2), 497–511.

89. Williams, J. R., (1995). The EPIC model. In: Singh, V. P., (ed.), *Computer Models of Watershed Hydrology* (pp. 909–1000). Colorado: Water Resources Publisher.

90. Xinchun, C., Wu, P., Wang, Y., & Zhao, X., (2014). Water footprint of grain product in irrigated farmland of China. *Water Resource Management, 28*, 2213–2227.

91. Yan, D., Wang, D., & Yang, L., (2007). Long-term effect of chemical fertilizer, straw, and manure on labile organic matter fractions in a paddy soil. *Biology and Fertility of Soils, 44*(1), 93–101.

92. Zhang, G. S., Zhang, X. X., & Hu, X. B., (2013). Runoff and soil erosion as affected by plastic mulch patterns in vegetable field at Dianchi Lake's catchment, China. *Agricultural Water Management, 122*, 20–27.

93. Zhang, H., Wang, X., You, M., & Liu, C., (1999). Water-yield relations and water-use efficiency of winter wheat in the North China Plain. *Irrigation Science*, *19*, 37–45.

94. Zhang, J., Sui, X., Li, B., Su, B., Li, J., & Zhou, D., (1998). An improved water-use efficiency for winter wheat grown under reduced irrigation. *Field Crops Research*, *59*(2), 91–98.

95. Zhao, J. H., Young, K. H., & Herrnstein, R. M., (2003). Variability of Sagittarius A*: Flares at 1 millimeter. *The Astrophysical Journal Letters*, *586*(1), L29–L31.

96. Zhou, G., Wei, X., Wu, Y., & Liu, S., (2011). Quantifying the hydrological responses to climate change in an intact forested small watershed in Southern China. *Global Change Biology*, *17*(12), 3736–3746.

97. Zhou, L., & Hoekstra, A. Y., (2017). The effect of different agricultural management practices on irrigation efficiency, water use efficiency, and green and blue water footprint. *Frontiers of Agricultural Science and Engineering*, *4*(2), 185–194.

98. Zwart, S. J., & Bastiaanssen, W. G. M., (2007). SEBAL for detecting spatial variation of water productivity and scope for improvement in eight irrigated wheat systems. *Agricultural Water Management*, *89*, 287–296.

99. Zwart, S. J., & Bastiaanssen, W. G. M., (2010). Global benchmark map of water productivity for rainfed and irrigated wheat. *Agricultural Water Management*, *97*, 1617–1627.

100. Zwart, S. J., & Bastiaanssen, W. G. M., (2010). WATPRO: A remote sensing-based model for mapping water productivity of wheat. *Agricultural Water Management*, *97*, 1628–1636.

CHAPTER 2

HEAVY METAL ACCUMULATION IN CROPS: STATUS, SOURCES, RISKS, AND MANAGEMENT STRATEGIES

MAYURI CHABUKDHARA, SANJAY K. GUPTA, and
ARVIND K. NEMA

ABSTRACT

This chapter assesses the current state of research in the field of heavy metals contaminations in soils, crops, and vegetables based on literature review. Potential sources of these metals, associated health risks, and remediation measures, have also been examined. Studies on heavy metals sources showed that in addition to natural sources, levels of heavy metals in soils, accumulations in crops and vegetables have increased due to mining and other related activities, industrial activities, wastewater (WW) irrigation, and applications of agrochemicals. Based on reported literature, health risks due to dietary intake of crops and vegetables grown around mining areas showed higher risks compared to others indicating that farming practices in and around may not be safe. Irrigation with WW may be considered after proper treatment to decrease the impact of heavy metals accumulations in soils and subsequently in crops and vegetables. Several research studies have been done in the laboratory or pilot scale to remediate heavy metal contaminated soils using different techniques. However, studies on successful applications in the field are very limited. More detailed research studies are needed on combinations of different remediation techniques to remove heavy metals from the soils. In addition, monitoring of metals in food crops should be performed regularly to ensure the quality of food items and to take up adequate mitigation measures.

2.1 INTRODUCTION

Food security is a worldwide issue and the provision of adequate and safe food is an effective health intervention. By 2050, around 60% more food would be needed to meet the demand by 9.5 billion global populations [1]. Food crops and vegetables are basic sources of nutrition and human health. They contain good amounts of proteins, fats, carbohydrates, vitamins, and minerals [55, 97, 117]. Human exposure to heavy metals via contaminated food chain is one of the most important and major pathways of exposure. Therefore, human health is threatened due to heavy metal contamination in food crops [11]. Metal contaminations have gained attention due to its toxicity, non-biodegradability, bioaccumulation, and bio-magnification capacity in the food chain [64].

Heavy metals can cause severe carcinogenic and non-carcinogenic health problems [22, 59, 66, 90, 132]; and Pb and Cd contaminated soil and vegetables can decrease our life expectancy [68]. Further, exposure to elevated levels of heavy metals in the soil, and consumption of metal-contaminated fruits and vegetables have caused the high occurrence of upper gastrointestinal cancer rates [57, 136]. Decreased immunity, malnutrition-related disabilities, intrauterine growth retardation, psychological dysfunctions, and gastrointestinal cancer are caused due to deficiency of various vital nutrients in the body caused by ingestion of heavy metals [136]. Increased awareness of the risk associated with metal contamination of the food chain has led to strict regulations of toxic metals in food items [111].

The contaminated food chain is among the dominant pathway of human exposure to heavy metals [27]. Bioaccumulation of metals in edible and non-edible parts of food crops through contaminated soil can affect food quality and safety [91]; and cause health problems to biota including humans [41, 140]. In China, metal pollution of foodstuffs caused a financial loss of around 3.2 billion US$ [70]. However, most part of the farmland in Europe can be considered safe for growing food crops in terms of heavy metals and only 6.24% of the land required local assessment and eventual remediation [135].

Because of rapid population and industrial growth, levels of heavy metals have increased in soils including agricultural or farmland soils; and thus, in crops and vegetables that are grown in such soils. Due to the risk associated with heavy metals and its exposure *via* consumption of food crops and vegetables, a better understanding of the accumulation of metals in different species of crops and vegetables and their potential sources will be required. This would also help to adopt the best appropriate agriculture practices.

This chapter discusses the status of heavy metal contaminations in food crops and vegetables based on a literature review, their potential sources, and associated health risks. This chapter also discusses various management strategies to reduce the impact of metal contamination in food crops.

2.2   HEAVY METALS IN AGRICULTURAL SOILS: STATUS AND SOURCES

Several studies on heavy metal contamination of agricultural soils have linked this issue to rapid urbanization, industrialization, and reuse of industrial and domestic wastewater (WW) for irrigation [70, 81, 91, 113]. In addition to the use of agrochemicals, atmospheric deposition and transportation emissions, mining activities, and WW irrigation are also main sources of heavy metal enrichment in soils [14, 93, 95, 138]. Multivariate analysis of agricultural soils in Spain (European Mediterranean) showed that parent rocks could be possible sources for Mn, Co, Cr, Fe, Ni, and Zn; while Cd, Cu, and Pb may have anthropogenic origin [88]. Peris et al. [104] found that Co, Fe, Mn, and Ni may have lithogenic origin while Cd, Cu, Pb, and Zn may have anthropic origin in Spain. In Southern Poland of Central Europe, Cd, and Pb in soils ranged between 0.5 to 68.5 and 3.41 to 2470 mg/kg, respectively [36]. Effects of pH, Fe content, soil extractable silicon on Cd and As availability in paddy fields was reported in Guangdong Province of China [150, 152].

Soil metal solubility and their accessibility by plants are influenced by pH value and the percentage of clay content in the case of Central Greece [44]. Cd greatly exceeded the guideline value in terraced paddy wetland soils in the Yunnan Plateau of China [13]. In the rice cropping system, Cd contamination in rice and soil is of serious concern [114]. Except for Si and Zr in paddy soils in the Khorat Basin of Thailand, other elements showed relatively lower concentrations [54]. Hotspots of pollution by heavy metals in the soil-rice system in eastern China were identified [72]. The contents of Cu, Zn, Ni, Cr, Pb, Cd, As, and Hg in agricultural soils of Guangdong in China ranged between 5.08–105.60, 8.43–169.50, 2.29–57.46, 20.36–137.20, 0.02–0.67, 0.40–28.87 and 0.01–1.01 mg/kg, respectively; and sources of Cr, Cu, Zn, Ni, and As were attributed to lithogenic origin but Pb, Hg, and Cd were of anthropic origin [74].

Concentrations of metals in agricultural soils reported around the world are shown in Table 2.1.

2.2.1 HEAVY METAL CONTAMINATIONS IN SOILS DUE TO WASTEWATER (WW) IRRIGATIONS AND AGROCHEMICALS

Intensive agricultural practices using agrochemicals and WW irrigation are one of the major sources of heavy metals in agricultural soils [24, 28, 107]. Industrial or municipal WWs are used for irrigation purposes due to ease of access and due to a shortage of freshwater [98]. According to Kachenko et al. [63], globally about 1500 Bm^3 of WW is generated and polluted water is being used for irrigation of around 20 million hectares (Mha) of agricultural land. Raw and untreated sewage is used for irrigating one-tenth of the world's irrigated crops [107]. These effluents lead to the accumulation of metals in waste irrigated and amended soils [129]. Increased level of metals (such as Fe, Zn, Ni, and Pb) was observed in soils receiving sewage irrigation for 10 years [113].

In the Varanasi city of India, heavy metals in soils ranged (mg/kg): Cu: 2.55–203.45, Cd: 0.55–8.85, Zn: 14.23–387.78, Cr: 13.40–679.89, Mn: 0.36–339.36, Pb: 0.46–44.50, and Ni: 2.0–34.45 during different months [125]. The Pb, Cu, and Ni exceeded the safe limits in irrigation WW in Titagarh, West Bengal-India [46]. In comparison to groundwater irrigated soils, irrigation with WW can cause the accumulation of toxic metals in soils [83]. Cu, Ni, Cd, and Zn exhibited pollution risk in the farmland of Southern Tehran in Iran due to the use of urban and industrial WW [51]. Levels of Cu, Zn, and Ni were high in the soils irrigated with treated WW from tanneries and the level of Cr was very high (30.5–311.5 mg/kg) in soil using tannery WW with Cr [130] (Figure 2.1).

FIGURE 2.1   Wastewater drains near agricultural sites in Ghaziabad city of India.

TABLE 2.1 Literature Review on Metals (in mg/kg) in Agricultural Soils in Different Regions Around the World

Soil	Heavy Metals, mg/kg									References
	Cu	Cr	Pb	Cd	Zn	Mn	Ni	As	Hg	
Alicante, Spain	23	27	23	0.3	53	295	21	—	—	[88]
Beijing, China	22.4	—	20.4	0.136	69.8	—	—	7.85	0.073	[130]
Durgapur, India (WW irrigated)	—	55	896	6	—	87	—	—	—	[47]
Enyigba, Nigeria	14.2–104.6	17–116	60.5–8324.7	0.1–2.5	56–778	231–2143	12.6–40.9	0.6–8.8	—	[94]
Ghaziabad, India	17.8	27.25	29.8	0.425	87	306.6	53.75	—	—	[27]
Kampala, Uganda			30–64.6	0.8–1.4	78.4–265.6					[92]
Malaysia	0.37–47.3	1.1–60.9	0.85–65.8	0.01–0.32	2.9–92.0		0.4–41.3	0.28–56.7	0.002–0.362	[154]
Taegu, Korea	63.5	—	146	2.31	393	—	—	64.4	—	[67]
Tehran, Iran	36.1	67.96	16.46	0.77	218	—	36.92	—	—	[51]
Titagarh, India	89.98	148.41	130.45	30.72	217.08	—	103.67	—	—	[46]
Varanasi, India	ND–1.54	BDL–0.82	BDL–0.37	BDL–0.23	BDL–1.25	BDL–1.08	BDL–0.37	—	—	[126]
Wuxi, China	40.4	58.6	46.7	0.14	112.9	—	—	14.3	—	[155]
Yangzhou, China	33.9	77.2	35.7	0.3	98.1	—	38.5	10.2	—	[53]

Regular application of large quantities of fertilizers in soils is a common practice to provide essential nutrients to enhance crop growth and yield [142]; and a significant amount of Hg, Cd, Pb, U, and Cr were found in final products of phosphate fertilizers that were ready for marketing [34]. Large accumulations of metals in the soil and plants were observed on the commercial farm compared to those in cooperative farms due to the continuing application of manure and fertilizers [101]. Soil heavy metal pollution due to the application of phosphate-based fertilizers causes serious concern [6]. Their applications caused increased levels of Cd, As, Cr, and Pb in soils and lowered the soil pH that caused increased soil metal leaching process [7]. In addition, several extensively used pesticides in the agricultural sector in the past decades contained a significant amount of metals [142].

The Cd, Cu, and Zn in soils were due to the application of fertilizers and manure in Shunyi, Beijing in China [79]. High Cd and As concentration is attributed to over-application of manure, herbicide, pesticide, and phosphate fertilizers [60, 96]. Cd speciation and complexation is influenced by fertilizer applications that affect the Cd transport and Cd uptake in plants [141]. Cd level in Kermanshah of Iran was 1.39 ± 0.09 mg/kg of soil and 1.57 ± 0.09 mg/kg of soil prior to fertilization and post-harvesting period, respectively [10]. In the United Kingdom, 10% of the used insecticides and fungicides contained Cu, Pb, Hg, Mn, and Zn [143].

2.2.2   HEAVY METALS IN SOILS: MINING OR SMELTING ACTIVITIES

Mining and non-ferrous metals smelting activities are one of the most significant sources of worldwide pollution by heavy metals [3, 12, 16, 31, 73, 115, 133]. Near a Pb-smelter in Lastenia of Argentina, elevated levels of Pb, Cd, Cu, and Zn were reported in urban soils [40]. Similarly, elevated levels of Cd, Cu, Pb, and Zn were found near a Cu-smelting plant of Montana, USA [23]. In another study, high levels of metals were observed in the mine dumpsites of a Cu-W Mine in Korea, with an average of 1.95, 419, 4.4, and 1030 mg/kg of soil for Cu, Zn, Cd, and Pb, respectively [61]. Smelting activities had led to significant contamination of local soils with Cd (3.88 mg/kg) and Zn (403.9 mg/kg) near a zinc-smelter in the northeast part of China [69].

Around Nanning smelter of China, Cd, and Pb were found in soils and vegetables [31]. The average As and Sb contents in soils around the abandoned mining area in ZlataIdka of Slovakia were 892 mg/kg and 818 mg/kg, respectively [112]. The paddy soil was highly contaminated with Cu, Zn, Pb, and Cd around Dabaoshan mine in South China [158]. Mining activities in Korea showed relatively higher levels of As in paddy soil and polished rice than the worldwide average and it was concluded that consumption of contaminated rice can pose health threats to local residents [67].

In another study, Cd concentrations of soils exceeded the safe limit by 8 to 20 times [124] near Daye smelter in Huangshi of Central China while Cu and Pb were a little higher than the safe limit near the smelter [144]. Similarly, in agricultural soils of Northern France (previously mined area), Pb was 15 times higher, and Cd was 20 times higher than the agricultural reference values, respectively [106]. In soils of China (Hunan Province), Pb, Cd, and Zn contents were approximately 3.7, 7.9, and 1.6 times, respectively, and higher than the corresponding background values [78].

Similarly, very high levels of heavy metals were found in agricultural soils close to non-ferrous metal mine area in Jiangxi Province of China; and the average level of metals (mg/kg) were [76]: As (33.99), Cd (1.22), Cr (70.28), Cu (138.4), Mn (468.7), Ni (32.2), Pb (125.3) and Zn (171.48), respectively. The highest levels of metals levels near the metal mining and smelting operations in South Africa (mg/kg) were [37]: 1450 for As, 8980 for Cu, 4640 for Pb kg, and 2620 for Zn [37]. Near Pb-Zn mines in Enyigba of Nigeria, in 87% of the soil samples, Pb exceeded the Canadian Environmental Quality Guideline (CCME) value and European Union (EU) level, while Zn exceeded in 31% of the samples set by CCME [94].

2.2.3   HEAVY METALS IN AGRICULTURAL SOILS DUE TO INDUSTRIES OR TRAFFIC

Li et al. [70] assessed levels of metal contamination in agricultural soils near a petrochemical complex to examine the effect of industrialization on these soils. Levels of Pb, Cu, Ni, and Cr in soil (mg/kg of soil) were: 73.6, 11.8, 58.9, and 158.6 mg/kg in soils of upper Assam and were mostly influenced by oil drilling sites and refineries [121]. Pb and Cd contents in urban soils ranged between 20.1–96.2 and 0.20–0.95 mg/kg in gardens of Salamanca in Spain [119]. In this study, Pb levels were related to the traffic density and Cd was related to the distance from roadways and age of the garden. Hg contamination in agricultural soils in Shunyi, Beijing of China was attributed to atmospheric deposition from Beijing city [79].

The concentrations of metals in soil was related to its distance from roads along highways in Kampala City of Uganda and thus leafy vegetables were recommended to be grown at least 30 meters away from roads in high-traffic urban areas [92]. When compared to residential areas with low traffic load, more industrialized or commercialized areas with high traffic load showed a higher deposition rate of metals, such as Cu, Zn, and Cd [126]. The study carried out in Ghaziabad of India showed that in addition to soil metal uptake by plants, the atmospheric deposition could be an important source of elevated concentrations of metals in vegetation [27].

2.3 ACCUMULATION OF HEAVY METALS IN AGRICULTURAL CROPS

Cd level in rice was as high as 4.4 mg/kg near the mine in Chenzhou city of China [75]. In urban gardens of Salamanca in Spain, the vegetables showed contaminations by Pb (4.17 and 52.7 mg/kg) and Cd (0.35 and 3.05 mg/kg) [119]. The vegetables collected from the industrial area in Lagos of Nigeria also showed contamination due to heavy metals, such as Cd, Cu, and Ni [153]. The 95.8%, 68.8%, 10.4%, and 95.8% of vegetable samples from agricultural soils in the Hunan Province of China exceeded the maximum safe limit for As, Cd, Ni, and Pb, respectively [78]. Prolonged use of WW caused high levels of heavy metals in vegetables [125, 127].

Arora et al. [9] assessed the levels of four metals in vegetables irrigated with water from different sources and substantial build-up of heavy metals was observed in WW irrigated vegetables. The concentrations of Pb, Zn, Cd, Cr, and Ni in all WW irrigated vegetables were beyond the safe limits in Tatagarh, West Bengal of India [46]. Plants grown in WW-irrigated soils exceeded the safe limits of State Environmental Protection Administration (SEPA), = of China, and the WHO; and showed significantly higher accumulation than those grown in the reference soils [64]. In Peshawar of Pakistan, the effects of industrial effluents were observed and concentration of metals in WW irrigated food-crops showed very high accumulations [58].

In WW irrigated sites in tropical urban areas of India, Cd, Pb, and Ni values exceeded the "safe limits" of Indian and WHO/FAO standards of all vegetables and cereals under study [128]. Further, metal accumulation in WW irrigated vegetables was higher than the control in Durgapur industrial area-West Bengal of India [47]. The Pb, Cd, Zn, and Ni in vegetables were several folds higher than the safe limits in and around Ghaziabad industrial area of India [27]. Vegetables grown in urban and peri-urban sites of Ghazi-abad district are shown in Figure 2.2.

FIGURE 2.2   Vegetables grown in (A) urban and (B) peri-urban sites in Ghaziabad district.

Leafy vegetables can accumulate more heavy metals from the soil compared to tubers or bulbs, etc., [83] and can cause a health risk to consumers [2]. Based on a study on metal accumulation in wheat crops in Pakistan, increased levels of metals in crops in less industrialized areas suggested the geological origin of these metals [8]. Further, the roots showed higher accumulation of metals than the shoots; and the seeds and grains showed the minimum levels [8]. In a similar study, wild vegetables showed sufficient nutrients necessary for good health and bioaccumulation of Cr, Pb, and Ni in Bangladesh; and these are considered toxic in nature [123].

Yusuf et al. [153] concluded that vegetables collected from industrial areas showed higher metal accumulation compared to those collected from residential areas (Lagos of Nigeria). Mapanda et al. [85] reported that the highest levels of heavy metals (mg/kg) in vegetable leaves in Harare of Zimbabwe were: 3.4 for Cu, 201 for Zn, 2.4 for Cd, 6.3 for Ni, 5.4 for Pb and 6.6 for Cr, respectively. Improved food quality assurance system is prereq-uisite in terms of heavy metal contamination [85]. Elevated concentrations of metals were found in corn plants in zinc-smelting areas of Guizhou in China; and Pb and Cd in corn plants exceeded the national safe limits either in totally or partially [21]. Mean As-concentration was highest in potato in Arsenic-affected area of West Bengal in India [20].

Metal accumulations in plants may be influenced by several physico-chemical and biological factors. Based on a study near Pb-Zn mine, total concentrations of metals in soil and the pH influenced the metal content in plants [62]. Factors causing metals accumulation in vegetables are: Heavy metal concentrations in soil, climate, and deposition from atmosphere, soil nature, and maturity level of the plants at the harvesting time [5, 19]. Uptake by roots from the soil and atmospheric deposition in vegetable leaves are dominant pathways for the accumulation of trace elements in plants [137]. Atmospheric deposition is the main source of most heavy metals in the leaves of vegetables in the dry tropicals region of India [99] (Table 2.2).

2.4 HEALTH RISKS DUE TO CONSUMPTION OF METAL-CONTAMINATED CROPS AND VEGETABLES

Assessment studies on the consumption of metal-contaminated crops grown on metal-contaminated agricultural soils may pose health threats and health risks have been indicated in the literature. In Varanasi, Sharma et al. [126] reported health threats due to Cu and Cd contamination of vegetables sold in local markets. Health risk study in terms of values of target hazard quotient (THQ) in an experimental farm in Dubai showed that Cu and Zn were significantly lower than the safe level (THQ <1.0), while Cr exceeded it in all vegetables with lettuce showing far higher (THQ = 11.5) value [110].

Health risk index (HRI) indicated that spinach and Brassica posed a grave health risk due to Cd and Mn contamination [83]. Song et al. [131] found that As posed the highest risk to local inhabitants in Beijing of China; and the daily intake (DI) of As was 0.08 μg/kg body weight/day for adults. The THQ of rice around the Dabaoshan mine of China varied from 1.43–1.99 for Pb, 0.48–0.60 for Zn, 0.66–0.89 for Cu, and 2.61–6.25 for Cd [158].

Singh et al. [128] showed health risk for the local population of Varanasi in India due to the consumption of vegetables contaminated with Cd, Pb, and Ni. Chary et al. [28] found that HQ (hazard quotient) was high for Zn in leafy vegetables, such as spinach, and Amaranthus. Consumption of market vegetables contaminated with Pb and Cd can pose a health risk to the local inhabitants in terms of HRI values in Ghaziabad [26]. Hazard index (HI) were above the safe limit (HI>1) due to contamination of soil and rice in Changshu of east China [50]. Health risks were observed due to the consumption of rice contaminated with Cd in Youxian, Hunan of China [139], and near mining area [74].

Heavy metal contamination in soils and food crops irrigated with WW exceeded the safe limits set by SEPA and WHO, in Beijing of China; and thus, the health risk was increased [64]. Pb showed HRI > 1 for both adults and children in Nigeria due to the contamination of cassava tuber and lemongrass [94].

TABLE 2.2 Literature Review on Heavy Metals in Cereal Crops and Vegetables Around the World

Cereal	Cu	Cr	Pb	Cd	Zn	Ni	As	References
Amaranthus, India	1.4	2.4	2.9	—	8	3.1	—	[28]
Brinjal, India	0.7	1.1	3	—	4.5	3.1	—	[28]
Brinjal, Malaysia	0.82	0.025	0.007	0.021	1.7	0.13	0.089	[154]
Cabbage, China	0.53	0.05	0.08	0.03	2	0.04	0.12	[53]
Cabbage, Malaysia	3.6	0.06	0.021	0.013	3	0.12	0.079	[154]
Carrot, Egypt	1.5	—	0.18	0.01	8.03	—	—	[111]
Cauliflower, India	35.7	—	1.56	2.57	63.63	—	—	[127]
Cauliflower, India	15.7	86.8	31.04	13.8	96.5	59.3	—	[46]
Lady's finger, Malaysia	0.79	0.039	0.0061	0.026	5.6	0.18	0.12	[154]
Potato, Pakistan	0.10	0.15	0.16	0.08	—	0.8	—	[102]
Potato, Egypt	0.83	—	0.08	0.02	7.16	—	—	[111]
Potato, Greece	0.1–0.14	0.09–0.11	0.05–0.08	0.003–0.013	0.18–0.22	0.06–0.1	—	[44]
Raddish, Pakistan	2.53–8.72	1.36–2.44	0.26–0.65	0.25–1.30	8.45–37	0.47–1.31	—	[149]
Radish, India	5.96	—	—	—	22.5	—	—	[9]
Rice, China	28.1	78	57.6	17.8	139.1	62.7	—	[46]
Rice, China	3.3 (2.8–4.5)	199 (62–424)	355 (167–745)	34.5 (3.6–69.7)		476 (201–818)	—	[41]
	4.9	0.8	2.5	0.03	32	—	—	[77]
Rice, Korea	4.69	—	0.804	0.174	16.8	—	0.247	[67]
Rice, Malaysia	1.9	37	0.24	0.011	42	1.1	1.27	[154]

TABLE 2.2 *(Continued)*

Cereal	Cu	Cr	Pb	Cd	Zn	Ni	As	References
Rice, Vietnam	2.6	—	—	—	—	869	—	[105]
Sorghum grain, Hyderabad, India	2.2	0.21	2.0	0.38	47	1.5	0.14	[57]
Spinach, China	17.8	4.11	2.96	0.4	69.3	—	—	[77]
Spinach, Egypt	4.5	—	0.34	0.11	20.89	—	—	[111]
Spinach, India	27.6	—	1.44	1.96	57.56	—	—	[127]
	16.5	—	—	—	33.1	—	—	[9]
	34.5	96.3	49.8	14.6	154.2	69.2	—	[46]
	0.09	2.9	3.1	—	10	3.2	—	[28]
Spinach, Malaysia	0.59	0.17	0.034	0.023	9.6	0.17	0.081	[154]
Spinach, Pakistan	4.99–9.1	0.48–1.72	0.27–1.44	0.28–0.81	8.4–26.5	0.58–1.25	—	[149]
Vegetables								
Wheat, Yangzhong, China	4.1	0.24	0.21	0.06	22.4	0.15	0.15	[53]

Ni, Pb, Cd, and Cr contents in most of the sampled vegetables in Saudi Arabia were above the safe limits [15]. The total non-carcinogenic risks (HI) from all vegetables were less than the guideline value (= 1.00) in the industrial and urban city of Gujranwala in Pakistan [149]. The order of observed values of HI was *Brassica oleracea* < *Solanum tuberosum* > *Raphanus sativus* > *Benincasa hispida* > *Daucus carota* > *Spinecia oleracea* > *Brassica rapa* [149].

2.5 MANAGEMENT STRATEGIES AND CHALLENGES OF HEAVY METAL CONTAMINATION IN SOILS AND AGRICULTURAL CROPS

There has been growing attention on research related to environmental clean-up of elements through efficient remediation techniques [48]. Both source control and enhanced remediation of metal-contaminated soil is required [156]. Huge investments have been made to remediate the contaminated soils in European countries [108]. Among different remediation measures, soil replacement as a physical remediation process involves the dilution of the pollutant concentration by replacing the polluted soil with clean soil [145]; and vegetable and soil quality was improved after soil replacement [35].

The use of soil amendments have gained wide attention due to their cost-effectiveness and impact and they could reduce trace element mobility through direct or indirect adsorption processes and by precipitation of their different forms, such as oxides, carbonates, hydroxides, hydroxy carbonates or phosphates [1]. Many inorganic and organic products are being investigated for its potential to be used as amendments for restricting the trace elements mobility in polluted soils [1, 18, 39, 52, 71, 103, 109].

Soil washing technologies using different agents have also been used for the elimination of heavy metals. Among these, several biosurfactants (microbial surface-active metabolites) have been effective and potential alternative agent for treating soils contaminated with metals [122] due to many advantages associated with these agents, such as less toxicity [120]; cost-effectiveness [30, 87]; eco-friendly and biodegradable nature, and a high ability to bind metal ions [29, 42, 84]. The 37% Cu, 7.5% Zn, and 33.2% Ni were removed using 5% of rhamnolipid [32]. *Candida sphaerica* derived biosurfactant could remove Fe (95%), Zn (90%) and Pb (79%) [82].

Acidic solution or chelating agents that are inorganic in nature can remove certain metals *via* mass transfer into aqueous phase by desorption, extraction, and dissolution from soils [33, 147, 148]. Co-contamination of soils with petrochemicals and heavy metals can be remediated by using oxidation and extraction processes and thus can be used as a rapid and cost-effective method [146].

Phytoremediation plays a significant role in the removal of low-to-moderate levels of heavy metals in soils and is considered to be an attractive, eco-friendly, economically viable, and energy-efficient, aesthetically pleasing, non-invasive technology [116]. Different categories of phytoremediation based on uptake mechanisms are phytoextraction, phytostabilization, and phytoevaporation. In phytoextraction, plants uptake heavy metals from soil. The success of the phytoextraction technique as a potential environmental cleaning method depends on the availability of metals and the absorption and accumulation capacity of plants [19].

The bioavailability and mobility of heavy metals in soils are decreased in the phytostabilization process [134], while phytovolatilization involves the conversion of heavy metals to vapors and its subsequent release into the atmosphere through transpiration process of plants [86, 118]. Plant-based genetic engineering has been in use for increasing the metal phytoextraction potential through a huge increase in absorption potential, xylem loading, and translocation to aerial shoots and increased tolerance [17, 43, 45, 65].

"IAPAR 61" black oat cultivar was efficient Cd, Cr, and Pb accumulator and addition of vermicompost improved the growth of black oat plants, but only the treatment with 25% of vermicompost showed promising results in the absorption of Cr and Pb, and the treatment with 50% in the absorption of Cd [49]. Based on a study on heavy metal-contaminated soils, nano-calcium silicate (NCS) was a suitable treatment for the gentle restriction of heavy metals mobilities in polluted soils [89].

Biochar has great potential for immobilizing heavy metals in soil [4], because of its highly porous structure, active functional groups, and generally high pH and cation exchange capacity [100]. The use of biochar can restrict the mobility of heavy metals to shoots in paddy plants and thus can reduce metal accumulations in rice [157]. Experiments carried out in pots showed that rice straw-based biochar was more efficient than bamboo-based biochar in decreasing the acid extractable form of metals, and the effect was more prominent with increased application of biochar to immobilize Pb, Cd, Cu, and Zn in soils [80].

In a study, greenhouse cultivation could decrease bioaccumulation in plants and the metal concentration thus could reduce associated health risks from edible vegetables [25]. Iron biogeochemistry processes (such as Fe^{2+} catalyzed recrystallization of iron (hydro) oxides, dissimilatory iron reduction, and Fe (II) oxidation) contribute to the immobilization of heavy metals [151].

2.6 SUMMARY

Due to rapid population growth, industrialization, and urbanization, heavy metals contaminations of the environment including soils have gained serious attention. In this chapter, the authors reviewed the status of heavy metals contaminations in (i) soils, crops, and vegetables; and (ii) their sources and associated risks. Studies on anthropogenic sources of heavy metals in soils (such as mining or smelting activities, WW irrigation, agrochemicals, industrial activities, and traffic) have been discussed. Increased heavy metals accumulation has been reported in vegetables irrigated with treated or untreated WW, and around mining areas. A review of studies on health risk assessment showed a relatively higher risk due to Pb and Cd contaminations in crops and vegetables. This chapter also summarizes various recent management and remediation measures to reduce the levels and impact of heavy metals contamination in soils. The results summarized in this chapter may provide insights to understand the importance of dietary exposure of heavy metals and to reduce contamination of soils, crops, and vegetables due to heavy metal contamination and their associated risk.

KEYWORDS

- agricultural soils
- anthropogenic sources
- heavy metals
- mining
- remediation
- wastewater irrigation

REFERENCES

1. Abad-Valle, P., Álvarez-Ayuso, E., Murciego, A., & Pellitero, E., (2016). Assessment of the use of sepiolite amendment to restore heavy metal polluted mine soil. *Geoderma*, *280*, 57–66.
2. Abdullahi, M. S., (2015). Soil contamination, remediation, and plants: Prospects and challenges. *Soil Remediation and Plants*, 525–546.
3. Adriano, D. C., (2001). *Trace Elements in Terrestrial Environments: Biogeochemistry, Bioavailability and Risks of Metals* (2nd edn., p. 317). New York: Springer-Verlag.

4. Ahmad, M., Rajapaksha, A. U., & Lim, J. E., (2014). Biochar as a sorbent for contaminant management in soil and water: A review. *Chemosphere, 99*, 19–33.

5. Ali, H., Khan, E., & Sajad, M. A., (2013). Phytoremediation of heavy metals: Concepts and applications. *Chemosphere, 91*, 869–881.

6. Alloway, B. J., (1990). *Heavy Metals in Soils* (p. 315). New York: John Wiley & Sons.

7. Alloway, B. J., (1995). *Soil Processes and the Behavior of Metals* (p. 238). New York: John Wiley & Sons.

8. Al-Othman, Z. A., Ali, R., Al-Othman, A. M., Ali, J., & Habila, M. A., (2013). Assessment of toxic metals in wheat crops grown on selected soils, irrigated by different water sources. *Arabian Journal of Chemistry*. Online: http://dx.doi.org/10.1016/j.arabjc.2012.04.006 (accessed on 20 June 2020).

9. Arora, M., Bala, K., Rani, S., Rani, A., Kaur, B., & Mittal, N., (2008). Heavy metal accumulation in vegetables irrigated with different water sources. *Food Chemistry, 111*(4), 811–815.

10. Atafar, Z., Mesdaghinia, A., Nouri, J., & Homaee, M., (2010). Effect of fertilizer application on soil heavy metal concentration. *Environmental Monitoring Assessment, 60*, 83–89.

11. Babu, A. G., Kim, J. D., & Oh, B. T., (2013). Enhancement of heavy metal phytoremediation by *Alnus firma* with endophytic *Bacillus thuringiensis* GDB-1. *Journal of Hazardous Materials, 2013*, 477–483.

12. Bacon, J. R., & Dinev, N. S., (2005). Isotopic characterization of lead in contaminated soils from the vicinity of a non-ferrous metal smelter near Plovdiv, Bulgaria. *Environmental Pollution, 134*, 247–255.

13. Bai, J., Xiao, R., Gong, A., Gao, H., & Huang, L., (2011). Assessment of heavy metal contamination of surface soils from typical paddy terrace wetlands on the Yunnan Plateau of China. *Physics and Chemistry of Earth, 36*, 447–450.

14. Bai, J., Zeng, X. B., Su, S. M., Duan, R., Wang, Y. N., & Gao, X., (2015). Heavy metal accumulation and source analysis in greenhouse soils of Gansu Province, China. *Environmental Science and Pollution Research, 22*, 5359–5369.

15. Balkhair, K. S., & Ashraf, M. A., (2016). Field accumulation risks of heavy metals in soil and vegetable crop irrigated with sewage water in western region of Saudi Arabia. *Saudi Journal of Biological Sciences, 23*, S32–S44.

16. Barcan, V., (2002). Nature and origin of multicomponent aerial emission of the copper-nickel smelter complex. *Environment International, 28*, 451–456.

17. Barrameda-Medina, Y., & Montesinos-Pereira, D., (2014). Comparative study of the toxic effect of Zn in *Lactuca sativa* and *Brassica oleracea* plants, part I: Growth, distribution, and accumulation of Zn and metabolism of carboxylates. *Environmental and Experimental Botany, 107*, 98–104.

18. Basta, N. T., & McGowen, S. L., (2004). Evaluation of chemical immobilization treatments for reducing heavy metal transport in a smelter-contaminated soil. *Environmental Pollution, 127*, 73–82.

19. Bhargava, A., Carmona, F. F., Bhagava, M., & Srivastava, S., (2012). Approaches for enhanced phytoextraction of heavy metals. *Journal of Environmental Management, 105*, 103–120.

20. Bhattacharya, P., Samal, A. C., Majumdar, J., & Santra, S. C., (2010). Arsenic contamination in rice, wheat, pulses, and vegetables: Study in an arsenic affected area of West Bengal, India. *Water Air Soil Pollution*. Online: doi: 10.1007/s11270-010-0361-9.

21. Bi, X., Feng, X., Yang, Y., Qui, G., & Li, G., (2006). Environmental contamination of heavy metals from zinc smelting areas in Hezhang County of western Guizhou, China. *Environment International*, *32*, 883–890.

22. Brewer, G. J., (2010). Copper toxicity in the general population. *Clinical Neurophysiology*, *121*, 459–460.

23. Burt, R., Wilson, M. A., Keck, T. J., & Dougherty, B. D., (2003). Trace element speciation in selected smelter-contaminated soils in Anaconda and Deer Lodge Valley, Montana, USA. *Advances in Environmental Research*, *8*, 51–67.

24. Cai, Q., Long, M. L., Zhu, M., Zhou, Q. Z., Zhang, L., & Liu, J., (2009). Food chain transfer of cadmium and lead to cattle in a lead-zinc smelter in Guizhou, China. *Environmental Pollution, 157*, 3078–3082.

25. Cao, C., Chen, X. P., Ma, Z. B., Jia, H. H., & Wang, J. J., (2016). Greenhouse cultivation mitigates metal-ingestion-associated health risks from vegetables in wastewater-irrigated agro-ecosystems. *Science of the Total Environment, 560*, 204–211.

26. Chabukdhara, M., Nema, A. K., & Gupta, S. K., (2012). Metal contamination in market-based vegetables in an industrial region, India. *Bulletin of Environmental Contamination and Toxicology*, *89*, 129–132.

27. Chabukdhara, M., Munjal, A., Nema, A. K., Gupta, S. K., & Kaushal, R. K., (2016). Heavy metal contamination in vegetables grown around peri-urban and urban-industrial clusters in Ghaziabad, India. *Human and Ecological Risk Assessment, 22*, 736–752.

28. Chary, N. S., Kamala, C. T., & Samuel, S. R. D., (2008). Assessing risk of heavy metals from consuming food grown on sewage irrigated soils and food chain transfer. *Ecotoxicology Environmental Safety*, *69*, 513–524.

29. Chen, W. J., Hsiao, L. C., & Chen, K. K. Y., (2008). Metal desorption from copper/nickel spiked kaolin as a soil component using plant-derived saponic biosurfactant. *Process Biochemistry, 43*, 488–498.

30. Chrzanowski, L., Lawniczak, L., & Czaczyk, K., (2012). Why do microorganisms produce rhamno lipids? *World Journal of Microbiology and Biotechnology, 28*, 401–419.

31. Cui, Y., Zhu, Y., Zhai, R., Chen, D., Huang, Y., & Qiu, Y., (2004). Transfer of metals from soil to vegetables in an area near a smelter in Nanning, China. *Environment International*, *30*, 785–791.

32. Dahrazma, B., & Mulligan, C. N., (2007). Investigation of the removal of heavy metals from sediments using rhamnolipid in a continuous flow configuration. *Chemosphere, 69*, 705–711.

33. Dermont, G., Bergeron, M., & Mercier, G., (2008). Soil washing for metal removal: A review of physical/chemical technologies and field applications. *Journal of Hazardous Materials*, *152*, 1–31.

34. Dissanayake, C. B., & Chandrajith, R., (2009). Phosphate mineral fertilizers, trace metals and human health. *Journal of National Science Foundation Sri Lanka, 37*, 153–165.

35. Douay, F., Pruvot, C., & Roussel, H., (2008). Contamination of urban soils in an area of Northern France polluted by dust emissions of two smelters. *Water Air Soil Pollution, 188*, 247–260.

36. Dziubanek, G., Piekut, A., Rusin, M., Baranowska, R., & Hajok, I., (2015). Contamination of food crops grown on soils with elevated heavy metals content. *Ecotoxicology and Environmental Safety, 118*, 183–189.

37. Ettler, V., Johan, Z., Křibek, B., & Veselovský, F., (2016). Composition and fate of mine-and smelter-derived particles in soils of humid subtropical and hot semi-arid areas. *Science of the Total Environment, 563–564*, 329–339.

38. FAO, (2013). *Statistical Yearbook: World Food and Agriculture* (p. 289). Food and Agriculture Organization of the United Nations (FAO), Rome. http://www.fao.org/docrep/018/i3107e/i3107e00.htm (accessed on 20 June 2020).

39. Farrell, M., & Jones, D. L., (2010). Use of composts in the remediation of heavy metal contaminated soil. *Journal of Hazardous Materials, 175*, 575–582.

40. Fernandez-Turiel, J. L., & Aceñolaza, P., (2001). Assessment of a smelter impact area using surface soils and plants. *Environmental Geochemistry and Health, 23*, 65–78.

41. Fu, J. J., Zhou, Q. F., Liu, J. M., Liu, W., Wang, T., Zhang, Q. H., & Jiang, G. B., (2008). High levels of heavy metals in rice (*Oryza sativa L.*) from a typical E-waste recycling area in southeast China and its potential risk to human health. *Chemosphere, 71*, 1269–1275.

42. Gao, L., Kano, N., Sato, Y., Li, C., Zhang, S., & Imaizumi, H., (2012). Behavior and distribution of heavy metals including rare earth elements, thorium, and uranium in sludge from industry water treatment plant and recovery method of metals by biosurfactants application. *Bioinorganic Chemistry and Applications, 10*, 1687–2479.

43. Gisbert, C., Ros, R., & De Haro, A., (2003). Plant genetically modified that accumulates Pb is especially promising for phytoremediation. *Biochemical and Biophysical Research Communications, 303*, 440–445.

44. Golia, E. E., Dimirkou, A., & Mitsios, I. K., (2008). Influence of some soil parameters on heavy metals accumulation by vegetables grown in agricultural soils of different soil orders. *Bulletin of Environmental Contamination and Toxicology, 81*, 80–84.

45. Guo, G., Wu, F., Xie, F., & Zhang, R., (2012). Spatial distribution and pollution assessment of heavy metals in urban soils from southwest China. *Journal of Environmental Science, 24*, 410–418.

46. Gupta, N., Khan, D. K., & Santra, S. C., (2008). Assessment of heavy metal contamination in vegetables grown in wastewater-irrigated areas of Titagarh, West Bengal, India. *Bulletin of Environmental Contamination Toxicology, 80*, 115–118.

47. Gupta, S., Satpati, S., Nayek, S., & Garai, D., (2010). Effect of wastewater irrigation on vegetables in relation to bioaccumulation of heavy metals and biochemical changes. *Environmental Monitoring Assessment, 165,* 169–177.

48. Gusiatin, Z. M., & Klimiuk, E., (2012). Metal (Cu, Cd, and Zn) removal and stabilization during multiple soil washing by saponin. *Chemosphere, 86*, 383–391.

49. Hoehne, L., De Lima, C. V. S., & Martini, M. C., (2016). Addition of vermicompost to heavy metal-contaminated soil increases the ability of black oat (*Avenastrigosa Schreb*) plants to remove Cd, Cr, and Pb. *Water Air Soil Pollution*, 227–443.

50. Hang, X., Wang, H., Zhou, J., Ma, C., Du, C., & Chen, X., (2009). Risk assessment of potentially toxic element pollution in soils and rice (*Oryza sativa*) in a typical area of the Yangtze River Delta. *Environmental Pollution, 157*, 2542–2549.

51. Hani, A., & Pazira, E., (2011). Heavy metals assessment and identification of their sources in agricultural soils of Southern Tehran, Iran. *Environmental Monitoring and Assessment, 176*(1–4), 677–691.

52. Huang, M., Zhu, Y., Li, Z., Huang, B., Luo, N., Liu, C., & Zeng, G., (2016). Compost as a soil amendment to remediate heavy metal-contaminated agricultural soil: Mechanisms, efficacy, problems, and strategies. *Water Air Soil Pollution, 227*, 359–366.

53. Huang, S. S., Liao, Q. L., Hua, M., & Wu, X. M., (2007). Survey of heavy metal pollution and assessment of agricultural soil in Yangzhong district, of Jiangsu province. China. *Chemosphere, 67*, 2148–2155.

54. Inboonchuay, T., & Suddhiprakarn, A., (2016). Amounts and associations of heavy metals in paddy soils of the Khorat Basin in Thailand. *Geoderma, 7*, 120–131.

55. Ivey, M. L., Lejeune, J. T., & Miller, S. A., (2012). Vegetable producers' perceptions of food safety hazards in the Midwestern USA. *Food Control, 26*, 453–465.

56. Iyengar, V., & Nair, P., (2000). Global outlook on nutrition and the environment: Meeting the challenges of the next millennium. *Science of the Total Environment, 249*, 331–346.

57. Jamali, M. K., Kazi, T. G., Arain, M. B., Afridi, H. I., Jalbani, N., Memon, A. R., & Shah, A., (2007). Heavy metals from soil and domestic sewage sludge and their transfer to Sorghum plants. *Environmental Chemistry Letter, 5*, 209–218.

58. Jan, F. A., Ishaq, M., Khan, S., & Ihsanullah, I., (2010). Comparative study of human health risks via consumption of food crops grown on wastewater irrigated soil (Peshawar) and relatively clean water irrigated soil. *Journal of Hazardous Materials, 179*, 612–621.

59. Jarup, L., (2003). Hazards of heavy metal contamination. *Brazilian Medical Bulletin, 68*, 167–182.

60. Ju, X. T., Kou, C. L., Christie, P., Dou, Z. X., & Zhang, F. S., (2007). Changes in the soil environment from excessive application of fertilizers and manures to two contrasting intensive cropping systems on the North China Plain. *Environmental Pollution, 145*, 497–506. Online: doi: 10.1016/j.envpol.2006.04.017.

61. Jung, M. C., (2008). Heavy metal concentrations in soils and factors affecting metal uptake by plants near a Korean Cu-W Mine. *Sensors, 8*, 2413–2423.

62. Jung, M. C., & Thornton, I., (1996). Heavy metal contamination of soils and plants near lead-zinc mine, Korea. *Applied Geochemistry, 11*, 53–59.

63. Kachenko, A. G., & Singh, B., (2006). Heavy metals contamination in vegetables grown in urban and metal smelter contaminated sites in Australia. *Water Air Soil Pollution, 169*, 101–123.

64. Khan, S., Cao, Q., Zheng, Y. M., Huang, Y. Z., & Zhu, Y. G., (2008). Health risk of heavy metals in contaminated soils and food crops irrigated with wastewater in Beijing, China. *Environmental Pollution, 152*, 686–692.

65. Khoudi, H., Maatar, Y., Brini, F., & Fourati, A., (2013). Phytoremediation potential of *Arabidopsis thaliana*, expressing ectopically a vacuolar proton pump, for the industrial waste phosphor-gypsum. *Environmental Science and Pollution Research, 20*, 270–280.

66. Knight, C., Kaiser, G. C., Robothum, L. H., & Witter, J. V., (1997). Heavy metals in surface water and stream sediments in Jamaica. *Environmental Geochemistry and Health, 19*, 63–66.

67. Kwon, J. C., Nejad, J. D., & Jung, M. C., (2017). Arsenic and heavy metals in paddy soil and polished rice contaminated by mining activities in Korea. *Catena, 148*, 92–100.

68. Lăcătuşu, R., Răuţă, C., Cârstea, S., & Ghelase, I., (1996). Soil-plant-man relationships in heavy metal polluted areas in Romania. *Applied Geochemistry, 11*(1/2), 105–107.

69. Li, B., Wang, Y., Jiang, Y., Li, G., Cui, J., Wang, Y., Zhang, H., et al., (2016). The accumulation and health risk of heavy metals in vegetables around a zinc smelter in northeastern China. *Environmental Science and Pollution Research* (p. 8). Online: doi: 10.1007/s11356-016-7342-5.

70. Li, G., Qiu, J., & Yin, C., (2009). Study on calculating losses of cropland degradation. *Chinese Agricultural Science Bulletin, 25*, 230–235 (in Mandarin).

71. Li, J., & Xu, Y., (2015). Immobilization of Cd in a paddy soil using moisture management and amendment. *Chemosphere, 122*, 131–136.

72. Li, W., Xu, B., Song, Q., Liu, X., Xu, J., & Brookes, P. C., (2014). The identification of hotspots of heavy metal pollution in soil-rice systems at a regional scale in eastern China. *Science of the Total Environment, 472*, 407–420.

73. Lim, H. S., Lee, J. S., Chon, H. T., & Sager, M., (2008). Heavy metal contamination and health risk assessment near the abandoned Songcheon Au-Ag mine in Korea. *Journal of Geochemical Exploration, 96*, 223–230.

74. Limei, C., Lanchun, H., & Yongzhang, Z., (2010). Heavy metal concentrations of agricultural soils and vegetables from Dongguan, Guangdong. *Journal of Geographical Science, 20*(1), 121–134.

75. Limei, Z., Xiaoyong, L., Tongbin, C., & Xiulan, Y., (2008). Regional assessment of cadmium pollution in agricultural lands and the potential health risk related to intensive mining activities: A case study in Chenzhou city. China. *Journal of Environmental Sciences, 20*, 696–703.

76. Liu, G., Tao, L., Liu, X., Hou, J., Wang, A., & Li, R., (2013). Heavy metal speciation and pollution of agricultural soils along Jishui River in non-ferrous metal mine area in Jiangxi Province, China. *Journal of Geochemical Exploration, 132*, 156–163.

77. Liu, W. H., Zhao, J. Z., & Ouyang, Z. Y., (2005). Impacts of sewage irrigation on heavy metal distribution and contamination in Beijing, China. *Environment International, 31*, 805–812.

78. Lixia, W., Zhaohui, G., Xiyuan, X., & Tongbin, C., (2008). Heavy metal pollution of soils and vegetables in the midstream and downstream of the Xiangjiang River, Hunan province. *Journal of Geographical Science, 18*, 353–362.

79. Lu, A. X., Wang, J. H., Qin, X. Y., Wang, K. Y., Han, P., & Zhang, S. Z., (2012). Multivariate and geostatistical analyses of the spatial distribution and origin of heavy metals in the agricultural soils in Shunyi, Beijing, China. *Science of the Total Environment, 425*, 66–74.

80. Lu, K., Yang, X., Gielen, G., Bolan, N., & Ok, Y. S. S., (2017). Effect of bamboo and rice straw biochars on the mobility and redistribution of heavy metals (Cd, Cu, Pb, and Zn) in contaminated soil. *Journal of Environmental Management, 186*, 285–292.

81. Lucho-Constantino, C. A., & Alvarez-Suárez, M., (2005). Multivariate analysis of the accumulation and fractionation of major and trace elements in agricultural soils in Hidalgo State, Mexico irrigated with raw wastewater. *Environmental International, 31*(3), 313–323.

82. Luna, J. M., Rufino, R. D., & Sarubbo, L. A., (2016). Biosurfactant from *Candida sphaerica* UCP0995 exhibiting heavy metal remediation properties. *Process Safety and Environmental Protection, 102*, 558–566.

83. Mahmood, A., & Malik, R. N., (2014). Human health risk assessment of heavy metals *via* consumption of contaminated vegetables collected from different irrigation sources in Lahore, Pakistan. *Arabian Journal of Chemistry, 7*, 91–99.

84. Maity, J. P., Huang, Y. M., & Hsu, C. M., (2012). Removal of Cu, Pb, and Zn by foam fractionation and a soil washing process from contaminated industrial soils using soapberry-derived saponin: A comparative effectiveness assessment. *Chemosphere, 92*, 1286–1293.

85. Mapanda, F., Mangwayana, E. N., Nyamangara, J., & Giller, K. E., (2005). The effect of long-term irrigation using wastewater on heavy metal contents of soils under vegetables in Harare, Zimbabwe. *Agriculture Ecosystem and Environment, 107*, 151–165.

86. Marques, A. P. G. C., & Oliveira, R. S., (2007). *Solanum nigrum* grown in contaminated soil: Effect of arbuscular mycorrhizal fungi on zinc accumulation and histo localization. *Environmental Pollution, 145*, 691–699.

87. Masli, P., & Maier, R. M., (2000). Rhamnolipid enhanced mineralization of phenanthrene in organic-metal co-contaminated soils. *Bioremediation Journal, 4*, 295–308.

88. Micó, C., Recatalá, L., Peris, M., & Sánchez, J., (2006). Assessing heavy metal sources in agricultural soils of European Mediterranean area by multivariate analysis. *Chemosphere, 65*, 863–872.

89. Mohammed, S. A. S., & Mughal, A. A. B., (2016). Efficacy of nano calcium silicate (NCS) treatment on tropical soils in encapsulating heavy metal ions: Leaching studies validation. *Innovative Infrastructure Solutions*, 1–21.

90. Mortada, W. I., Sobh, M. A., El-Defrawy, M. M., & Farahat, S. E., (2001). Study of lead exposure from automobile exhaust as a risk for nephrotoxicity among traffic policemen. *American Journal of Nephrology, 21*, 274–279.

91. Muchuweti, M., Birkett, J. W., & Chinyanga, E., (2006). Heavy metal content of vegetables irrigated with mixtures of wastewater and sewage sludge in Zimbabwe: Implication for human health. *Agricultural Ecosystem and Environment, 112*, 41–48.

92. Nabulo, G., Oryem-Origa, H., & Diamond, M., (2006). Assessment of lead, cadmium, and zinc contamination of roadside soils, surface films, and vegetables in Kampala City, Uganda. *Environmental Research, 101*, 42–52.

93. Nicholson, F., Smith, S., Alloway, B., Carlton-Smith, C., & Chambers, B., (2003). An inventory of heavy metals inputs to agricultural soils in England and Wales. *Science of the Total Environment, 311*, 205–219.

94. Obiora, S. C., Chukwu, A., & Davies, T. C., (2016). Heavy metals and health risk assessment of arable soils and food crops around Pb-Zn mining localities in Enyigba, Southeastern Nigeria. *Journal of African Earth Science, 116*, 182–189.

95. Ok, Y. S., Usman, A. R. A., & Lee, S. S., (2011). Effect of rapeseed residue on cadmium and lead availability and uptake by rice plants in heavy metal contaminated paddy soil. *Chemosphere, 85*, 677–682.

96. O'Neill, P., (1995). *Arsenic: Heavy Metals in Soils* (2nd edn., pp. 105–121). Glasgow: Blackie Academic & Professional.

97. Onwordi, C. T., Ogungbade, A. M., & Wusu, A. D., (2009). The proximate and mineral composition of three leafy vegetables commonly consumed in Lagos, Nigeria. *African Journal of Pure and Applied Chemistry, 3*(6), 102–107.

98. Ozturk, M., & Gucel, S. S., (2011). Overview of the possibilities for wastewater utilization for agriculture in Turkey. *Israel Journal of Plant Science, 59*, 223–234.

99. Pandey, J., & Pandey, U., (2009). Accumulation of heavy metals in dietary vegetables and cultivated soil horizon in organic farming system in relation to atmospheric deposition in a seasonally dry tropical region of India. *Environmental Monitoring Assessment, 148*, 61–74.

100. Park, J. H., Choppala, G. K., Bolan, N. S., Chung, J. W., & Chuasavathi, T., (2011). Biochar reduces the bioavailability and phytotoxicity of heavy metals. *Plant Soil, 348*, 439–451.

101. Parkpian, P., Leong, S. T., Laortanakul, P., & Thunthaisong, N., (2003). Regional monitoring of lead and cadmium contamination in a tropical grazing land site, Thailand. *Environmental Monitoring and Assessment, 85*(2), 157–173. doi: 10.1023/A:1023638012736.

102. Parveen, Z., Khuhro, M. I., & Rafiq, N., (2003). Market basket survey for lead, cadmium, copper, chromium, nickel and zinc in fruits and vegetables. *Bulletin of Environmental Contamination and Toxicology, 71*, 1260–1264.

103. Pérez-de-Mora, A., Madrid, F., Cabrera, F., & Madejón, E., (2006). Amendments and plant cover influence on trace element pools in a contaminated soil. *Geoderma, 139*, 1–10.

104. Peris, M., Recatalá, L., Micó, C., Sánchez, R., & Sánchez, J., (2008). Increasing the knowledge of heavy metal contents in agricultural soils of the European Mediterranean Region. *Water, Air, and Soil Pollution, 192*, 25–37.

105. Phuong, T. D., Chuong, P. V., & Khiem, D. T., (1999). Elemental content of Vietnamese rice, Part 1: Sampling, analysis and comparison with previous studies. *Analyst, 124*, 553–560.

106. Pruvot, C., Douay, F., Hervé, F., & Waterlot, C., (2006). Heavy metals in soil, crops, and grass as a source of human exposure in the former mining areas. *Journal of Soils and Sediments, 6*, 215–220.

107. Qadir, M., Sharma, B. R., & Bruggeman, A., (2007). Nonconventional water resources and opportunities for water augmentation to achieve food security in water scarce countries. *Agricultural Water Management, 87*, 2–22.

108. Qian, S. Q., & Liu, Z., (2000). Overview of development in the soil-remediation technologies. *Chemical Industrial and Engineering Process, 4*, 10–20.

109. Querol, X., Alastuey, A., & Moreno, N., (2006). Immobilization of heavy metals in polluted soils by the addition of zeolitic material synthesized from coal fly ash. *Chemosphere, 62*, 171–180.

110. Qureshi, A. S., Hussain, M. I., Ismail, S., & Khan, Q. M., (2016). Evaluating heavy metal accumulation and potential health risks in vegetables irrigated with treated wastewater. *Chemosphere, 163*, 54–61.

111. Radwan, M. A., & Salama, A. K., (2006). Market basket survey for some heavy metals in Egyptian fruits and vegetables. *Food and Chemical Toxicology, 44*, 1273–1278.

112. Rapant, S., Dietzová, Z., & Cicmanová, S., (2006). Environmental and health risk assessment in abandoned mining area of Zlataldka, Slovakia. *Environmental Geology, 51*, 387–397.

113. Rattan, R. K., Datta, S. P., Chhonkar, P. K., Suribabu, K., & Singh, A. K., (2005). Long-term impact of irrigation with sewage effluents on heavy metal content in soils, crops and groundwater, a case study. *Agriculture Ecosystem and Environment, 109*, 310–322.

114. Reeves, P. G., & Chaney, R. L., (2001). Mineral nutrient status of female rats affects the absorption and organ distribution of cadmium from sunflower kernels (*Helianthus annuus* L.). *Environmental Research, 85*, 215–225.

115. Rieuwerts, J., & Farago, M., (1996). Heavy metal pollution in the vicinity of a secondary lead smelter in the Czech Republic. *Applied Geochemistry, 11*, 17–23.

116. Sabir, M., Waraich, E. A., & Hakeem, K. R., (2015). *Phytoremediation, Soil Remediation and Plants* (p. 316). New York: Elsevier Inc. doi: 10.1016/B978-0-12-799937-1.00004-8.

117. Saikia, P., & Deka, D. C., (2013). Mineral content of some wild green leafy vegetables of North-East India. *Journal of Chemical and Pharmaceutical Research, 5*(3), 117–121.

118. Sakakibara, M., Ohmori, Y., Ha, N. T. H., Sano, S., & Sera, K., (2011). Phytoremediation of heavy metal-contaminated water and sediment by *Eleocharis acicularis*. *Clean Soil Air Water, 39*, 735–741.

119. Sánchez-Camazano, M., Sánchez-Martín, M. J., & Lorenzo, L. F., (1994). Lead and cadmium in soils and vegetables from urban gardens of Salamanca (Spain). *Science of the Total Environment, 146, 147*, 163–168.

120. Sandrin, T. R., & Maier, R. M., (2000). Impact of metals on the biodegradation of pollutants. *Environmental Health Perspectives, 111*, 1093–1100.

121. Sarma, H., Islam, N. F., Borgohain, P., Sarma, A., & Prasad, M. N. V., (2016). Localization of polycyclic aromatic hydrocarbons and heavy metals in surface soil of Asia's oldest oil and gas drilling site in Assam, north-east India: Implications for the bio-economy. *Emerging Contaminants, 2*, 119–127.

122. Sarubbo, L. A., Rocha, J. R. B., & Luna, J. M., (2015). Some aspects of heavy metals contamination remediation and role of biosurfactants. *Chemical Ecology, 31*, 707–723.

123. Satter, M. M. A., Khan, M. M., & Jabin, S. A., (2016). Nutritional quality and safety aspects of wild vegetables consume in Bangladesh. *Asian Pacific Journal of Tropical Biomedicine, 6*, 125–131.

124. SEPA, (1995). *Environmental Quality Standards for Soils* (p. 98). State Environmental Protection Administration (SEPA), China; Report GB15618-1995.

125. Sharma, R. K., Agrawal, M., & Marshall, F. M., (2007). Heavy metals contamination of soil and vegetables in suburban areas of Varanasi, India. *Ecotoxicology and Environmental Safety, 66*, 258–266.

126. Sharma, R. K., Agrawal, M., & Marshall, F. M., (2008). Atmospheric deposition of heavy metals (Cu, Zn, Cd, and Pb) in Varanasi City, India. *Environmental Monitoring and Assessment, 142*, 269–278.

127. Sharma, R. K., Agrawal, M., & Marshall, F. M., (2009). Heavy metals in vegetables collected from production and market sites of a tropical urban area of India. *Food and Chemical Toxicology, 47*, 583–591.

128. Singh, A., Sharma, R. K., Agrawal, M., & Marshall, F. M., (2010). Health risk assessment of heavy metals via dietary intake of foodstuffs from the wastewater irrigated site of a dry tropical area of India. *Food and Chemical Toxicology, 48*, 611–619.

129. Singh, K. P., Mohan, D., Sinha, S., & Dalwani, R., (2004). Impact assessment of treated/untreated wastewater toxicants discharged by sewage treatment plants on health, agricultural, and environmental quality in the wastewater disposal area. *Chemosphere, 55*, 227–255.

130. Sinha, S., Gupta, A. K., Bhatt, K., Pandey, K., Rai, U. N., & Singh, K. P., (2006). Distribution of metals in the edible plants grown at Jajmau, Kanpur (India) receiving treated tannery wastewater: Relation with physicochemical properties of the soil. *Environmental Monitoring and Assessment, 115*, 1–22.

131. Song, B., Mei, L., Chen, T., Zheng, Y., Xie, Y., Li, X., & Gao, D., (2009). Assessing the health risk of heavy metals in vegetables to the general population in Beijing, China. *Journal of Environmental Science, 21*, 1702–1709.

132. Steenland, K., & Boffetta, P., (2000). Lead and cancer in humans: Where are we now? *American Journal of Industrial Medicine, 38*, 295–299.

133. Sterckeman, T., Douay, F., Proix, N., & Fourier, H., (2000). Vertical distribution of Cd, Pb, and Zn in soils near smelters in the north of France. *Environmental Pollution, 107*, 377–389.

134. Sylvain, B., Mikael, M. H., & Florie, M., (2016). Phyto stabilization of As, Sb and Pb by two willow species (*S. viminalis and S. purpurea*) on former mine techno sols. *Catena, 136*, 44–52.

135. Tóth, G., Hermann, T., Da Silva, M. R., & Montanarella, L., (2016). Heavy metals in agricultural soils of the European Union with implications for food safety. *Environment International, 88*, 299–309.

136. Türkdoğan, M. K., Kilicel, F., Kara, K., Tuncer, I., & Uygan, I., (2003). Heavy metals in soil, vegetables, and fruits in the endemic upper gastrointestinal cancer region of Turkey. *Environmental Toxicology Pharmacology, 13*, 175–179.

137. Voutsa, D., Grimanis, A., & Samara, C., (1996). Trace elements in vegetables grown in an industrial area in relation to soil and air particulate matter. *Environmental Pollution, 94*, 325–335.

138. Wagner, A., & Kaupenjohann, M., (2014). Suitability of biochars (pyro-and hydrochars) for metal immobilization on former sewage-field soils. *European Journal of Soil Science, 65*, 139–148.

139. Wang, M., Chen, W., & Peng, C., (2016). Risk assessment of Cd polluted paddy soils in the industrial and township areas in Hunan, Southern China. *Chemosphere, 144*, 346–351.

140. Wang, X. P., Shan, X. Q., Zhang, S. Z., & Wen, B., (2004). A model for evaluation of the phyto availability of trace elements to vegetables under the field conditions. *Chemosphere, 55*, 811–822.

141. Wångstrand, H., Eriksson, J., & Oborn, I., (2007). Cadmium concentration in winter wheat as affected by nitrogen fertilization. *European Journal of Agronomy, 26*, 209–214.

142. Wijayawardhana, D., Herath, V., & Weerasinghe, A., (2016). Heavy metal pollution in Sri Lanka with special reference to agriculture: A review of current research evidences. *Rajarata University Journal, 4*, 52–66.

143. Wuana, R. A., & Okieimen, F. E., (2011). Heavy metals in contaminated soils: Review of sources, chemistry, risks, and best available strategies for remediation (Review article). *International Scholarly Research Network, 2011*, 1–20.

144. Yan, S., Ling, Q. C., & Bao, Z. Y., (2007). Metals contamination in soils and vegetables in metal smelter contaminated sites in Huangshi, China. *Bulletin of Environmental Contamination and Toxicology, 79*, 361–366.

145. Yao, Z., Li, J., Xie, H., & Yu, C., (2012). Review on remediation technologies of soil contaminated by heavy Metals. *Procedia Environmental Sciences, 16*, 722–729.

146. Yousaf, B., Liu, G., Wang, R., & Imtiaz, M., (2016). Bioavailability evaluation, uptake of heavy metals and potential health risks via dietary exposure in urban-industrial areas. *Environmental Science Pollution Research, 23*, 22443–22453.

147. Yu, H. Y., Ding, X., Li, F., Wang, X., & Zhang, S., (2016). The availabilities of arsenic and cadmium in rice paddy fields from a mining area: The role of soil extractable and plant silicon. *Environmental Pollution, 215*, 258–265.

148. Yu, H. Y., Liu, C., Zhu, J., Li, F., Deng, D. M., Wang, Q., & Liu, C., (2016). Cadmium availability in rice paddy fields from a mining area: The effects of soil properties highlighting iron fractions and pH value. *Environmental Pollution, 209*, 38–45.

149. Yoo, I. C., & Lee, C., (2017). Simultaneous application of chemical oxidation and extraction processes is effective at remediating soil co-contaminated with petroleum and heavy metals. *Journal of Environmental Management, 2017*, 314–319.

150. Yoo, J. C., Kim, E. J., Yang, J. S., & Baek, K., (2015). Step-wise extraction of metals from dredged marine sediments. *Separation Science and Technology, 50*, 536–544.

151. Yoo, J. C., Lee, C. D., Yang, J. S., & Baek, K., (2013). Extraction characteristics of heavy metals from marine sediments. *Chemical Engineering Journal, 228*, 688–699.

152. Yu, H. Y., Li, F. B., Liu, C. S., Huang, W., Liu, T. X., & Yu, W. M., (2015). Iron redox cycling coupled to transformation and immobilization of heavy metals: Implications for paddy rice safety in the red soil of South China. *Advances in Agronomy, 137*, 279–317.

153. Yusuf, A. A., Arowolo, T. A., & Bamgbose, O., (2003). Cadmium, copper, and nickel levels in vegetables from industrial and residential areas of Lagos City, Nigeria. *Food and Chemical Toxicology, 41*, 375–378.

154. Zarcinas, B. A., Ishak, C. F., McLaughlin, M. J., & Cozens, G., (2004). Heavy metals in soils and crops in Southeast Asia 1. Peninsular Malaysia. *Environmental Geochemistry and Health, 26*, 343–357.

155. Zhao, Y. F., Shi, X. Z., Huang, B., & Yu, D. S., (2007). Spatial distribution of heavy metals in agricultural soils of an industry-based peri-urban area in Wuxi, China. *Pedosphere, 17*, 44–51.

156. Zhou, D. M., Hao, X. Z., & Xue, Y., (2004). Advances in remediation technologies of contaminated soils. *Ecology and Environmental Sciences, 13*(2), 234–242.

157. Zhu, Q., Wu, J., Wang, L., & Zhang, X., (2015). Effect of biochar on heavy metal speciation of paddy soil. *Water Air Soil Pollution, 226*, 429–434; doi: 10.1007/s11270-015-2680-3.

158. Zhuang, P., McBride, M. B., Xia, H., Li, N., & Li, Z., (2009). Health risk from heavy metals via consumption of food crops near Dabaoshan mine in South China. *Science of the Total Environment, 407*, 1551–1561.

CHAPTER 3

WASTEWATER USE IN AGRICULTURE: AN EMERGING ISSUE AND CURRENT AND FUTURE TRENDS

VINOD KUMAR TRIPATHI and ASHISH KUMAR

ABSTRACT

Water, an important natural resource needs to be conserved and managed to meet the increasing demand of water for drinking, agriculture, hydropower generation, and ecological balance, because a continuous increase in the population is putting tremendous pressure on water resources. Water is limited and mostly water is confined in oceans, polar regions, and glaciers, which is not available for human use. In the quest of jobs, people are migrating from rural areas to urban areas. Per capita requirement of water is more in urban areas and these areas generate a greater volume of wastewater (WW). Collectively, a larger amount of freshwater is being diverted for domestic, commercial, and industrial use. Sewage water is being disposed of in rivers without treating making environment and soil polluted.

In the present study, a proper alternative for sewage water has been suggested as the use of WW for irrigation purposes. Various sources of pollution like organic compounds, domestic, heavy metals, and environmental fate of pollutants in WW irrigated soils have been discussed. Major positive impacts of WW reuse as irrigation on crop yield, soil quality (physical structure, soil microbial activity), and negative impacts on soil like pollution by pathogenic microbial pollution and its effect have been presented.

3.1 INTRODUCTION

Water is vital to the existence of all living organisms as it is required for drinking, food production, energy production, manufacturing, and ecological balance. Although nearly 75% of the earth's surface is covered with water, yet the availability of freshwater is very low due to confinement of water in oceans, polar regions, and glaciers [61]. Only less than 1% of global water is available as freshwater for human use, because most of freshwater is in deep substrata [57].

In the view of continuous population growth at a very faster pace, the world population will have grown 1.5 times by 2050, and the urban population will have grown three times, with almost 50% total population living in cities by the year 2025 [22]. Owing to continuous growth in the human population, water required for household, commercial, industrial, and agricultural purposes is increasing greatly. Land and water are limited resources but continuous increase in population consequently forces the people to displace from rural to urban areas as cities attract the growth of industries and jobs, they bring together both the human and the entrepreneurial resources for productive uses of technologies. A larger amount of freshwater is being diverted for domestic, commercial, and industrial use [3, 43, 52].

Most cities and towns have no sewerage and sewage treatment plants (STPs). Commonly sewage is either directly released to natural water bodies or open fields, which causes environmental pollution. Bhardwaj [4] estimated that about 38,254 million liters per day (MLD) of wastewater (WW) is generated in urban areas comprising of Class I cities and Class II towns (with the population more than 50,000) accounting more than 70% of the total urban population. Whereas present WW treatment capacity is about 11,787 MLD (31% of WW generation) in both classes and it is projected that WW generation in, urban areas may cross 120 billion liters per day in 2051 and that rural India will also generate not less than 50 billion liters per day [4].

Nearly 7 million commercially available chemicals are being routinely dumped into the sewage line after use [16]. At present only, 269 STPs are in India, out of which only 231 are operational, which shows only 21% generated WW can be treated [12]. This indicates two main inevitable problems: the shortage of water and sewage overload.

Globally, 70% of the total water withdrawn is used in the irrigation sector [19]. Many river basins of the world are approaching towards closure. The desertification caused by climate change in large areas has put tremendous pressure on the availability of water throughout the world [25]. Among many problems being faced in India, freshwater scarcity is on the war footing. Irrigation using freshwater limits the freshwater availability, which makes water recycling necessary for irrigation in dry zones [18, 30]. The scarcity of water for irrigation makes the farmers in urban and peri-urban areas use WW for irrigation purposes. Since around 3000 B.C, WW is being used for irrigation as it contains both water and nutrients that improve soil fertility along with prevention to contaminate rivers [15].

Use of WW makes the availability of nutrients like carbon, nitrogen, phosphorus, potassium, magnesium, and boron, molybdenum, selenium, and copper to the plant and improves the soil fertility. These nutrients improve the productivity of the crop but at the same time have many harmful impacts. Although, many farmers not aware of consequences of use of polluted water often use WW for irrigation due to unavailability of other sources of water, yet in some cases, farmers are familiar with consequences they will be facing due to presence of associated pathogens, heavy metals and other undesirable constituents depending on the source [52]. Moreover, in many countries, farmers, consumers, and government agencies are not aware of the potential impacts of irrigation with WW. Therefore, it is required that soon interim solutions and risk management are needed to prevent adverse environmental and health impacts of irrigation with WW [48, 64].

At present, some countries are using WW in agriculture as a convenient environmental strategy [1, 42]. The ecosystem possesses self-purification quality that maintains the quality of WW supplied for irrigation purposes. Photolysis and biodegradation promote the dissipation of pollutants in the soil while the adsorption force makes organic pollutants confine within the solid matrix onto the soil particles [66]. The depuration of WW before reusing it for irrigation is the most reasonable option to avoid soil pollution by WW reuse. WW treatment is used to remove nitrogen, carbon, phosphorus, and minerals thus reducing the quality of effluents (as fertilizers), consequently impacting crop yields.

To meet the water quality standards for the specific use, it is obvious to treatWW by passing through various treatment stages. For the crop like fodder, primary treatment can be satisfactory, while for unrestricted crops secondary treatment is required [56, 63]. Generally, developing countries use treated water for irrigation, while in low-income countries WW is commonly used for irrigation with untreated or partially treated WW [33, 35]. In China, Mexico, India, and Pakistan, untreated WW has generally been reused for irrigation purposes [36]. The WHO estimated that nearly 20 million-ha, the land is being irrigated without treating WW throughout the world [64]. Also, in some cities, up to 80% of the vegetables locally-consumed are produced using WW for irrigation [21]. For proper planning and management of WW for irrigation, scientific knowledge about the impact on the environment is important.

This chapter focuses on the review of status and future trends of WW generation, its treatment. It was shown that how WW reuse can benefit soil/ water and farmers and at the same time its negative impact ecosystem was also assessed. It can provide understanding about which type of approaches for risk reduction and management measures should be adopted by institutions and policymakers to improve WW management and minimize adverse impacts on irrigators and produce consumer health.

3.2   IMPACT OF WASTEWATER (WW) REUSE IN AGRICULTURE

It is very important to know the benefits and negative impacts of WW reuse in irrigation. In this chapter, the focus has been given to provide the assistant to the planner to define the desired limits for WW reuse based on quality parameters to make the ecology balanced.

3.2.1   POSITIVE IMPACTS OF WW REUSE IN IRRIGATION

WW (treated and untreated) is widely being used in agriculture as it provides nutrients and all the moisture necessary for crop growth. Based on the local condition, benefits can vary and the overview is discussed in this chapter.

3.2.2   IMPROVEMENT IN CROP YIELD

WW is a water source being produced continuously and is available year-round. It is desirable to select crops to be sown year-round, mainly those having high profitability and water requirement. Most crops give higher than potential yields with WW irrigation, reduce the need for chemical fertilizers, resulting in net-cost savings to farmers; and it also stabilizes soil nutrients. When the nitrogen delivered exceeds the desired limit, vegetative growth increases but ripening and maturity time exceeded the optimum period, which results in low yield. During the dry season, irrigation with treated and untreated WW has made possible vegetable production and also it increases yield and selling prices by three to five folds compared to the monsoon season [44].

Countries like Pakistan, Ghana, and Senegal, rural, and urban farmers make use of WW to irrigate crops resulting in three to six harvests per year [27, 62]. Organic nitrogen converts into nitrates at a higher rate in the presence of WW microorganisms [21]. A small amount of phosphorus is found in WW therefore continuous use of WW does not pose any negative impact on the environment [14, 29]. Potassium demand cannot be met by sewage water as it contains low concentrations of potassium.

3.2.3 IMPROVEMENT IN SOIL QUALITY

Soil, a media for plant growth, provides habitat for flourishing the crops [11]. WW as irrigation not only provides nutrients to the soil but also improves soil quality. Improvement in soil quality has been evidenced in soils irrigated with WW, such as enhancement of physicochemical and biological characteristics of the soil.

3.2.4 IMPROVEMENT OF THE PHYSICAL PROPERTIES OF SOIL

Irrigation with WW increases soil porosity and water infiltration but decreases erodibility [5]. By adding the organic material and some inorganic compounds in the soil, micro-aggregates are formed and macro-aggregates form when these micro-aggregates join. Otherwise, particulate organic matter plays an important role as a macro-aggregates form around it, while exudates due to soil microorganisms serve as binding material, which increases the stability of micro and macro-aggregates [20]. Both elements supplied by WW can improve the physical structure of soil through an increase in the stability of soil aggregates. For instance, the stability of soil aggregates increases with the use of WW [45]. WW contains calcium and magnesium in abundance thus increasing the formation of micro-aggregates. Micro and macro-aggregates form when insoluble calcium and magnesium carbonates are present in arid soils and soils with low organic matter contents [8]. Furthermore, when the sodium concentration increases in soil, aggregates break up due to dispersion by calcium [2].

In the formation of macro-aggregates, polysaccharides act as binding agents [34]. The consistency of soil aggregates increases by the introduction of certain chemicals like surfactants, lipids, and hydrocarbons into the soil via WW [8]. The physical structure of WW irrigated soil is also improved by agricultural activities because some crops (i.e., maize, alfalfa, and leguminous plants) can improve the physical structure of the soil. Phenolic compounds present in municipal and industrial WW produce similar effects to those of corn wastes [9]. The method of irrigation in use has a significant role in the degree of compaction after irrigation.

3.2.5   INCREASE OF SOIL MICROBIAL ACTIVITY

Irrigating with WW increases in the microbial activity of the soil. The nutritional condition of soil microorganisms improves when irrigated with WW for long periods because of a decrease in C/N ratio (carbon/nitrogen ratio) by 45% [24]. When the soil is irrigated with the WW for 100 years, the population of copiotrophic, oligotrophic bacteria, actinomycetes, and fungi is increased by 234%, 217%, 234%, 206%, respectively, compared with those populations found in non-irrigated soils. Soil enzymatic activity did not change for twenty years after WW irrigation had ceased, while microbial activity changes after a few days when not irrigated [38].

Rhizospheric activities are accelerated due to the increase in the populations of fungi, bacteria, and actinomycetes in the irrigated soil, which lead to an improvement in the formation and stability of soil aggregates, increment in the growth of plants, higher stabilization rates of organic matter, and increment in depuration rate. Proliferation of soil (micro- and macro-) fauna essential for soil formation and development of plants is supported by the transformation of carbon and nitrogen by soil microorganisms. The highest N_2O production was found in soils irrigated with treated WW [40, 54]. Planning of agricultural systems based on WW reuse requires a thorough knowledge about role soil microorganisms in soil development as well as purification of WW.

3.2.6   IMPROVEMENT OF SOIL PERFORMANCE AS A WW TREATMENT SYSTEM

Soils can purify WW to tertiary treatment level when WW is applied into soils and as the water gets infiltrated, the purification process intensified. Soil maintains the quality of surface and groundwater bodies through this mechanism. To convince policymakers, it is important to explore the potential of this mechanism for WW depuration to simultaneously solve problems of water scarcity and negative impacts on the environment. The extent to which this natural system can work depends on local conditions and types of pollutants as it varies from almost nonexistent to very high. WW applied to the soil reduces pathogenic microorganisms by six to seven log units for bacteria while 100% for helminths and other protozoa. Levels of recalcitrant compounds in WW (like phosphorus, nitrogen, and metals) are reduced by 70–95% of P_2O_5, 20–90% of N, 20–70% of heavy metals, respectively, while total organic carbon is reduced by 90%.

The bio-degradation of carbonaceous substances at a very high rate can be achieved by WW application to the soil under controlled conditions [7]. In the infiltrated water, total organic carbon is reduced from 80–200 ppm to 1–5 ppm [53]. Generally, the pH value of the soil increases when it is irrigated with treated/untreated WW. However, in a few cases, soil pH tends to decrease after the application of WW.

3.2.7   NEGATIVE IMPACTS OF WASTEWATER (WW) REUSE IN IRRIGATION

Soils continuously irrigated with WW are receiving newly synthesized substances because of increasing industrial development. These substances may have several negative impacts on the effectiveness of the soil by destroying the physical structure of the soil, disturbing natural cycles occurring within the soil, and poisoning the degrading microorganisms.

3.2.8   SOIL POLLUTION BY PATHOGENIC MICROBIAL AGENTS

Irrigation with WW has a prominent effect on the daily life of farmers and their families, people living nearby to irrigated fields, crop handlers, and product consumers as such irrigation is composed of bacteria, protozoa, and viruses obtained from human and animal excreta, which makes it vector for intestinal infections. Owing to the high exposure of farmers to WW, they are highly affected segments of the community from which children and elders are at higher vulnerable conditions [64]. Its exposure may be of the direct or indirect type.

3.2.9   EFFECTS CAUSED BY MICROBIAL POLLUTION IN SOIL

WW reuse has many disasters effects not only on human health but also on animal health, where WW is used without treatment for irrigation and drinking water is of low quality, it does not mean that treated WW reuse cannot be harmful for human health. Cholera, caused by bacterium Vibrio cholera, is one of the infectious diseases closely linked to WW reuse in poor countries. The most common viral infections are viral enteritis (caused by rotaviruses) and Hepatitis A [55]. Skin diseases (such as dermatitis (eczema)) and nail problems such as koilonychia (spoon-formed nails) in farmers have also been reported [6, 37]. Health and growth problems have also been observed in cattle that consume forage produced by WW irrigation and drink the WW.

3.2.10 MICROBIAL AGENTS IN WW IRRIGATED SOILS

The soil samples taken from area irrigated with untreated WW in West Bengal-India have 69% of *Ascaris lumbricoides* (hookworm) and *Trichuris trichiura* (human whipworm) [31]. Antibiotic-resistant pathogens (ARPs) and antibiotic resistance genes (ARGs) movement in the soil by means of WW are evolving issues. The ARPs reaching the soil through WW may migrate to groundwater sources if conditions permit others it survives on the soil surface. ARGs may reach aquifers by infiltration of WW or surface water by runoff. The occurrence of ARGs had a positive relationship with the time [13].

3.2.11 MICROBIAL POLLUTION IN CROPS

WW reused as irrigation can pollute edible parts of plants by direct contact with WW. Severity depends on the water quality, the irrigation method, quantity applied to the soil, and crop type. Microbial contamination of crops can also occur during washing, packing, transportation, and marketing. With subsurface drip irrigation (SDI), microbial pollution can be reduced to a minimum even when using WW [49]. Fruits from trees are rarely polluted when irrigation is provided using sprinklers. *E. coli* (non-pathogenic fecal coliform indicator) can survive in soil for nearly a month, whereas the pathogenic strain of *E. coli* O157:H7 survives for 14 days only in spinach leaves [50].

3.2.12 FATE OF PATHOGENIC MICROORGANISMS IN SOIL

Bacteria can migrate up to the extent of 830 meters, while viruses can mobilize up to 408 m as viruses survive for a short period than the bacteria [39, 58]. Pathogen transportation also depends on climatic conditions, soil texture, the interconnection of pores, preferential paths within the soil matrix. For example, pathogens has longer life-span in frozen soils and thus they can mobilize for longer distances compared to those in tropical and desert soils [28]

3.2.13 EFFECT OF ORGANIC POLLUTANTS ON DOMESTIC WW

Owing to antibiotic resistance by pathogenic microorganisms due to the occurrence of antibiotics in WW, it has become growing concern for scientific community in recent years [41, 60]. Damages in liver, gills, and kidney caused by ibuprofen, naproxen, diclofenac, paracetamol, and ketoprofen have been reported in aquatic species [23]. Non-steroidal anti-inflammatory drug diclofenac causes visceral gout in vultures [59]. Lessening in the development of human embryo cells due to chronic exposure has been traced to anti-inflammatory drugs [51].

The presence of psychotropic agents in WW affects the behavior of some fish species [47]. It is a fact that WW is the primary source for allowing OPECs (organic pollutants of emerging concerns) to reach soil, yet not enough attention has been given in this type of research. Top 30 cm soil has highest concentrations of OPECs. In the areas irrigated with untreated WW for eight decades, the concentration of the pharmaceuticals naproxen, ibuprofen, and carbamazepine varied from 0.25 to 6.48 µg/kg of soil (on dry weight (DW) basis) in Phaeozem and Leptosol soils [17]. Average concentration of triclosan, estrone, nonphenols, and plasticizers was 1.8 µg/kg, 2.5 µg/kg, 14–80 µg/kg and 140–2610 µg/kg, respectively [10, 65]. Hu et al. [32] reported that a concentration of di-2 (ethylhexyl) phthalate was up to 7110 µg/kg.

3.2.14 ENVIRONMENTAL FATE OF ORGANIC POLLUTANTS IN WW IRRIGATED SOILS

It is governed by the physicochemical properties of both the pollutant and the irrigated soil along with climatic conditions at the site. In WW irrigated systems, dissolved, and particulate organic matter tend to accumulate on the topsoil. The heteroatoms in organic molecules can affect environmental fate in soil. Photodegradation of organic pollutants significantly affects by physical structure of the soil as it defines the depth of solar radiation that can penetrate the soil. Rhodococcusrhodochorus bacteria [26] can degrade carbamazepine down to levels of 15% of its initial concentration in soil, while fungi Trametes versicolor can degrade naproxen [46]. Knowledge about assimilation of OPECs by plants in WW irrigated soils is still limited; therefore, assessment of health risk is a must for proper management of risks rising from WW irrigated soil.

3.2.15   REMOVAL OF HEAVY METALS FROM WASTEWATER (WW)

WW contains chemical elements having atomic weights from 63.5 to 200 called heavy metals. These elements are toxic in nature if their concentration is more than the prescribed limit recommended by various pollution agencies. They can also cause environmental pollutions. To utilize WW, it is necessary to reduce the concentrations of heavy metals below the permissible limits.

The technologies available to reduce the concentration of heavy metals in WW are electrochemical methods, membrane filtration, adsorption, ion-exchange, chemical precipitation, etc. It is not necessary that all the techniques will be effective for a single type of heavy metal. Out of these methods, membrane filtration, adsorption, and ion exchange techniques are extensively adopted in WW treatment process. Traditionally heavy metals are being removed by the process of precipitation. This method is effective when the concentration of heavy metals is high. This method is economical, but sludge is generated as a byproduct through this process. The sludge should be disposed-off by adopting proper process.

3.3   SCOPE FOR FURTHER RESEARCH ON WASTEWATER (WW) USE IN AGRICULTURE

Given the scarcity of water and erratic nature of rainfall, it has become very important to go for alternate sources of water to be used for irrigation of the crop. In line with this quest, WW reuse irrigation has come up as an emerging source of irrigation. This section presents some perspectives for further studies as:

- On a long-term basis, improvement in soil properties to produce food should be checked. Monitoring studies should be carried out for OPECs and pathogens in soils irrigated with WW. It can identify a compound-based solution for each site.
- Studies on the effect of climate change, extreme rainfall events on the migration of contaminants in soil should be carried out. The effect of temperature on the biodegradation of organic pollutants in the soil should be checked.

- There is a need for the development and implementation of cost-effective WW treatment systems to remove organic, inorganic, and biological pollution without reducing the content of organic matter in the water. Advanced primary treatment systems may represent a plausible strategy in such cases.
- The development and validation of environmentally-friendly analytical techniques are needed for the determination of OPECs in soils.
- The fate of pollutants that accumulated in the soil during continuous input via WW should be determined.
- The study of the environmental fate of emerging contaminants at laboratory and field-scale using different model molecules in soil should be carried out.

3.4 SUMMARY

The reuse of WW irrigation can help to identify a compound-based solution for each site. Quantitative evaluation of microbiological risk is required for the proper implementation of agricultural reuse of WW. Therefore, improvement in the risk estimation model is the next milestone to check the negative effect on the human and to make the safe reuse of residual water.

KEYWORDS

- **antibiotic resistance genes**
- **biological pollutant**
- **irrigation**
- **sewage treatment plants**
- **sewage water**
- **wastewater**

REFERENCES

1. Anderson, J., (2003). The Environmental benefits of water recycling and reuse. *Water Science and Technology: Water Supply, 3*(4), 1–10.
2. Armstrong, A. S. B., & Tanton, T. W., (1992). Gypsum application to aggregated saline-sodic clay topsoils. *Journal of Soil Science, 43*(2), 249–260.

3. Asano, B. T. E., & Mubofu, E. B., (1999). Heavy metals in edible green vegetables grown along the sites of the Sinza and Msimbazi Rivers in Dares Salaam, Tanzania. *Food Chemistry, 66*, 63–66.

4. Bhardwaj, R. M., (2005). Status of wastewater generation and treatment in India. *Paper Presented at Inter-Secretariat Working Group on Environment Statistics (IWG–Env) Joint Work Session on Water Statistics* (p. 7). International Working Group on Environment (IWG-Env.), Vienna-Austria.

5. Boix-Fayos, C., Calvo-Cases, A., Imeson, A. C., Soriano-Soto, M. D., & Tiemessen, I. R., (1998). Spatial and short-term variations in runoff, soil aggregation and other soil properties along a Mediterranean climatological gradient. *Catena, 33*(2), 123–138.

6. Bos, R., Carr, R., & Keraita, B., (2010). Assessing and mitigating wastewater related health risks in low-income countries: An introduction. In: Drechsel, P., Scott, C. A., Raschid-Sally, L., Redwood, M., & Bahri, A., (eds.), *WW Irrigation and Health: Assessing and Mitigating Risk in Low-Income Countries* (pp. 29–51). London/Ottawa/Colombo: Earthscan/IDRC/IWMI.

7. Bouwer, H., (1991). Groundwater recharge with sewage effluent. *Water Science and Technology, 23*(10–12), 2099–2108.

8. Bronick, C. J., & Lal, R., (2005). Soil structure and management. *Geoderma, 124*(1/2), 3–22.

9. Brunetti, G., Senesi, N., & Plaza, C., (2007). Effects on amendments with treated and untreated olive oil mill WW on soil properties, soil humic substances and wheat yield. *Geoderma,138*(1/2), 144–152.

10. Chen, F., Ying, G. G., Kong, L. X., Wang, L., Zhao, J. L., Zhou, L. J., & Zhang, L. J., (2011). Distribution and accumulation of endocrine-disrupting chemicals and pharmaceuticals in wastewater irrigated soils in Hebei, China. *Environmental Pollution, 159*(6), 1490–1498.

11. Coleman, D. C., Crossley, J. D. A., & Hendrix, P. F., (2004). *Fundamentals of Soil Ecology* (2ndedn., p. 408). London: Academic Press.

12. CPCB, (2009). *Status of Water Supply, Wastewater Generation and Treatment in Class-I Cities and Class-II Towns of India* (p. 93). Control of Urban Pollution Series: CUPS/70/2009-10. Central Pollution Control Board (CPCB), Ministry of Environment and Forest, Government of India, New Delhi.

13. Dalkmann, P., Broszat, M., Siebe, C., Willaschek, E., Sakinc, T., Huebner, J., Amelung, W., et al., (2012). Accumulation of pharmaceuticals, enterococcus, and resistance genes in soils irrigated with wastewater for zero to 100 years in Central Mexico. *PLoS One,7*(9), E-article ID: 45397.

14. Degens, B., Schipper, L., Claydon, J., Russell, J., & Yeates, G., (2000). Irrigation of an allophanic soil with dairy factory effluent for 22 years: Responses of nutrient storage and soil biota. *Australian Journal of Soil Research, 38*(1), 25–35.

15. DeFeo, G., Mays, L. W., & Angelakis, A. N., (2011). Water and wastewater management technologies in the ancient Greece and Roman civilizations. In: Wilder, P., (ed.), *Treatise in Water Science* (pp. 3–22). Oxford: Elsevier.

16. Diao, X., Jensen, J., & Hansen, A. D., (2007). Toxicity of the anthelmintic abamectin to four species of soil invertebrates. *Environmental Pollution, 148*(2), 514–519.

17. Duran-Alvarez, J. C., Becerril-Bravo, E., Castro, V. S., Jiménez, B., & Gibson, R., (2009). The analysis of a group of acidic pharmaceuticals, carbamazepine, and potential

endocrine disrupting compounds in WW irrigated soils by gas chromatography-mass spectrometry.*Talanta, 78*(3), 1159–1166.

18. Duran-Alvarez, J. C., & Jiménez-Cisneros, B., (2014). Beneficial and negative impacts on soil by the reuse of treated/untreated municipal wastewater for irrigation-a review of the current knowledge and future perspectives. Chapter 5; In: Hernandez, M. C., (ed.), *Environmental Risk Assessment of Soil Contamination* (pp. 137–197). http://www.intechopen.com/books/environmental-risk-assessment-of-soilcontamination/beneficial-and-negative-impacts-on-soil-by-the-reuse-of-treated-untreated-municipal-WW-for-a (accessed on 20 June 2020).

19. Earth Trends: Environmental Information, (2019). Online: http://www.wri.org/project/earthtrends (accessed on 20 June 2020).

20. Edwards, A. P., & Bremner, J. M., (1967). Micro-aggregates in soil. *European Journal of Soil Science, 18*(1), 64–73.

21. Ensink, J. H. J., Simmons, R. W., & Van, D. H. W., (2004). WW use in Pakistan: The cases of Haroonabad and Faisalabad. In: Scott, C. A., Faruqui, N. I., & Raschid-Sally, L., (eds.), *Use in Irrigated Agriculture: Confronting the Livelihood and Environ-Mental Realities* (pp. 91–99). WWWallingford: CAB International.

22. EPA, (1992). *Manual: Guidelines for Water Reuse.* Report EPA/625/R-92/004 (NTIS 93-222180). US Environmental Protection Agency (EPA), Washington, DC.

23. Fent, K., Weston, A. A., & Caminada, D., (2006). Ecotoxicology of human pharmaceuticals. *Aquatic Toxicology, 76*(2), 122–159.

24. Filip, Z., Kanazawa, S., & Berthelin, J., (1999). Characterization of effects of a long-term wastewater irrigation on soil quality by microbiological and biochemical parameters. *Journal of Plant Nutrition and Soil Science,162*(4), 409–413.

25. Food and Agriculture Organization of the United Nations (FAO), (2012). *Coping with Water Scarcity: An Action Framework for Agriculture and Food Security* (p. 100). Rome: FAO. http://www.fao.org/3/a-i3015e.pdf (accessed on 20 June 2020).

26. Gauthier, H., Yargeau, V., & Cooper, D. G., (2010). Biodegradation of pharmaceuticals by *Rhodococcusrhodochrous* and *Aspergillus niger* by co-metabolism. *Science of the Total Environment,408*(7), 1701–1706.

27. Gaye, M., & Niang, S., (2002). *Purification of Wastewater Reuse in Urban Agriculture. Research Studies* (p. 110). Dakar: ENDA–TM.

28. Gerba, C. P., Melnick, J. L., & Wallis, C., (1997). Fate of wastewater bacteria and viruses in soil. *Journal of the Irrigation and Drainage Division, 101*(3), 157–174.

29. Girovich, M., (1996). *Biosolids Treatment and Management: Processes for Beneficial Use* (p. 449). New York: Marcel Dekker Inc.

30. Goyal, M. R., (2015). *Book Series: Research Advances in Sustainable Micro Irrigation* (Vols. 1–10). Oakville, ON, Ca: Apple Academic Press Inc.

31. Gupta, N., Khan, D. K., & Santra, S. C., (2009). Prevalence of intestinal *helminth* eggs on vegetables grown in wastewater irrigated areas of Titagarh, West Bengal, India. *Food Control, 20*(10), 942–945.

32. Hu, X. Y., Wen, B., & Shan, X. Q., (2003). Survey of phthalate pollution in arable soils in China. *Journal of Environmental Monitoring,5*(4), 649–653.

33. Hussain, I., Raschid-Sally, L., Hanjra, M. A., Marikar, F., & Van, D. H. W., (2002). *Wastewater Use in Agriculture: Review of Impacts and Methodological Issues in Valuing Impacts* (pp. 1–18). Working Paper 37; Colombo, Sri Lanka: International Water Management Institute.

34. Jastrow, J. D., (1996). Soil aggregate formation and the accrual of particulate and mineral associated organic matter. *Soil Biology and Biochemistry*, *28*(4/5), 665–676.

35. Jimenez, B., (1993). Wastewater reuse to increase soil productivity. *Water Science and Technology*, *32*(12), 173–180.

36. Jimenez, B., & Asano, T., (2008). Water reclamation and reuse around the world. In: Jimenez, B., & Asano, T., (eds.), *Water Reuse-An international Survey of Current Practice, Issues and Needs* (pp. 3–27). London: IWA Publishing.

37. Jimenez, B., Drechsel, P., Koné, D., Bahri, A., Raschid-Sally, L., & Qadir, M., (2010). General wastewater, sludge, and excreta use situation. In: Drechsel, P., Scott, C. A., Raschid-Sally, L., Redwood, M., & Bahri, A., (eds.), *WW Irrigation and Health: Assessing and Mitigating Risk in Low-Income Countries* (pp. 3–29). London/Ottawa/Colombo: Earthscan/IDRC/IWMI.

38. Kay, B. D., (1998). Soil structure and organic carbon: A review. In: Lal, R., Kimble, J. M., Fol-Lett, R. F., & Stewart, B. A., (eds.), *Soil Processes and the Carbon Cycle* (pp. 169–197). Boca Raton: CRC Press.

39. Keswick, B. H., & Gerba, C. P., (1980). Viruses in groundwater. *Environmental Science and Technology*, *14*(11), 1290–1297.

40. Kim, D. Y., & Burger, J. A., (1997). Nitrogen transformations and soil processes in a wastewater irrigated, mature Appalachian hardwood forest. *Forest Ecology and Management*, *90*(1), 1–11.

41. Kummerer, K., (2005). Significance of antibiotics in the environment. *Journal of Antimicrobial Chemotherapy*, *52*(1), 5–7.

42. Lazarova, V., Levine, B., Sack, J., Cirelli, G., Jeffrey, P., Muntau, H., Salgot, M., & Brissaud, F., (2001). Role of water reuse for enhancing integrated water management in Europe and Mediterranean countries. *Water Science and Technology*, *43*(10), 25–33.

43. Lazarova, V., & Bahri, A., (2005). *Water Reuse for Irrigation: Agriculture, Landscapes, and Turf Grass* (p. 336). CRC Press, Boca Raton, USA.

44. Leach, L., Enfield, C., & Harlin, C., (1980). *Summary of Long-Term Rapid Infiltration System Studies* (p. 4). EPA Report EPA–600/2-80-165. Oklahoma: U.S. Environment Protection Agency (EPA).

45. Martens, D. A., & Frankenberger, W. T., (1992). Modification of infiltration rates in an organic-amended irrigated. *Agronomy Journal*, *84*(4), 707–717.

46. McLain, J. E. T., & Williams, C. F., (2010). Development of antibiotic resistance in bacteria of soils irrigated with reclaimed WW. *Proceedings of the 5th National Decennial Irrigation Conference by ASABE* (p. 8).

47. Mennigen, J. A., Stroud, P., Zamora, J. M., Moon, T. W., & Trudeau, V. L., (2011). Pharmaceuticals as neuro endocrine disruptors: Lessons learned from fish on prozac, Part B. *Journal of Toxicology and Environmental Health*, *14*(5/7), 387–412.

48. Obuobie, E., Keraita, B., Danso, G., Amoah, P., Cofie, O. O., Raschid-Sally, L., & Drechsel, P., (2006). *Irrigated Urban Vegetable Production in Ghana: Characteristics, Benefits, and Risks* (p. 150). Report IWMI-RUAF-IDRC-CPWF; Accra, Ghana: IWMI. Available at www.cityfarmer.org/GhanaIrrigateVegis.html (accessed on 20 June 2020).

49. Oron, G., DeMalach, Y., Hoffman, Z., & Manor, Y., (1992). Effect of effluent quality and application method on agricultural productivity and environmental control. *Water Science and Technology*, *26*(7/8), 1593–1601.

50. Patel, J., Millner, P., Nou, X., & Sharma, M., (2010). Persistence of enterohaemorrhagic and non-pathogenic E. coli on spinach leaves and in rhizosphere soil. *Journal of Applied Microbiology*, *108*(5), 1789–1796.

51. Pomati, F., Castiglioni, S., Zuccato, E., Fanelli, R., Vigetti, D., Rossetti, C., & Calamari, D., (2006). Effects of a complex mixture of therapeutic drugs at environmental levels on human embryonic cells. *Environmental Science and Technology*, *40*(7), 2442–2447.

52. Qadir, M., Sharma, B. R., Bruggeman, A., Choukr-Allah, R., & Karajeh, F., (2007). Nonconventional water resources and opportunities for water augmentation to achieve food security in water scarce countries. *Agricultural Water Management*, *87*, 2–22.

53. Quanrud, D. M., Hafer, J., Karpiscak, M. M., Zhang, J., Lansey, K. E., & Arnold, R. G., (2003). Fate of organics during soil-aquifer treatment: Sustainability of removals in the field. *Water Research*, *37*(14), 3401–3411.

54. Russell, J. M., Cooper, R. N., & Lindsey, S. B., (1993). Soil denitrification rates at wastewater irrigation sites receiving primary-treated and anaerobically treated meat-processing effluent. *Bioresource Technology*, *43*(1), 41–46.

55. Seymour, I. J., & Appleton, H., (2001). Food borne viruses and fresh produce. *Journal of Applied Microbiology, 91*(5), 759–773.

56. Shelef, G., & Azov, Y., (1996). The Coming era of intensive wastewater reuse in the Mediterranean region. *Water Science and Technology*, *33*(10/11), 115–125.

57. Shiklomanov, I. A., (1998). *World Water Resources: A New Appraisal and Assessment for the 21st Century* (p. 110). St Petersburg, UNESCO.

58. Stewart, L. W., & Reneau, R. B., (1981). Spatial and temporal variation of fecal coliform movement surrounding septic tank-soil absorption systems in two Atlantic coastal plain soils. *Journal of Environmental Quality, 10*(4), 528–531.

59. Swan, G. E., Cuthbert, R., Quevedo, M., Green, R. E., Pain, D. J., Bartels, P., Cunningham, A. A., & Duncan, N., (2006). Toxicity of diclofenac to gyps vultures. *Biology Letters*, *2*(2), 279–282.

60. Thiele-Bruhn, S., (2003). Pharmaceutical antibiotic compounds in soils: A review. *Journal of Plant Nutrition and Soil Science, 166*(2), 145–167.

61. United States Geological Survey (USGS), (2013). The USGS water science school. *The World's Water-Distribution of Earth's Water*. http://ga.water.usgs.gov/edu/earth-wherewater.html (accessed on 20 June 2020).

62. Van, D. H. W., Ul-Hassan, M., & Ensink, J. H. J., (2002). *Urban Wastewater: A Valuable Resource for Agriculture* (p. 29). Report 63; Colombo: International Water Management Institute, Research.

63. WHO, (1989). *Guidelines for the Safe Use of Wastewater, Excreta and Grey Water*, (Vol. 2, p. 114). Wastewater Use in Agriculture; World Health Organization (WHO), Geneva, Switzerland.

64. World Health Organization, (2006). *WHO Guidelines for the Safe Use of Wastewater, Excreta and Grey Water* (p. 90). Geneva: WHO.

65. Xu, J., Chen, W., Wu, L., Green, R., & Chang, A. C., (2009). Leach ability of some emerging contaminants in reclaimed municipal wastewater-irrigated turf grass fields. *Environmental Toxicology and Chemistry, 28*(9), 1842–1850.

66. Yamamoto, H., Nakamura, Y., & Morigushi, S., (2009). Persistence and partitioning of eight selected pharmaceuticals in the aquatic environment: Laboratory photolysis, biodegradation, and sorption experiments. *Water Research*, *43*(2), 351–362.

Part II
Wastewater Management with Biological Systems

CHAPTER 4

MICROORGANISM-BASED BIOLOGICAL AGENTS IN WASTEWATER TREATMENT: POTENTIAL USE AND BENEFITS IN AGRICULTURE

DJADOUNI FATIMA

ABSTRACT

The wastewater (WW) from domestic/hospital/industrial/municipal/agricultural sites affects adversely the human/animal/marine life, water quality, soil pollution, balance in biodiversity, the spread of microbes/pathogens/bacteria/viruses, and world economy, etc. Bacteria in WW cause many diseases, such as inflammation of the intestine, small intestine ulcers, cholera, typhoid, respiratory diseases, fever, and jaundice; while viruses cause intestinal infections, meningitis, paralysis, jaundice, respiratory diseases, and unusual heart disease. This chapter focuses on the potential use and benefits of microorganisms-based biological agents in WW treatment and agriculture.

4.1 USE OF TREATED WASTEWATER (WW) IN IRRIGATION

Drought, population growth, low water availability, and deterioration of freshwater quality by pollution resulting from a high quantity of waste released into the environment have led to the use of wastewater (WW) in irrigation in water-scarce countries, developing, and the industrialized countries. Recycled WW is a high-quality product that releases nutrients, organic matter, and fertilizers into the soil to promote plant growth and crop production [9, 16].

The WW management contributes to the irrigation of crops, recycling of nutrients, reduces artificial fertilizer costs, minimizes toxic pollutants in soil and plants, and reduces the pathogenic microorganisms in the environment [17, 18].

Ignoring WW management leads to two principle water quality impacts, namely: chemical contamination and microbial pollution. Pollution of healthy, undesirable chemical constituents, salinity effects on soil, crop contamination, microbial pathogens, and waterborne diseases are most environmental and health risks associated with direct or indirect reuse of untreated WW for irrigation [19, 24, 37].

This chapter focuses on the potential use and benefits of microorganism-based biological agents in WW treatment and agriculture.

4.2 MICROORGANISMS IN WASTEWATER (WW)

Bacteria, protozoa, metazoan, filamentous bacteria, algae, fungi, and helminths are major groups of organisms found in WW. WW contains large quantities of bacteria, the majority of which are of fecal origins, such as *Escherichia coli*, *Salmonella*, *Klebsiella*, and *Enterococci* [8, 32].

A wide range of opportunistic pathogenic bacteria [22, 23, 36] can be detected in WW. Among these, bacteria are pathogens that are generally responsible for intestinal infections, such as *Salmonella* or certain types of *E. coli* entero-pathogens (Table 4.1). *Mycobacterium* spp., *Legionella* spp. and *Leptospira* are most bacterial pathogens, which cause non-enteric illnesses [22].

TABLE 4.1 Some Wastewater Microorganisms and Their Effects on Human Health

Microorganisms	Diseases
Bacteria	
Arcobacter spp.	Bacillary dysentery, cholera, typhoid, and
Campylobacter spp.	paratyphoid, epidemic hepatitis, meningitis, gastroenteritis, vomiting, colitis, diarrhea
Clostridium perfringens	(Enterotoxin), enteritis, erythema,
Clostridium tetani	tetanus, inflammation of the urinary
Pseudomonas aeruginosa	tract, pus-accumulation, respiratory tract inflammation, local infection, endocarditis,
E. coli	otitis, and pneumonia.
Enterococci spp.	
Helicobacter spp.	
Klebsiella pneumoniae	

TABLE 4.1 *(Continued)*

Microorganisms	Diseases
Legionella spp.	
Leptospira spp.	
Proteus spp.	
Salmonella typhi	
Salmonella typhimurium	
Salmonella paratyphi	
Shigella dysenteriae	
Staphylococcus aureus	
Streptococcus spp.	
Vibrio cholerae	
Yersinia enterocolitica	
Fungi	
Aspergillus spp.	Broncho-pulmonary mycosis, mycosis of
A. fumigatus	the nails, otitis, granuloma, and mycosis of
A. Niger	the skin.
Trichophyton spp.	
Nematodes	
Ascaris lumbricoides	Infect the small intestine, lungs, intestine,
Anclyostoma duodenale	and internal bodies.
Toxocara canis	
Protozoa	
Cryptosporidium spp.	Giardiasis infects liver, intestine, biliary
Entamoeba histolytica	gall bladder, and internal organs (liver,
Giardia lamblia (Lamblia intestinalis)	brain, heart), cryptosporidiosis.
Sarcocystis spp.	
Toxoplasma gondii	
Tapeworms	
Echinococcus granulosus	Infect intestine, liver, and lungs.
Taenia saginata	
Taenia solium	
Virus	
Adenovirus	Poliomyelitis, fever, respiratory diseases,
Coronavirus	meningitis, myocarditis, hepatitis,
Coxsackie	encephalitis, eye inflammation, and cold.
Echo	

TABLE 4.1   *(Continued)*

Microorganisms	Diseases
Enterovirus	
Hepatite A, B, and C	
Polio	
Yeast	
Candida albicans	Disorders of the mucous membranes
Candida crusei	(mouth), lungs, meningoencephalitis, and granuloma.
Cryptococcus neoformans	

4.3   IRRIGATION WITH WASTEWATER (WW): DYNAMICS OF MICROORGANISM IN SOIL

WW irrigation has high influence on soil parameters and quality, such as heavy metal (cadmium, copper, lead, selenium, zinc, sodium) contents, nutrient types, and organic matter and therefore affect the microbiological parameters (microbial biomass, microbial activities, and enzyme activities), dynamics of pathogen and indicator microorganism content in soil [5, 44].

The activated sludge process is composed of aerobic and anaerobic microorganisms, such as bacteria, archaea, fungi, and protists. It is capable of degrading organic compounds, including petroleum products, toluene, benzopyrene, and neutralizing chemical pollutants such as toxicants and xenobiotics [39].

The organisms used in the activated sludge process include a specific variety of bacterial strains. These strains use oxygen to develop. They work in synergy to effectively degrade sludge, bioremediation, and decontamination water from heavy metal (iron, sulfur, ammonium ferrous sulfate, and ferrous sulfate, sulfate) [21, 30, 33]. Five major groups of microorganisms are generally found in the aeration basin of the activated sludge process:

- **Algae and Fungi-Fungi:** It is present with pH changes and older sludge.
- **Bacteria:** Aerobic bacteria remove organic nutrients.
- **Filamentous Bacteria:** Bulking sludge (poor settling and turbid effluent).
- **Metazoa:** Dominate longer age systems including lagoons.
- **Protozoa:** Remove and digests dispersed bacteria and suspended particles.

Some examples of bacteria, which are commonly used in the activated sludge process, are *Bacillus* and *Pseudomonas* species [26, 27, 29, 43]. *Trichoderma harzianum*, *Bacillus subtilis* QST 713, *B. pumilus* QST 2808, *B. megaterium* ATCC 14581 and *Pseudomonas chlororaphis* MA 342 are bacteria that act through a fungistatic and fungi toxic action, but also through stimulation of the plant's defense, decomposition of cellulose and other organic waste components; and produces bioactive substances that evolve the defenses against biotic stress [3, 4, 7, 25, 38].

There are also ubiquitous microorganisms of water environments (storage basins) capable of causing non-enteric infections, such as bacteria of the genera *Pseudomonas, Aeromonas, Campylobacter, Legionella, Mycobacterium, Bacillus, Streptomyces*, and *Leptospira*. Important roles of these are: (i) improved nutrient uptake by the plant; (ii) improved bioavailability of nutrients in the soil; (iii) stimulation of the degradation of organic matter; (iv) produced a variety of biocontrol agents and bioactive compounds with antibacterial and antifungal properties [10, 11, 13–15, 20].

4.4 WASTEWATER (WW) VERSUS AGRICULTURAL CROPS

One of the most important vegetables irrigated with sewage water include lettuce, mint, parsley, potatoes, cucumber, squash, pumpkin, watercress, and radish, it is noted that WW contains high levels of nutrients required for agricultural crop production and yield, such as organic matter, inorganic compounds, micronutrients, nitrogen, phosphorus, and microorganisms [12, 31, 46].

WW nutrients will act as the sole source of soil fertilizers, which contain dissolved minerals, phosphorus, nitrogen, sodium, potassium, iron, calcium, and compounds (such as fats, sugars, and protein) and these substances are used by soil microorganisms as foods and source of energy for the synthesis of cell components and to maintain life processes [34, 44].

Treated WW is used to the irrigation of crops that are not intended for direct human consumption (restricted irrigation), and thus covers irrigation of industrial crops (e.g., cotton, sisal, and sunflower); crops processed prior to consumption (e.g., wheat, barley, oats), and fruit trees, fodder crops, and pastures. Unrestricted irrigation, on the other hand, refers to all crops grown for direct human consumption, including those eaten raw (e.g., lettuce, salads, cucumber, etc.), and irrigation of sports fields and public parks [44, 45].

4.5 BIOLOGICAL PROCESSES FOR URBAN WASTEWATER (WW) TREATMENT

These processes are based on the use of microorganisms for the conversion of organic contaminants into less toxic compounds (mineralization into carbon dioxide, water, and inorganic salts in specific bioreactors). Several studies describe the biological treatment of contaminated water by various chemical pollutants such as perchlorates, bromates, and polycyclic aromatic hydrocarbons [40, 45].

These processes have disadvantages, such as the addition of additional energy and the pumping of the water to be treated. Moreover, these processes are generally not applicable to high concentrations of pollutants of high toxicity or very low biodegradability [45].

4.5.1 MAIN PURIFYING BACTERIA

Biological purification due to the action of the bacteria allows the decomposition of organic matter by nitrification in an aerobic zone, denitrification in the absence of oxygen, and possibly an anaerobic zone at depth, which ensures the digestion of other organic compounds phosphates and sulfates).

4.5.1.1 CARBONATE-REDUCING BACTERIA

They are anaerobic bacteria capable of oxidizing organic compounds, which are predominantly presented by methanogen producing methane CH_4, generating bacteria, such as *Methanococcus*, *Methanosarcina*, sporulated *Methanobacillus,* and non-sporulated *Methanobacterium* [41].

4.5.1.2 NITRIFYING AND DENITRIFYING BACTERIA

Following nitrifying and denitrifying bacteria are participate in biological purification by reducing nitrogen accumulation in WW, elimination of nitrate by denitrification and decreasing the eutrophication of sewage water ecosystems [2, 35]:

1. *Achromobacter alcalinigens;*

2. *Azospirillum brasilense;*
3. *B. azotoformans;*
4. *B. licheniformis;*
5. *Ch. Lividum;*
6. *Chromobacterium violaceum;*
7. *Corynebacterium nephridia;*
8. *Halobacterium marismortui;*
9. *Kingella denitrificans;*
10. *N. flavescens;*
11. *N. mucosa;*
12. *N. subflava;*
13. *Neisseria sicca;*
14. *Nirozomonas;*
15. *Nitrobacter;*
16. *Paracoccus halodenitrificans;*
17. *Peudomonas;*
18. *Pr. Cidipropionici;*
19. *Propionibacterium pentosaceum;*
20. *Sp. psychrophilum;*
21. *Spirillum lipoferumi;*
22. *Thiobacillus denitrificans;* etc.

4.5.1.3 *DEPHOSPHATOUS BACTERIA ACINETOBACTER*

The following *Dephosphatous Bacteria Acinetobacter* are main *Acinetobacter* species that are dominant in sewage and raw WW, and play important role in activated sludge:

1. *A. baylyi;*
2. *A. bouvetii;*
3. *A. gerneri;*
4. *A. grimontii;*
5. *A. johnsonii;*
6. *A. junii;*
7. *A. kyonggiensis;*
8. *A. lwoffii;*
9. *A. pakistanensis;*
10. *A. rudis;*
11. *A. tandoii;*

12. *A. tjernbergiae;*
13. *A. towneri;*
14. *Acinetobacter baumannii.*

Dephosphatation process is carried out by alternating the anaerobic/ aerobic sequences by modifying the enzymatic equilibrium and subsequently induced phases of phosphorus accumulation. Under anaerobic condition, the bacterium releases phosphorus, and then, as the oxygen concentration rises, it reabsorbs it. The assimilation of polyphosphate by the bacterium can be used either as a reserve of energy or as phosphorus [1, 6, 47].

4.5.1.4   SULFATE-REDUCING BACTERIA

Sulfate-reducing bacteria (*Desulfovibrio desulfiricans, Desulfobacterium autotrophicum,* and *Desulfobulbus propionicus*) are anaerobic heterotrophs and autotrophs bacteria, which play an interesting role in organic matter mineralization, biocorrosion, and sulfur-cycle based WW treatment process to allow the reduction of sulfate to sulfide and decomposition of WW sediments [28, 42].

4.6   SUMMARY

The use of treated WW in agriculture can be beneficial to the ecosystems and agricultural production if it is carefully planned and managed. However, WW negatively affects humans and the surrounding environment; as they contribute to the spread of microbes and pathogens, which are harmful to human health and increase the likelihood of being infected by incurable diseases. Bacteria in WW cause many diseases, inflammation of the intestine, small intestine ulcers, cholera, typhoid, respiratory diseases, fever, and jaundice, while viruses cause intestinal infections, meningitis, paralysis, jaundice, respiratory diseases, and unusual heart disease. The modesty of human infection is common with diarrhea and amoeba and epidemics of hepatitis and other diseases. It dissolves oxygen in water by microbes, leading to the death of aquatic organisms, and the appearance of rot in water, and the spread of odors. Imbalance of biodiversity, pollution of soil is on the run and is reaching agricultural land.

KEYWORDS

- biological control
- environmental stress
- indigenous microorganisms
- irrigation wastewater
- microbial treatment of wastewater
- pathogens microorganisms
- soil profile
- wastewater

REFERENCES

1. Abbas, S., Ahmed, I., Kudo, T., Iida, T., Ali, G. M., & Fujiwara, T., (2014). Heavy metal-tolerant and psychrotolerant bacterium *Acinetobacter pakistanensis* sp. nov. isolated from a textile dyeing wastewater treatment pond. *Pakistan Journal of Agricultural Sciences, 51*, 593–606.
2. Abeliovich, A., (1987). Nitrifying bacteria in wastewater reservoirs. *Applied and Environmental Microbiology, 53*(4), 754–760.
3. Abhilash, M., Shiva, R. D. M., Mohan, B. K., Nataraja, S., & Krishnappa, M., (2010). Isolation and molecular characterization of *Bacillus megaterium* isolated from different agro-climatic zones of Karnataka and its effect on seed germination and plant growth of *Sesamum indicum*. *Research Journal of Pharmaceutical, Biological, and Chemical Sciences, 1*(3), 3–10.
4. Abriouel, H., Franz, C. M. A. P., Ben, O. N., & Gálvez, A., (2011). Diversity and applications of *Bacillus* bacteriocins. *FEMS Microbiol. Review, 35*, 201–232.
5. Agnieszka, C. K., & Magdalena, Z., (2016). Bacterial communities in full-scale wastewater treatment systems. *World Journal of Microbiology and Biotechnology, 32*, 66–69.
6. Al Atrouni, A., Joly-Guillou, M. L., Monzer, H., & Kempf, M., (2016). Reservoirs of non *baumannii Acinetobacter* species. *Frontiers in Microbiology, 7*, 49–53.
7. Amalraj, E. L., Maiyappan, S., & John, P. A., (2012). *In vivo* and *In vitro* studies of *Bacillus megaterium* var. *phosphaticum* on nutrient mobilization, antagonism, and plant growth promoting traits. *Journal of Ecobiotechnology, 4*(1), 35–42.
8. Baduru, L. K., & Sai, G. D. V., (2015). Effective role of indigenous microorganisms for sustainable environment. *Biotechnology, 5*, 867–876.
9. Benoît, P. C., Weixiao, Q., Huijuan, L., Beat, M., & Michael, B., (2012). Sources and pathways of nutrients in the semi-arid region of Beijing-Tianjin, China. *Environmental Science and Technology, 46*(10), 5294–5301.

10. Chakraborty, U., Chakraborty, B., & Basnet, M., (2006). Plant growth promotion and induction of resistance in *Camellia sinensis* by *Bacillus megaterium*. *Journal of Basic Microbiology*, *46*, 186–195.

11. Chun, H. J., Fang, W., Zhen, Y. Y., Ping, X., Hong, J. K., Hong, W. L., Yi, Y. Y., & Jian, H. G., (2015). Study on screening and antagonistic mechanisms of *Bacillus amyloliquefaciens* 54 against bacterial fruit blotch (BFB) caused by *Acidovorax avenae* subsp. *Citrulli*. *Microbiological Research*, *170*, 95–104.

12. Deepak, B., Ansari, M. W., Ranjan, K. S., & Narendra, T. B., (2014). Biofertilizers function as key player in sustainable agriculture by improving soil fertility, plant tolerance, and crop productivity. *Microbial Cell Factories*, *13*, 66–70.

13. Devi, K. K., & Natarajan, K. A., (2015). Isolation and characterization of a bioflocculant from *Bacillus megaterium* for turbidity and arsenic removal. *Minerals and Metallurgical Processing*, *32*(4), 222–229.

14. Djadouni, F. M., (2007). *Production and Isolation of Bacteriocins from Bacilli Isolates* (p. 118). MSc Thesis; Alexandria University, Alexandria-Egypt.

15. Djadouni, F., & Madani, Z., (2016). Recent advances and beneficial roles of *Bacillus megaterium* in agricultures and other fields. *International Journal of Biology, Pharmacy and Applied Sciences*, *5*(12), 3160–3173.

16. FAO, (2008). *Water at a Glance: The Relationship Between Water, Agriculture, Food Security and Poverty* (p. 18). Water Development and Management Unit, Food and Agriculture Organization (FAO) of the United Nations, Rome, Italy.

17. FAO, (2010). *The Wealth of Waste: The Economics of Wastewater Use in Agriculture* (p. 35). Water Development and Management Unit, Food and Agriculture Organization of the United Nations, Rome, Italy.

18. FAO, (2002). *Agricultural Drainage Water Management in Arid and Semi-Arid Areas* (p. 61). FAO Irrigation and Drainage; Rome, Italy.

19. FAO, (2010). *Gender Differences in Assets* (p. 121). Document *rédigé par l'équipe charge du rapport sur la situation mondiale de l'alimentation et de l'agriculture* (Document prepared by the report team on the state of food and agriculture). Food and Agriculture Organization of the United Nations, Rome, Italy.

20. Gopinath, S. M., Ismail, S. M., & Ashalatha, A. G., (2015). Bioremediation of lubricant oil pollution in water by *Bacillus megaterium*. *International Journal of Innovative Research in Science, Engineering, and Technology*, *4*(8), 6773–6780.

21. Jayakumar, P., & Natarajan, S., (2012). Microbial diversity of vermin compost bacteria that exhibit useful agricultural traits and waste management potential. *Springer PLus*, *1*, 26–30.

22. João, P. S. C., (2010). Water microbiology: Bacterial pathogens and water. *International Journal of Environmental Research and Public Health*, *7*(10), 3657–3703.

23. Keiser, J., & Utzinger, J., (2005). Food-borne trematodiasis: An emerging public health problem. *Emerging Infectious Diseases*, *11*(10), 1503–1510.

24. Khaled, S., & Balkhair, S., (2016). Microbial contamination of vegetable crop and soil profile in arid regions under controlled application of domestic wastewater. *Journal of Biological Sciences*, *23*(1), S83–S92.

25. Khan, J. A., (2011). Biodegradation of azo dye by moderately halotolerant *Bacillus megaterium* and study of enzyme azoreductase involved in degradation. *Advanced Biotechnology*, *10*, 21–27.

26. Kildea, S., Ransbotyn, V., Khan, M. R., Fagan, B., Leonard, G., Mullins, E., & Doohan, F. M., (2008). *Bacillus megaterium* shows potential for the biocontrol of *Septoria tritici* blotch of wheat. *Biological Control, 47*, 37–45.

27. Kumar, A., Prakash, A., & Johri, B. N., (2011). *Bacillus* as PGPR in crop ecosystem: Chapter 2. In: Maheshwari, D. K., (ed.), *Bacteria in Agrobiology: Crop Ecosystems* (p. 37–59). Springer-Verlag Berlin Heidelberg.

28. 28. Lau, G. N., Sharma, K. R., Chen, G. H., & Van, L. M. C. M., (2006). Integration of sulfate reduction, autotrophic denitrification and nitrification to achieve low-cost excess sludge minimization for Hong Kong sewage. *Water Science and Technology, 53*(3), 227–235.

29. Léon, M., Yaryura, P. M., Montecchia, M. S., Hernández, A. I., Correa, O. S., Pucheu, N. L., Kerber, N. L., & Garćia, A. F., (2009). Antifungal activity of selected indigenous *Pseudomonas* and *Bacillus* from the soybean rhizosphere. *International Journal of Microbiology, 10*, 9–13.

30. Liang, L., Zhigang, Z., Xiaoli, H., Xue, D., Changan, W., Jinnan, L., Liansheng, W., & Qiyou, X., (2016). Isolation, identification, and optimization of culture conditions of a bioflocculant-producing bacterium *Bacillus megaterium* SP1 and its application in aquaculture wastewater treatment. *BioMed Research International, 25*, 9–14.

31. Lopez-Bucio, J., Campos-Cuevas, J. C., Hernandez-Calderon, E., Velasquez-Becerra, C., Farias-Rodriguez, R., Macias-Rodriguez, L. I., & Valencia-Cantero, E., (2007). *Bacillus megaterium* rhizobacteria promote growth and alter root-system architecture through an auxin-and ethylene-independent signaling mechanism in *Arabidopsis thaliana*. *Molecular Plant Microbe Interactions, 20*, 207–217.

32. Michael, H. G., (2006). *Wastewater Bacteria* (p. 251). John Wiley & Sons, Inc., Hoboken, New Jersey [Published simultaneously in Canada.

33. Narmadha, D., & Mary, S. K., (2012). Treatment of domestic wastewater using natural flocculants. *Pakistan Journal of Zoology, 45*(6), 1655–1662.

34. Richardson, A. E., Barea, J. M., McNeill, A. M., & Prigent-Combaret, C., (2009). Acquisition of phosphorus and nitrogen in the rhizosphere and plant growth promotion by microorganisms. *Plant Soil, 321*, 305–339.

35. Roger, K., (1982). Denitrification. *Microbiological Reviews, 46*(1), 43–70.

36. Samuel, B., (1996). *Medical Microbiology* (4th edn., p. 310). The University of Texas Medical Branch, Galveston, Texas; ISBN-10: 0-9631172-1-1.

37. Shakir, E., Zahraw, Z., & Al-Obaidy, A. H., (2016). Environmental and health risks associated with reuse of wastewater for irrigation. *Egyptian Journal of Petroleum, 26*(1), 95–102.

38. Shumaila, S., Saira, A., & Abdul, R., (2013). Characterization of cellulose degrading bacterium, *Bacillus megaterium* S3, isolated from indigenous environment. *Pakistan Journal of Zoology, 45*(6), 1655–1662.

39. Stottmeister, U., Wiebner, A., Kuschk, P., Kappelmeyer, U., Kästner, M., Bederski, O., Müller, R. A., & Moormann, H., (2003). Effects of plants and microorganisms in constructed wetlands for wastewater treatment. *Biotechnology Advances, 22*(1/2), 3–117.

40. Swiontek, M. B., Urszula, J., Aleksandra, B., & Maciej, W., (2014). Chitinolytic microorganisms and their possible application in environmental protection. *Current Microbiology, 68*, 71–81.

41. Tingting, Z., & Dittrich, M., (2016). Carbonate precipitation through microbial activities in natural environment, and their potential in biotechnology: A review. *Frontiers in Bioengineering and Biotechnology*, *4*, 4–10.

42. Tsukasa, I., Satoshi, O., Hisashi, S., & Yoshimasa, W., (2002). Successional development of sulfate-reducing bacterial populations and their activities in a wastewater biofilm growing under microaerophilic conditions. *Applied and Environmental Microbiology*, *68*(3), 1392–1402.

43. Wang, Y. H., Dong, B., Mao, Y. B., & Yan, Y. S., (2010). Bioflocculant producing *Pseudomonas alcaligenes*: Optimal culture and application. *The Chinese Journal of Environmental Science and Technology*, *33*(3), 68–71.

44. WHO, (2006). WHO guidelines for the safe use of wastewater, excreta, and grey water. In: *Wastewater Use in Agriculture* (3rd edn., p. 95). World Health Organization (WHO), Geneva, Switzerland.

45. WHOA, (2005). *Regional Overview of Wastewater Management and Reuse in the Eastern Mediterranean Region* (p. 67). World Health Organization, Regional Office for the Eastern Mediterranean Regional, California Environmental Health Association, Cairo, Egypt.

46. Yazdani, M., Bahmanyar, M. A., Pirdashti, H., & Esmaili, M. A., (2009). Effect of phosphate solubilization microorganisms (PSM) and plant growth promoting rhizobacteria (PGPR) on yield and yield components of Corn (*Zea mays* L.). *Proceedings of World Academy of Science Engineering and Technology*, *37*, 90–92.

47. Zhang, Z., Li, H., Zhu, J., et al., (2011). Improvement strategy on enhanced biological phosphorus removal for municipal wastewater treatment plants: Full-scale operating parameters, sludge activities, and microbial features. *Bioresources Technology*, *102*, 4646–4653.

CHAPTER 5

BIOSORPTION OF HEAVY METALS BY CYANOBACTERIA FROM TEXTILE EFFLUENTS

B. JEBERLIN PRABINA, K. KUMAR, and N. O. GOPAL

ABSTRACT

A diverse group of microorganisms including cyanobacteria can accumulate heavy metals in their cells. The metabolism of cyanobacteria is highly flexible and is adapted to diverse environmental conditions. Cyanobacteria can remove the pollutants, heavy metals, and toxic organic compounds that are present in industrial effluents since they possess the ability to excrete polysaccharides. Textile industry being one of the largest consumers of water, the effluent generated from the textile industry is a major cause for water pollution that is loaded with heavy metals and dyes. This chapter focuses on the potential of cyanobacterial isolates *Anabaena*-TE1 and *Nostoc*-TE1 in remediating the textile effluents. The bioassay carried out in sunflower indicates the absence of stress enzymes, peroxidase, and catalase proving the potential of cyanobacteria in removing the heavy metals from the polluted water.

5.1 INTRODUCTION

Environmental pollution is one of the most alarming dangers that challenges mankind today and has become an international issue. The vast amount of industrial wastes and drains generated due to urbanization and industrialization are discharged into the environment and affect the life support systems, thus unbalancing the structural and functionality of the ecosystems. Thus, it is necessary for mankind to direct his intellectual power not only to technological development, but also towards the protection of natural life-support systems. China stands first in the textile industry and the Indian textile industry is the second largest in the world employing about 20 million persons. In India, there are 1,200 medium to large scale textile mills. Twenty percent of these mills are located in Coimbatore, Tamil Nadu. The annual water consumption for various processes by these textile industries is 829.8 million m^3 with a discharge of 637.30 million m^3 [65].

The textile industry in India guzzles double the accepted amount of consumption of water (200 to 250 m^3/ton of cotton cloth), while the global best is less than 100 m^3/ton of cotton cloth. A major factor in the textile industry is the obsolete technology, which permits recycling and reuse of process water in an economic manner. The chemical processes are expensive for treating a large volume of wastewater (WW). As an alternate, microbes-based processes have become an ideal choice as they are eco-friendly and cost-effective. The cyanobacteria in the algal system could prove beneficial in treating the wastes in different ways, since they bring about oxygenation and mineralization [14, 60]. Algae need simple dissolved nutrients and are flexible in adaptation that makes them suitable for treatment of WW. The potential of the cyanobacteria to reduce the pollutants level in industrial WW has been studied [1, 25, 54, 66]. *Chlorella* sp. and *Scenedesmus* sp. could reduce BOD, COD, and high level of nitrogenous compounds of different industrial WW [63].

The ability of producing copious amounts of exo-polysaccharides (EPSs) is beneficial for the cyanobacteria as they could immobilize heavy metals. The carboxyl groups on algal cell biomass act as bio-scavenging sites for binding various ions [15]. Reports also indicate the participation of intracellular polyphosphates in metal sequestration [72]. The capacity of algae to reduce the heavy metals (*viz.*, iron, zinc, and copper) has been tested by Sengar et al. [55]. Various studies on the uptake of heavy metal ions by the biomass of cyanobacteria have been performed. This chapter focuses on the role of cyanobacterial EPSs in binding the heavy metals.

5.2 REVIEW OF LITERATURE

5.2.1 THE CYANOBACTERIA AND GREEN ALGAE

Cyanobacteria (commonly known as blue-green algae) are an ancient and widespread group of microorganisms belonging to the eubacteria. They are found in almost all environments on earth. Some are found on the surface of glaciers, inside rocks, in hot deserts, and in hot springs. They have been found in fossil form in rocks as old as 3.5 billion years. Cyanobacteria come in many different forms, such as unicellular or form filaments of cells connected to each other. Traditionally, taxonomists divide cyanobacteria into five major groups based on their mode of cell division, whether or not they form filaments, and whether those filaments are able to differentiate into heterocysts. These groups are [8]:

1. **Chroococcales:** Unicellular reproducing by binary fission or budding.
2. **Pleurocapsales:** Unicellular divides by multiple fission.
3. **Oscillatoriales:** Filamentous, non-heterocyst forming, and dividing in one plane.
4. **Nostocales:** Filamentous heterocyst forming and dividing in one plane.
5. **Stigonematales:** Filamentous, heterocyst forming, and dividing in more than one plane.

The green algae are commonly known as chlorophytes. They commonly occur in near-shore marine environments, rocky shorelines of eutrophic lakes, snow, and form extensive coatings on terrestrial surfaces including mud, rocks, wood, and tree bark. Based on flagellation and other characteristics, green algae are divided into five major groups [17]: Prasinophyceae, Ulvophyceae, Trebouxiophyceae, Chlorophyceae, and Charophyceae.

Donna Johnson, Co-ordinator of Aquatic Species Programme of Solar Energy Research Institute, Golden Colorado (USA), opined that algae can produce anything, but the trick is to find an organism that makes what we want. Thus, algae find its application in a wide area right from the production of bio-chemicals, as a protein source to treat WW. The applications of algae as bio-scavenger of pollutants relevant to this work are reviewed in this chapter.

5.2.2   ALGAE IS AN IDEAL WASTE REMOVER

Algae have often been considered as an ideal waste remover for effluents because of their requirement for dissolved forms of nutrients, which are major components of WW. Another useful characteristic feature of algae is that they produce extracellular organic material, which can bind with dissolved metals, thereby reducing or eliminating metal toxicity. The cell surfaces of algae are so made that they provide sites for the dissolved organic materials and metals to get adsorbed. Each of these properties of algae makes them potentially useful for the removal of a wide variety of waste products in WW [49]. The cell wall polysaccharides of several species of algae affect ion exchange reactions with polyvalent metals. For example, Fucoidin from *Ascophyllum nodosum* can bind lead [42].

Carrageenan from *Euchema striatum* and *Euchema spinosum* strongly bound lead and cadmium [69]. Various solution parameters such as pH, the nature, and concentration of organic solutes and ionic solutes can affect algal-metal interactions. Biological surfaces contain a variety of surface functional sites that consist of carboxylic, amino, sulfhydryl, and other groups [11] wherein solutes may specifically or non-specifically adsorb to surfaces [29]. Biological surfaces are further complicated by the fact that some sites are thought to be physiologically active and aid in intracellular transport, while others is physiologically inactive. Algae have adaptive mechanisms to detoxify metals. Jones et al. [26] observed involvement of plasmalemma enzymes in detoxification process.

Hunstman and Sunda [23] discussed another adaptation mechanism, which include changes in cell wall permeability to metals and compartmentalization of trace metals into inert intracellular sites. It is clear from earlier works that algae have the appropriate biochemical pathways to make them potentially useful for the reclamation of WW.

5.2.3 PHYSICOCHEMICAL NATURE OF EFFLUENTS: TEXTILE AND DYEING FACTORIES

The physical, chemical, and biological nature of the effluents varies and the color of WW is mainly due to the presence of dyes [5]. The physicochemical properties of textile and dyeing factory effluent are given in Table 5.1.

TABLE 5.1 Physicochemical Properties of Effluents from Textile and Dyeing Factories

Characteristics	Range	References
Bicarbonates	9.00 me L^{-1}	[46]
Biological oxygen demand (BOD)	400–800 mg L^{-1}	[21]
Calcium	0.5–2.0 mg L^{-1}	[21]
Carbonates	4–16 meq L^{-1}	[21]
Chemical oxygen demand (COD)	1160–1760 mg L^{-1}	[30]
Chlorides	300–570 mg L^{-1}	[30]
Cobalt	0.001 mg L^{-1}	[46]

TABLE 4.1 *(Continued)*

Characteristics	Range	References
Color	Blue	[44]
Copper	0.072 mg L^{-1}	[46]
Dissolved oxygen (DO)	1.2–1.9 mg L^{-1}	[44]
Electrical conductivity (EC)	8.5–13.9 dSm^{-1}	[21]
Heavy metals		
Hexavalent chromium	0.2–0.6 mg L^{-1}	[30]
Iron	0.106 mg L^{-1}	[46]
Lead	1.342 mg L^{-1}	[46]
Magnesium	20–48 meq L^{-1}	[46]
Manganese	0.147 mg L^{-1}	[46]
Nickel	0.112 mg L^{-1}	[46]
Oil and grease	900 mg L^{-1}	[12]
pH	9–11	[2]
	10–12	[21]
Phosphates	20.0–25.0 mg L^{-1}	[30]
Potassium	1.28 meq L^{-1}	[46]
Residual sodium carbonate (RSC)	11–90 meq L^{-1}	[2]
Sodium	600–3500 mg L^{-1}	[2]
Sodium absorption ratio (SAR)	19.72 meq L^{-1}	[46]
Sulfates	660–1600 mg L^{-1}	[30]
Total chromium	5–20 mg L^{-1}	[30]
Total dissolved solids (TDS)	5120–7180 mg L^{-1}	[30]
Total nitrogen	45 mg L^{-1}	[30]
Total suspended solids (TSS)	1550–1950 mg L^{-1}	[30]
Zinc	0.108 mg L^{-1}	[46]

5.2.3 IMPACTS OF TEXTILE EFFLUENTS ON ENVIRONMENT

5.2.3.1 IMPACT ON IRRIGATION WATER

The textile mill effluents are discharged into the river stream and can adversely affected the water quality rendering the downstream water not suitable for drinking, bathing, irrigation, and fish culture [38]. Similar observations were reported by Gupta and Jain [21]. Kihage and Magnusson [28] reported untreated textile effluent when discharged into the Noyyal River deteriorated the water quality. Mohan-Rao [37] reported that the effluent discharged from textile and other industries influence the biological systems of water resources. The release of industrial effluents into the water bodies increased the BOD and COD and the same was reported for river water [68]. The high BOD and COD could deplete the dissolved oxygen (DO) content of the river water thereby creating an anaerobic condition in the river bed, thus affecting the aquatic life. The entry of hydrogen sulfide, ammonia, and chloride into the water bodies through the textile mill effluent was found to be highly toxic to fish [36].

5.2.3.2 IMPACT ON GROUND WATER

Apart from river water, the pollution of underground water due to the disposal of effluent into the river [47] and on open land [48] has also been reported. Seepage of the effluent into wells situated on banks resulted in contamination with the increase in EC and SAR [21]. Similarly, the well water in adjoining areas of rivers showed increased pH, EC, chloride, sulfate, and calcium [43]. The concentration of sodium and chloride was increased with a proportional increase in total dissolved solids (TDS) and EC in well water [27].

5.2.3.3 IMPACT ON CROP PLANTS

Industrial WW is tested for its suitability for irrigation both by scientists and industries. Somashekar et al. [58] reported that effluent from textile mills inhibited the germination and plant height of jowar, bajra, and paddy even at 25% concentration. The raw dyeing factory effluent at various concentrations significantly reduced the germination of seed and vigor index of paddy, finger millet, cowpea, soybean, and maize [44]. However, Swaminathan et al. [62] showed that the diluted dyeing factory effluent favored the groundnut seed germination, hypocotyl development, and seed vigor. And the diluted effluent increased the chlorophyll and protein content of the seedlings. Vijayarengan et al. [71] found that green gram seeds (ADT-3) soaked in textile mill effluent at pH 9.15, with high BOD, COD, chloride, sulfate, and sodium showed 90% germination. The textile effluent was also used for the establishment of tree species of economic value by adopting amendments [2].

5.2.4   BIO-SCAVENGING OF POLLUTANTS BY ALGAE

During recent years, few studies have dealt with the uptake and accumulation of heavy metals and other pollutants by green algae and cyanobacteria. Algae have appropriate biochemical pathways for the uptake of nitrogen and phosphorus from the environment. The ability to use different forms of nitrogen (such as gaseous nitrogen, nitrate, nitrite or ammonium) appears to be general among algae. Salomonson [52] purified and characterized nitrate reductase from *Chlorella* and from the diatom, *Thalassiosira*. Nitrite reductase has been studied in *Anabaena, Dunaliella* [19], and *Chlorella* [73]. Light seems to influence the uptake and assimilation of nitrogen by stimulating the uptake and reduction of nitrate nitrite and hydroxylamine. This could be attributed to direct photo-reduction of the nitrogen compounds to the supply of energy *via* photo-phosphorylation or by an indirect effect of photosynthesis providing carbon skeletons to accept reduced nitrogen [18].

The major form, in which the microalgal cells assimilate phosphorus, is inorganic phosphorus ($H_2PO_4^-$ or HPO_4^{2-}). Organic phosphate might be used as a primary source of phosphorus, but it must be hydrolyzed by extracellular enzymes, such as phospho-esterases or phosphatases [50]. Phosphate uptake by algae is an energy-dependent reaction and is stimulated by light [56]. Phosphate in the cell is channeled into polyphosphate and various organic compounds and nucleotides, such as ATP, nucleic acids, phospholipids, and sugar phosphates.

Assadi [4] used different algal species (*viz., Phagus, Euglena, Scenedesmus, Chlamydomonas,* and *Chlorella*) for the purification of textile WW and reported 81 and 78% reduction of BOD and COD content, respectively. Tam and Wong [63] reported the reduction of BOD, COD, and high levels of nitrogenous compounds by algal cultures in different industrial WW.

Algae can accumulate trace metals from the aqueous environment and it is an economic process for the reclamation of WW [13] that they bioconcentrate in cell surfaces. Hassett et al. [22] observed removal of 99% of trace metals and 85% of trace organics present in waste effluent from mining operations by algal systems. Smith [57] noticed intracellular metal binding in *Nostoc muscorum* and described that it was closely related to the presence of polyphosphate granules within the cells. The maximum metal binding capacity of *Chroococcus paris* has been determined and is 53, 110, and 65 mg g^{-1} of the dry algal weight of cadmium, copper, and zinc, respectively [31]. Schencher and Driscoll [53] reported removal of copper and lead within an equilibrium period of 8 h in a pH range of 7.5–8.9 by *Nostoc muscorum*. Cell surfaces of microalgae show strong adsorption of metal ions including lead, copper, nickel, zinc, cadmium, silver, and mercury.

Greene et al. [20] applied lyophilized preparations of whole cells of *Chlorella vulgaris* complexed to silica gel and observed adsorption of copper. *Phaeodactylum tricornutum* released extracellular products, which complexed with metals, such as, lead [26]. Sengar et al. [55] used mixed algal cultures and found complete removal of Fe^{2+}, Zn^{2+}, and Cu^{2+} on 30th day of growth of cyanobacteria. Immobilized cyanobacterial systems showed higher rate of metal uptake in comparison to free cells, which indicates potential application of immobilized algal cells as a continuous system for heavy metal recovery [41]. Uptake of metals in algae varied with metal concentration, pH values, and growth conditions [10]. Taneja and Fatma [64] observed direct correlation between copper uptake and amount of copper present in the surrounding medium by *Anacystis nidulans*. Jha et al. [24] observed dramatic response by *Westiellopsis prolifica* and *Anabaena* sp. to salt stress by excess production of polysaccharides. Therefore, tress induced polysaccharide production is advantageous.

5.3   MATERIALS AND METHODS

5.3.1   *ISOLATION, PURIFICATION, AND MAINTENANCE OF CYANOBACTERIAL CULTURES FROM EFFLUENT SAMPLES*

Representative effluent sample was serially diluted to 10^{-3} and plated in N-free BG-11 medium for the isolation of cyanobacteria. The plates were incubated in algal culture room fitted with white fluorescence light of 3000 lux maintained at a temperature of 23°C for 21 days. Well-developed colonies were observed under microscope and purified for further studies.

The algal isolates were purified by adopting the method described by Castenholz [7]. The blue-green algal culture for purification was homogenized in nitrogen-free BG-11 medium and purified by the streak plate method. A loopful of the homogenized culture was streaked over N-free BG-11 solid medium containing the antibiotic cycloheximide at 100 ppm concentration. The Petri plates were incubated under cool white fluorescent lamps (3000 lux) at a temperature of 23°C under a 16:8 h light: dark photoperiod for 21 days. Well-developed single colonies were observed under binocular microscope for the respective characteristic features.

5.3.2 MOLECULAR IDENTIFICATION OF THE CYANOBACTERIAL ISOLATES

The cyanobacterial isolates were identified morphologically based on the cyanobacterial taxonomy handbook [3]. The molecular level identification of the cyanobacterial isolates was done using PCR by comparing the DNA molecular markers generated for the cyanobacterial isolates with two STRR (short tandemly repeated repetitive) primers (CRA22 CCGCAGCCAA, CRA23 GCGATCCCA) with those obtained for standard cyanobacterial cultures, obtained from The Centre for Conservation and Utilization of Blue Green Algae (CCUBGA), Indian Agricultural Research Institute, New Delhi.

The total genomic DNA from the cyanobacterial cultures was isolated by the standard CTAB (cetyl trimethyl ammonium bromide) method described by Melody [34] with slight modifications. The DNA pellet obtained was dissolved in 50μL of TE buffer and stored at- 20°C for further use. Quantification of DNA was done with the method described by Brunk et al. [6] using fluorometer at 260 nm and expressed at $ng\mu L^{-1}$. Amplification of genomic DNA with the STRR primers was carried out with 20.0 μL of the reaction mixture (25 $ng\mu L^{-1}$ DNA-3.0 μL, 10 mM dNTP's-1.0 μL, Primer-1.2 μL, 10X Assay buffer-2.0 μL, Taq polymerase-0.2 μL, Sterile distilled water-12.6 μL). The thermal cycler was programmed as follows:

Profile 1:	95°C for 6 minutes	Initial denaturation
Profile 2:	94°C for 1 minute	Denaturation
Profile 3:	56°C for 1 minute	Annealing
Profile 4:	65°C for 5 minutes	Extension
Profile 5:	65°C for 16 minutes	Final extension
Profile 6:	4°C forever	

Note: Profiles 2, 3, and 4 were programmed for 35 cycles.

The PCR products thus obtained were separated on 1.5% horizontal agarose gels with 10 μL of each sample. Ten μL of the PCR product was mixed with 5 μL of tracking dye and loaded into the well. Electrophoresis was carried out with a 1X TBE buffer. After the separation of the PCR products with 1.5% agarose gel, it was viewed and photographed using Alpha Imager TM1200 documentation and analysis system. The DNA fingerprinting pattern obtained from the standard cultures was compared with those obtained for the isolates and identified. Each lane was scored for the presence or absence of a specific band in the STRR profiles. The cluster analysis was carried out using NTSYS-pc version 1.7 [51]

5.3.3   ESTIMATION OF HEAVY METAL REMOVAL BY CYANOBACTERIAL EXO-POLYSACCHARIDES (EPSS)

The EPSs produced by the cyanobacterial isolates grown in nutrient medium and in raw textile effluent were estimated.

One milliliter of the cyanobacterial suspension was collected and centrifuged at 10,000 rpm for 15 min. The supernatant was collected and acidified with 1% acetic acid. To this, twice the volume of cool (4°C) 95% ethanol was added and kept at 4°C overnight for the precipitation of crude polysaccharide. Then, centrifuged at 8000 rpm for 15 min. and the crude precipitate was washed with 100% ethanol and finally with acetone. The acetone residues were removed in vacuum and the crude extract was used for estimation of EPS production [61].

The EPS produced was estimated using Anthrone reagent test [39]. To 0.2 mL of crude extract in a boiling tube, 1.8 mL of distilled water and 4 mL of anthrone reagent (0.2% anthrone in 95% sulfuric acid) was added and mixed gently by shaking. The tubes were kept in boiling water bath for ten minutes and cooled to room temperature. The absorbance was measured at 620 nm in Beckman DU-64 spectrophotometer against anthrone reagent as blank. A standard graph was prepared with various concentrations of glucose ranging from 200 to 1000 $\mu g\ mL^{-1}$ and polysaccharide production was calculated from the standard curve and expressed as μg of glucose mL^{-1} of the suspension culture.

Approximately 0.1 g of the crude polysaccharide extracted from effluent grown samples was digested with triacid (conc. $H_2SO_4:HNO_3:$ Perchloric acid = 2:9:1) in 2:1 ratio. The volume was made up to 50 mL with double distilled water and was kept overnight. It was analyzed in AAS (atomic absorption spectrophotometer).

5.3.4   EVALUATION OF TEXTILE WASTEWATER (WW) TREATMENT EFFICIENCY BY CYANOBACTERIAL ISOLATES THROUGH BIO-ASSAY

To validate and to evaluate the efficiency of cyanobacteria for treatment of textile effluent, a bioassay was carried out. The effect of treated and untreated textile effluent on seed germination and vigor index was studied with various treatments on sunflower (var. CO-4 (TNAU variety)) (raw and treated effluent 25–100%).

Plant samples were collected on 15th day of germination for peroxidase (EC 1.11.17) and catalase (EC 1.11.1.6) assay.

The protein extract for analyzing catalase and peroxidase was prepared by homogenizing one gram of plant sample in 1 mL of 0.1 M sodium phosphate buffer (pH 7.0) and centrifuged at 16,000 g for 20 min at 4°C. The method developed by Lowry et al. [33] was followed for the estimation of protein. Sample (50 µg protein) was loaded into 8% polyacrylamide gel for the analysis of peroxidase and catalase.

After electrophoresis, peroxidase isoforms were visualized by soaking the gels in 0.05% benzidine in the dark for 30 min (Sigma, USA) in acetate buffer (20 mM at pH of 4.2). For assessing the activity of catalase, the gel was incubated with 0.01% hydrogen peroxide for 20 minutes and the gel was immersed in the staining solution (potassium ferricyanide: 0.5 g, ferric chloride: 0.5 g and distilled water 100 mL). The regions of catalase activity appear as clear areas on a blue green background.

5.4 RESULTS AND DISCUSSION

5.4.1 MOLECULAR CHARACTERIZATION OF THE CYANOBACTERIAL ISOLATES

Polymerase chain reaction (PCR) technology had a significant impact in almost all areas of molecular biology and the modifications of this basic procedure did allow numerous assays for detecting variation at the nucleotide level. In this study, DNA fingerprinting method was tried for identification and to find the relation of the isolates with the standard cyanobacterial cultures. The STRR primers CRA-22 and CRA-23 generated almost similar finger printing for standard *Anabaena variabilis* and the textile effluent isolate *Anabaena*-TE1 and both are related by 93%. *Nostoc muscorum* and *Nostoc*-TE1 are also related to each other by 93%. Since the relationship was more than 80%, the isolates were identified as *Anabaena* sp. and *Nostoc*sp (Figure 5.1).

5.4.2 ACCUMULATION OF HEAVY METALS BY CYANOBACTERIAL ISOLATES FROM TEXTILE WASTEWATER (WW)

The ability of algae to bioconcentrate trace metals from aqueous environment is well-documented and algae are an inexpensive process for the reclamation

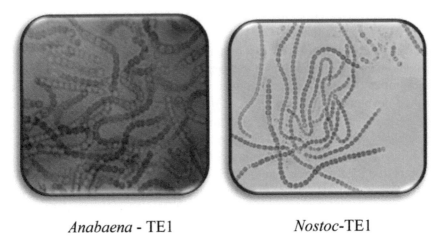

Anabaena - TE1   *Nostoc*-TE1

FIGURE 5.1   The cyanobacterial isolates from textile effluent (*Anabaena*-TE1 and *Nostoc*-TE1).

of WW [13]. Mierle and Stokes [35] suggested two-step process for metal uptake by algae: (i) an initial reaction between the metals and the surface of the algal cell that is rapid; (ii) followed by a slower metal accumulation step due to intracellular transport. Smith [57] described that the intracellular binding of metals in *Nostoc muscorum* is due to the presence of polyphosphate granules inside the cells. The EPSs released by the algae also bind significant amount of heavy metals [16].

In this research, the contribution of polysaccharides for accumulation of heavy metals was highly significant. The isolate *Anabaena*-TE1 cell biomass removed 94% chromium out of which 60.70% was removed by polysaccharide. Similarly, *Nostoc*-TE1 also removed 91% of chromium present in the raw textile effluent out of which 67.3% was accumulated by polysaccharides alone. Ozer et al. [40] found that the alga *Cladophora crispata* could remove chromium efficiently from WW.

The capacity of the algae to reduce the heavy metal pollution in river water has been studied by Senger et al. [55] and they found complete removal of Fe^{2+}, Zn^{2+}, and Cu^{2+} by mixed algal culture on 30th day of growth. In the present study by authors of this chapter, *Anabaena*-TE1 and *Nostoc*-TE1 removed considerable amounts of lead and copper on 21st day of incubation and the contribution of polysaccharides in heavy metal removal was highly appreciable, and ranged between 65–72%. The rest of the heavy metals might be transported intracellular and accumulated within the cell.

5.4.3   BIO-EFFICACY TEST WITH THE TREATED TEXTILE EFFLUENT

Bioassays validate the treatment efficiency. Comparative toxicological profile for sunflower at different dilutions with treated and untreated effluent was done. The bioassay carried out with the raw and treated effluent on sunflower revealed significant differences in germination percentage and vigor index among the treatments. The treated effluent at all levels showed higher germination of 90–100%. Highest vigor index of 2690 was recorded with 100% treated effluent. Application of treated effluent @ 75% recorded a vigor index of 2610 and was on par with seedlings grown in nutrient solution. Raw effluent at 100% affected the seedling growth and recorded the least germination percentage of 30 and a Vigor index of 210.

The expression of stress enzymes peroxidase and catalase showed significant differences between treated and untreated effluents. Regarding the enzyme peroxidase, three isoforms were expressed in seedlings grown with 75% raw effluent and with 100% raw effluent. The sample treated with 50% raw effluent showed two peroxidase isoforms and one peroxidase isoform in 25% raw effluent. No peroxidase activity was expressed in treated effluent at various levels and those grown in nutrient solution. The catalase activity was also noticed in samples treated with raw effluent and variation in intensity of expression was noticed. The samples with 75 and 100% raw effluent showed higher production of catalase compared with 25 and 50% treated samples (Figures 5.2 and 5.3).

The germination percentage and morphological characters (such as shoot and root lengths) were decreased gradually with the increase in the untreated effluent concentration. On the contrary, there was a gradual increase in the germination percentage and root and shoot lengths with an increase in the treated effluent concentration. Rajula and Padmadevi [45] recorded similar results in sunflower with automotive industry effluent amended with blue-green algae. The biometric and biochemical features of plants was on the decreasing side in plants exposed to higher concentration of the textile effluent compared to diluted samples in soybean [70] and Uma et al. [67] also recorded similar observations in black gram (*Vigna mungo*) on experimentation with textile-mill effluent.

Peroxidase and catalase expression is used as an indicator of stress condition in plant shoot tissues [32] and thought to be part of the non-specific resistance system of plants [59]. In the present study, the enzyme assays with untreated and treated effluent obviously revealed the toxicity of untreated textile effluent to sunflower.

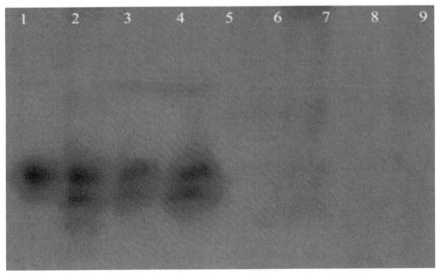

1 - 25% raw effluent   6 - 50% treated effluent
2 - 50% raw effluent   7 - 75% treated effluent
3 - 75% raw effluent   8 - 100% treated effluent
4 - 100% raw effluent   9 - Nutrient solution
5 - 25% treated effluent

FIGURE 5.2   Native-PAGE profile of peroxidase in plant samples grown in untreated and treated raw effluents.

With increased concentrations of untreated textile effluent, increased expression of catalase and peroxidase was observed. No peroxidase or catalase activity was observed in treated effluent grown samples indicating the non-toxicity of algal treated effluent. Cordova-Rosa et al. [9] also recorded enhanced peroxidase and catalase activity in rice, wheat, and soybean seedlings grown in textile effluent. Bioassay results have shown a significant reduction in the toxicity of textile effluents after treatment with mixed algae.

5.5   SUMMARY

The contamination of soils, air, and water on a global basis constitutes a major concern about aesthetics, health, and overall environmental well-being. The growing awareness of the magnitude of current environmental pollution demands vigorous efforts on the part of industry and research

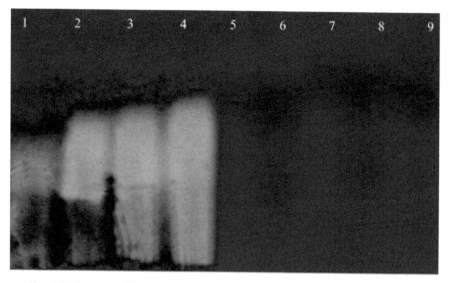

1 - 25% raw effluent   6 - 50% treated effluent
2 - 50% raw effluent   7 - 75% treated effluent
3 - 75% raw effluent   8 - 100% treated effluent
4 - 100% raw effluent   9 - Nutrient solution
5 - 25% treated effluent

FIGURE 5.3 Native-PAGE profile of catalase in plant samples grown in untreated and treated raw effluents.

community to find more effective measure to diminish and ameliorate pollution. The mechanical methods of treating WW are energy intensive; and because of the high cost incurred, WW treatments have not been widely applied in the industries. Bioremediation strategy has been accepted as a method for ameliorating and biostabilizing high concentration of pollutants and is eco-friendly, cost-effective, and resource generating. Microorganisms can remove metals from the surrounding environment with various mechanisms, either as metabolically mediated processes or as a passive adsorption of metals on the charged macromolecules of the cell envelope. Owing to the presence of many negative charges on the external cell layers, EPSs producing cyanobacteria have been considered very promising as chelating agents for the removal of positively charged heavy metal ions from an aqueous environment.

In this study, two native cyanobacterial isolates *viz.*, *Anabaena*-TE1, and *Nostoc*-TE1 from Denim textile effluent site in Coimbatore showed promising results for sorption of chromium, lead, and copper from the textile effluent. It was found that the role of the released polysaccharides in sorbing the heavy metals is significant and could pave way for a clean technology. The *Anabaena*-TE1 cell biomass removed 94%, 91%, and 94.2% of chromium, lead, and copper, respectively from the raw textile effluent. Out of this, 60.7% chromium, 66.4% lead, and 65% copper was removed by EPSs alone. *Nostoc*-TE1 accumulated a higher amount of copper in the cell biomass (94.2%) than the other two metals and the removal of the same by its EPSs was about 68.3%. Among the three metals removed by *Nostoc*-TE1, the contribution for metal accumulation by EPSs was more for lead (71.3%) followed by removal of 68.3% copper and 67.3% chromium. The ability of cyanobacteria to bio-concentrate heavy metals from the effluent is proved and EPSs, which are the most efficient fraction in metal sorption, could be easily separated from the cultures and could be immobilized for practical applications. The enzymes peroxidase and catalase were expressed in plant samples grown in untreated raw effluent indicating the response of plants on exposure to toxic compounds, whereas no activity was seen in treated effluent that indicates the algal treatment process efficiency.

KEYWORDS

- anabaena
- biological oxygen demand (BOD)
- bioscavenging
- chemical oxygen demand (COD)
- cyanobacteria
- electrophoresis
- exo-polysaccharide
- *Nostoc*
- polymerase chain reaction
- STRR primers
- textile effluent
- vigor index

REFERENCES

1. Adhikary, S. P., Bastia, A. K., & Tripathy, P. K., (1992). Growth response of the nitrogen fixing cyanobacteria *Westiellopsis prolifica* to fertilizer factory effluents. *Bulletin of Environmental Contamination and Toxicology, 49*, 137–144.

2. Aggarwal, P. K., & Kumar, P., (1990). Textile industrial effluents. Implication and possible use for afforestation in Western Rajasthan. *Annals of Arid Zone, 29*, 295–302.

3. Anand, N., (1989). *Hand Book of Blue Green Algae(of Rice Fields of South India)* (p. 232). Bishen Singh Mahendra Pal Singh, Dehradun.

4. Assadi, M., (1979). Application of stabilization pond for purification of textile wastewater in Esfahan. *Water, Air and Soil Pollution, 11*, 247–252.

5. Banat, I. M. P., Nigum, P., Singh, D., & Marchnat, R., (1996). Microbial decolorization of textile dye containing effluents: A review. *Bioresource Technology, 58*, 217–227.

6. Brunk, C. F., Jones, K. C., & James, T. W., (1979). Assay for nanogram quantities of DNA in cellular homogenates. *Analytical Biochemistry, 92*, 497–500.

7. Castenholz, R. W., (1988). Culturing of cyanobacteria. *Methods in Enzymology, 167*, 68–93.

8. Castenholz, R. W., & Waterbury, J. B., (1989). Oxygenic photosynthetic bacteria group II. Cyanobacteria. In: Staley, J. T., Bryant, M. P., Pfenning, N., & Holt, J. G., (eds.), *Bergey's Manual of Systematic Bacteriology* (Vol. 3, pp. 1710–1789). Williams and Wilkins, Baltimore-MD.

9. Chang, C., & Sibley, T. H., (1993). Accumulation and transfer of copper by *Oocystis pusilla. Bulletin of Environmental Contamination and Toxicology, 50*, 689–695.

10. Christ, R. H., Ober, H. K., Shank, N., & Nguyen, M., (1981). Copper and lead interactions with *Nostoc muscorum. Environmental Science and Technology, 15*, 12–12.

11. Cordova, R. E. V., Vargas, C., Souza-Sierra, M. M., Correa, A. X. R., & Radetski, C. M., (2003). Biomass growth, micronucleus induction, and antioxidant stress enzyme responses in *Vicia faba* exposed to cadmium in solution. *Journal of Environmental Toxicology and Chemistry, 22*, 645–649.

12. Dhabadgaonkar, S. M., & Patil, S. J., (1987). Oil and grease removal from textile mill wastewater by chemical treatment. *Proceedings of the 2nd National Conference of Environmental Engineers, 1987,* 25–30.

13. Dor, I., & Svi, B., (1980). Effect of heterotrophic bacteria on green algae growing in wastewater. In: Sheief, G., & Soeder, C. J., (eds.), *Algal Biomass* (pp. 421–429). Elsevier/North-Holland Biomedical Press.

14. Elnabaraway, M., & Welter, A. N., (1984). Utilization of algal cultures and assay for industry. In: Shubert, L. E., (ed.), *Algae as Ecological Indicators* (pp. 317–328). New York: Academic Press Inc.

15. Gardea-Torresdey, J. L., Becker-Hapak, M. K., Hosea, J. M., & Darnall, D. W., (1990). Effect of chemical modification of algal carboxyl groups on metal ion binding. *Environmental Science and Technology, 24*, 1372–1378.

16. Geesey, G. G., & Jang, L., (1990). Extracellular polymers for metal binding. In: Brierley, C., & Ehrlich, H., (eds.), *Microbial Mineral Recovery* (pp. 223–249.) McGraw Hill, New York.

17. Graham, L. E., & Wilcox, L. W., (2000). *Algae* (p. 397). Prentice-Hall, New Jersey.

18. Grant, B. R., (1967). The action of light on nitrate and nitrite assimilation by the marine chlorophyte *Dunaliella tertiolecta*. *Journal of General Microbiology*, *48*, 379–384.

19. Grant, B. R., (1970). Nitrite reductase in *Dunaliella tertiolecta*: Isolation and properties. *Plant Cell Physiology*, *11*, 55–60.

20. Greene, B., Henzl, M. T., Hosea, J. M., & Darnall, D. W., (1986). Elimination of bicarbonate interference in the binding of chromium (VI) in mill waters to freeze dried *Chlorella vulgaris*. *Biotechnology and Bioengineering*, *28*, 764–769.

21. Gupta, I. C., & Jain, B. L., (1992). Salinization and alkalization of ground waters polluted due to textile hand processing industries in Pali. *Current Agriculture*, *16*, 59–62.

22. Hassett, J. M., Jennett, J. C., & Smith, J. E., (1980). Heavy metal accumulation by algae. In: Baker, R. A., (ed.), *Contaminants and Sediments* (Vol. 2, pp. *409*–424). *Ann Arbor Science* Publishers Inc., Ann Arbor, MI.

23. Huntsman, S. A., & Sunda, W. G., (1980). The role of trace metals in regulating phytoplankton growth with emphasis on Fe, Mn, and Cu. In: Morris, I., (ed.), *The Physiological Ecology of Phytoplankton* (p. 285).University of California Press, Berkeley-CA.

24. Jha, M. N., Venkataraman, G. S., & Kaushik, B. D., (1987). Response of *Westiellopsis prolifica* and *Anabaena* sp. to salt stress. *MIRCEN Journal*, *3*, 307–317.

25. Jones, B. B., & Bishop, N. I., (1976). Growth rate and hydrogen production by cyanobacteria. *Plant Physiology*, *57*, 659.

26. Jones, G. J., Palenik, B. P., & Morel, F. M. M., (1987). Trace metal reduction by phytoplankton: The role of plasmalemma redox enzymes. *Journal of Phycology*, *23*, 237–242.

27. Khan, M. A., (2001). Pollution of water resources due to industrialization in arid zone of Rajasthan. *Journal of Environmental Sciences*, *13*, 218–223.

28. Kihage, M., & Magnusson, C., (1993). *Salinity Problems in Noyyal River Basin* (pp. 33–36). MSc Thesis; Institute of Technical Education, Kalmar, Sweden.

29. Kneip, T. J., & Lauer, G. P., (1973). Trace metal concentration factors in aquatic ecosystems. In: Cohen, E. M., Kneip, T. J., & Sweig, G., (eds.), *Chemical Analysis of the Environment and Other Modern Techniques* (pp. 43–47). New York: Plenum Press.

30. Kothandaraman, V., Aboo, E. K. M., & Sastry, C. A., (1976). Characteristics of waste from a textile mill. *Indian Journal of Environmental Health*, *18*, 99–112.

31. Les, A., & Walker, R. W., (1984). Toxicity and binding of copper, zinc and cadmium by the blue-green alga (*Chroococcus paris*). *Water Air and Soil Pollution*, *23*, 129–139.

32. Levitt, K., (1972). *Responses of Plants to Environmental Stress* (p. 697). Academic Press, New York.

33. Lowry, O. H., Rosebrought, N. J., Larr, A. I., & Randall, R. J., (1951). Protein measurement with folin-phenol reagent. *The Journal of Biological Chemistry*, *193*, 265–275.

34. Melody, S. C., (1997). *Plant Molecular Biology: A Laboratory Manual* (p. 239). Springer-Verlag, New York.

35. Mierle, G. M., & Stokes, P. M., (1976). Heavy metal tolerance and metal accumulation by plaktonic algae. In: Hemphill, D. D., (ed.), *Trace Substances in Environmental Health* (pp. 113–120). Columbia University, Missouri.

36. Mishra, S., Sharma, D., & Uma, M., (1990). Fish mortality as affected by effluents from Raza Textile Ltd., Rampur (UP). *International Journal of Ecology and Environmental Sciences*, *16*, 119–124.

37. Mohan-Rao, G. J., (1972). Dairy waste characteristic with reference to ISI standards. *Indian Journal of Environmental Health, 14,* 218–224.

38. Mohapatra, P. K., Patnaik, L. N., & Mishra, G., (1990). Pollution due to textile industry: A case study. *Environment Asia, 12,* 51–66.

39. Morris, D. L., (1948). Quantitative determination of carbohydrates with Dreywood' Santhrone reagent. *Science, 107,* 254–255.

40. Ozer, D., Aksu, Z., Kutsal, T., & Caglar, A., (1994). Adsorption isotherm of Pb and Cr on *Cladophora cripata*. *Journal of Environmental Technology, 15,* 439–448.

41. Pant, A., Srivastava, S. C., & Singh, S. P., (1992). Methyl mercury uptake by free and immobilized cyanobacteria. *Biometals, 5,* 229–234.

42. Paskins-Hurlburt, A., Tanaka, Y., & Skoryna, S. C., (1976). Isolation and metal binding properties of fucoidin. *Botanica Marina, 19,* 327–336.

43. Patel, S. S., & Srivastava, V. S., (1999). Impact of textile dying and printing industrial effluents on soil and ground water quality: A case study. *Indian Journal of Environmental Protection, 19,* 771–773.

44. Rajannan, G., (1987). *Studies on the Pollution of Bhavani River by Different Industrial Effluents, Their Treatment and Recycling for Crop Production* (p. 238). PhD Thesis; Tamil Nadu Agricultural University, Coimbatore.

45. Rajula, R., & Padmadevi, S. N., (2000). Effect of industrial effluents without and with BGA on the growth and biochemical contents of the seedlings of *Helianthus annuus* L. *Asian Journal of Microbiology Biotechnology and Environmental Sciences, 2*(3), 151–154.

46. Ramachandran, K., (1994). *Studies on the Effect of Dyeing Factory Effluent on Soils and Adjoining Ground Waters.* MSc (Ag.) Thesis; Tamil Nadu Agricultural University, Coimbatore.

47. Ramasami, V., & Rajaguru, P., (1991). Groundwater quality of Tiruppur. *Indian Journal of Environmental Health, 33,* 187–191.

48. Rastogi, R., & Gaumat, M. M., (1990). *Pollution of Ground Water* (Vol. 5, pp. 49–51). In Madurai city, U.P., Bhujal News.

49. Redalje, D. G., & Duerr, E. O., (2000). Algae as ideal waste remover: Biochemical pathways. In: Huntley, M. E., (ed.), *Biotreatment of Agricultural Wastewater* (pp. 92–99). CRC Press Inc., Boca Raton, Florida.

50. Reichardt, W., Overbeck, J., & Steubing, L., (1968). Free dissolved enzymes in lake waters. *Nature, 216,* 1345.

51. Rohlf, F. J., (1992). *NTSYS-PC: Numerical Taxonomy and Multivariate Analysis System, Version 1.7* (p. 138) Exeter Software: Setauket, NY.

52. Salomonson, L. P., (1979). Structure of *Chlorella* nitrate reductase. In: Hewitt, E. J., & Cutting, C. V., (eds.), *Nitrogen Assimilation of Plants* (pp. 199–205). Academic Press, New York.

53. Schecher, W. D., & Driscoll, C. T., (1985). Interactions of copper and lead with *Nostoc muscorum*. *Water, Air and Soil Pollution, 24,* 85–101.

54. Sengar, R. M. S., & Sharma, K. D., (1986). Role of algae on the assessment of pollution in river Yamuna. *The Journal of Indian Botanical Society, 66,* 325–334.

55. Sengar, R. M. S., Sharma, K. D., & Mittal, S., (1990). *In-vitro* studies for the removal of heavy metal of river water by algal treatment. *Geobios, 17,* 77–81.

56. Smith, F. A., (1966). Active phosphate uptake by *Nitellatranslucens*. *Biochimica et Biophysica Acta, 126,* 94–98.

57. Smith, W., (1983). The role of trace metals in regulating algal growth with emphasis on Fe, Mn and Cu. In: Morris, I., (ed.), *The Physiology Ecology of Algae* (pp. 285–299). University of California Press, Berkeley.

58. Somashekar, P. K., Gowda, M. T. G., Shettigar, S. L. N., & Srinath, K. P., (1984). Effect of industrial effluent on crop plants. *Indian Journal of Environmental Health, 26,* 136–146.

59. Spanu, P., & Bonfante-Fasolo, P., (1988). Cell bound peroxidase activity in roots of mycorrhizal *Allium porrum. New Phytologists, 109,* 11–19.

60. Subramanian, G., & Shanmugasundaram, S., (1986). Sewage utilization and waste recycling of cyanobacteria. *Indian Journal of Environmental Health, 28,* 250–253.

61. Sutherland, I. W., & Wilkinson, J. F., (1971). Chemical extraction methods of microbial cells. In: Norris, J. R., & Ribbons, D. W., (eds.), *Methods in Microbiology* (pp. 360–361). Academic Press, New York.

62. Swaminathan, K., & Vaidheeswaran, P., (1991). Effect of dyeing factory effluent on seed germination and seedling development of groundnut. *Journal of Environmental Biology, 12,* 353–358.

63. Tam, N. F. Y., & Wong, Y. S., (1989). Wastewater nutrient removal by *Chlorella pyrenodosa* and *Scenedesmus* sp. *Environmental Pollution, 58,* 19–34.

64. Taneja, L., & Fatma, T., (1998). Studies on cyanobacterium *Anacystis nidulans* with respect to biochemical constituents and copper uptake. In: Subramanian, G., Kaushik, B. D., & Venkataraman, G. S., (eds.), *Cyanobacterial Biotechnology* (pp. 437–445). Vijay Primlani Publishing Co. Pvt. Ltd., New Delhi.

65. The Center for Science and Environment, (2001). In: *CPCB in Water Quality in India* (Accessed on 20 June 2020).

66. Uma, L., & Subramanian, G., (1990). Effective use of cyanobacteria in effluent treatment. In: *Proceedings of the National Symposium on Cyanobacteria in Nitrogen Fixation* (pp. 437–443). Indian Agric. Res. Institute (IARI), New Delhi.

67. Uma, S., Balamurugan, V., & Vijayalakshmi, G. S., (2003). Impact of textile mill effluent on germination and seedling growth of *Vigna mungo. Journal of Ecotoxicology and Environmental Monitoring, 13*(1), 47–51.

68. Verma, S. R., Tyagi, A. K., & Daleta, R. C., (1974). Studies on characteristics and disposal problems of industrial effluent with reference to ISI standards, Part II. *Indian Journal of Environmental Health, 19,* 165–175.

69. Veroy, R. L., Montano, N., De Guzman, M. L. B., Laserna, E. C., & Cajipe, G. J. B., (1980). Studies on the binding of heavy metals to algal polysaccharides from Philippine seaweeds: Carrageenan and binding of lead and cadmium. *Botanica Marina, 23,* 59–64.

70. Vijayakumari, B., (2003). Impact of textile dyeing effluent on growth of soyabean (*Glycine max*). *Journal of Ecotoxicology and Environmental Monitoring, 13*(1), 59–64.

71. Vijayarengan, P., & Lakshmanachary, A. S., (1993). Effect of textile mill effluent on growth and development of green gram seedling. *Advances in Plant Sciences, 6,* 359–365.

72. Zhang, W., & Majidi, V., (1994). Monitoring the cellular response of *Stichococcus bacillaris* to exposure of several different metals using *in vivo*^{31}P NMR and other spectroscopic techniques. *Environmental Science and Technology, 28,* 1577–1581.

73. Zumpt, W. G., (1978). Ferredoxin: Nitrite oxidoreductase from *Chlorella*: Purification and properties. *Biochimica et Biophysica Acta,* 276–363.

CHAPTER 6

WATER QUALITY IMPROVEMENT: USE OF INDIGENOUS PLANT MATERIALS

S. SIVARANJANI and AMITAVA RAKSHIT

ABSTRACT

Nowadays, the use of seed materials and aquatic plants are receiving considerable attention for the effective use of water treatment process. The available technologies should be easy to implement and useful for the rural people. The treatment process uses natural plant materials; therefore, it does not generate any non-treatable waste from this process. The process should require little or no maintenance cost and easily operable. The plant materials as coagulants in future can be obtained from algae or chitosan, but the dosages should be the main focus during the water treatment.

6.1 INTRODUCTION

About 75% of our planet is covered by water, in the form of lakes, rivers, oceans, etc. Of this percentage, only (113 m³) a small portion is available to the living organisms on earth [19]. Around the world, only 0.5% of freshwater is available from the water resources present on earth, because water is not evenly distributed around the globe. About 10 countries have only 60% of the freshwater supply in the world. And about 1.8 billion persons still have problem in accessing the freshwater. Since 1940, there has been continuous increase in population thus increasing the demand for water consumption in the world. Most of the European cities, use of groundwater is common for daily use. In the same scenario, the old water storage systems also lose water through leaks and cracks. All of us must understand that water is very essential for all living organisms, but only few of us understand the risks of it. About 5,000/year children die due to use of polluted water and poor hygiene problems. The water-related diseases kill one child in every 15 seconds. Therefore, it is important to understand that investment of one dollar in water sanitation problem will result in economic growth of 8 dollars.

The 70% of the world's water resources is required for food production. Increase in the population requires more than 50% of freshwater to increase the food production. In many developing countries, use 90% of their freshwater is for irrigation purpose. Therefore, it is important to find new irrigation technologies, which will reduce irrigation losses and usage of freshwater for irrigation.

Nowadays, it is a serious problem to access the clean water for drinking purpose, because of poor land management systems. The polluted water from sewages, industrial discharge, and runoff from agricultural land severely affects the groundwater source. Therefore, these water resources must undergo treatment process, so that the consumers can get the clean water for drinking and domestic purposes [2]. Treatment of drinking water process is carried out by coagulation, sedimentation, filtration, disinfection process [13]. Coagulants play a major role in the treatment of wastewater (WW). Many coagulants for conventional treatment process include inorganic coagulants (such as: Aluminum sulfate, alum). Organic coagulants act as polyelectrolytes and are obtained from synthetic or plant-based materials [11, 32].

The residues present post-treatment process lead to many health problems. For example, aluminum is characterized as a poisoning factor for encephalopathy, and many research reports show the impact of aluminum on human health. It has been clearly documented that aluminum-based coagulants are linked to the development of neuro-degenerative illnesses, such as sessile dementia [17] and Alzheimer's diseases [23]. Synthetic polyelectrolytes have been questioned due to high level of toxicity [7].

The overuse of alum and chlorine in the WW treatment leads to the high cost of drinking water. Therefore, poor persons are forced to use the contaminated water. Also, the treated drinking water with chlorine and alum causes some carcinogenic diseases that have been reported in several studies [3]. However, the risks from treatment with these products are not significant compared with the inadequate disinfection. Therefore, it is necessary to find some natural coagulants for the treatment process to replace the chemical and organic coagulants. The natural coagulants are cheaper, easily available, eco-friendly, and safe to human health [6]. Today, there is considerable interest to develop natural coagulants for treatment of WW in many developing countries [12].

In domestic purposes, the coagulants are used in the form of powder or paste, containing 90% of substances other than polyelectrolytes. Apart from this, it may have effective coagulant properties [10]. Some of the plant materials that are identified for coagulation purposes are: *Moringa oleifera* [9, 16, 26], *Nirmali* (*Strychnos potatorum*) [31], and cactus [30].

Research reports confirm that natural polymers have been used for the treatment of various types of water and WW. Table 6.1 indicates selected natural coagulants, which have been used in the treatment process, based on literature survey. The most available coagulants are chemically modified and can remove 50 to 90% of turbidity. Also, the natural coagulants cause change in the dissolved organic carbon (DOC) level, thus showing concern to chlorination process [17, 18]. Because the biomaterials have relatively, lower efficiency of turbidity removal, the production, and extraction of these biomaterials as coagulants is expensive thus making them impractical for full-scale applications.

TABLE 6.1   Summary of Recently Published Literature on Coagulating Efficiency of Natural Coagulants

Coagulant	Optimum Dose (mg/L)	Turbidity Removal (%)	Change in DOC (Dissolved Organic Carbon)	References
1. Step purification of M. oleifera	2	97	32% increase	[24]
2. Step purification of M. oleifera	2	98.5	17% increase	[25]
Chestnut and acorn	0.5	70–80	1 mL/L	[27]
Grafted Plantago psyllium Mucilage	1.6	—	—	[29]
Hercofloc	9	50	—	[15]
Hibiscus esculentus seedpods	5	93–97.3	190% increase	[1]

TABLE 6.1   *(Continued)*

Coagulant	Optimum Dose (mg/L)	Turbidity Removal (%)	Change in DOC (Dissolved Organic Carbon)	References
M. oleifera seeds	50	90	450% increase	[14]
	6000	—	52% reduction	[4]
	100	—	64% reduction	[5]
Malva sylvestris mucilage	12	96.3–97.4	80% increase	[1]
Nalco 610	10	62	—	[15]
Purifloc C-31	50	50	34% reduction	[22]

Source: Reprinted with permission from Anastasakis et al., 2009. © Elsevier.

The possible replacements identified for treating WW are watermelon, tannins that are present in the bark and wood of trees like Acacia, Castanea, or Schinopsis. However, the use of tannin needs modification in chemical structure before being used in WW treatment [20, 21, 25]. Opuntia reduced 98% turbidity in the treated water. Heredia et al. [3] reported that tannin is a feasible source for treated water. The turbidity removal of natural coagulants depends mainly on the plant characteristics and their origin. KCl/NaNO$_3$ [16, 17] and NaCl [17, 24] are chemicals, which have been used in the extraction process.

This chapter explores technology of improving water quality using indigenous materials of plant origin that are used in selected rural areas in the world.

6.2   WASTEWATER (WW) TREATMENT BY USING NATURAL COAGULANT

A natural coagulant is a plant-based material, which can be used for the coagulation process in WW treatment to reduce the turbidity of the treated water. The objective of this chapter is to assess the possibility of using the following plant-based coagulants as an alternative to the chemical coagulants.

6.2.1   WATERMELON SEEDS AS POTENTIAL COAGULANT FOR WATER TREATMENT

The dosage of about 0.1 g/L at pH 7.0 of watermelon seeds was effective in the turbidity removal. The reduction in turbidity was below the value recommended by the World Health Organization (WHO) and the best color removal was also not up to the recommended value of WHO. When the watermelon seeds were used in combination with alum, there were unfavorable changes in the pH of the treated water, but the best color and turbidity removal was obtained. Therefore, watermelon seeds can be used as a natural coagulant for the water treatment.

6.2.2 NIRMALI SEEDS FOR WASTEWATER (WW) TREATMENT

Strychnos potatorum (Nirmali) is a moderate-sized deciduous tree, which is found in the Southern and Central regions of India, Sri Lanka, and Burma. The seeds have been mostly used for medicinal purposes. About 4000 years ago, Sanskrit writings indicate their use for water treatment. These seed extracts contain anionic polyelectrolyte, which destabilizes the particles present in water by means of interparticle bridging. The seeds also contain lipids, carbohydrates, and alkaloids, which can enhance the coagulant extraction capacity.

The mixture of polysaccharide fraction from *S. potatorum* seeds has galactomannan and galactan that can reduce the turbidity by about 80%. The specific coagulation mechanism of Nirmali seed extract has not been extensively investigated. The chemical studies may provide the information of about how many OH^- groups present in galactomann and galactan, but not on the abundant adsorption sites, which are mainly used by coagulant interparticle bridging effect. Therefore, further studies should be conducted to evaluate the coagulation effect in relation to purification of water [28].

6.2.3 M. OLEIFERA USED FOR WASTEWATER (WW) TREATMENT

Moringa oleifera (horseradish or drumstick tree) is a tropical plant found throughout India, Asia, Saharan Africa, and Latin America. The seeds of *M. oleifera* contain edible oil and some water-soluble substances that can be used for treating WW. Almost every part of the plant can be used for different health benefits. In less developed countries, it is most frequently used for food and medicinal purposes.

Rural-communities in Africa have mostly used the crude seed extract to clear turbidity of river water. *M. oleifera* seeds contain coagulant protein (cationic), which is helpful used for drinking water clarification or WW treatment. It is one of the effective coagulants used nowadays. The dimeric protein in these seeds has a molecular weight of about 6.5 to 14 kDa. However, the use of this crude extract has issues related to residual DOC, which makes the water not feasible for drinking. It is necessary for to purify the coagulants before its use in water treatment.

6.2.4 TANNIN AS A NATURAL COAGULANT FOR WATER TREATMENT

The polyphenol compounds are generally named as Tannins with a molecular weight ranging from hundreds to tens of thousands. It is used as a tanning agent in the leather industry. It has a negative effect on human health and limited use in water treatment. The tannin from Valonia (alga) is an excellent substitute to chemical coagulant. The chemical structure of tannin can influence the effectiveness of water treatment. The anionic of tannin illustrates the molecular interaction that is required needed for coagulation. The presence of more phenolic groups increases the coagulation capability.

The coagulant characteristics of the tannins from Acacia catechu were examined. The powdered material extract from the bark of acacia catechu was used to test the coagulant reduction rate and dosage. The turbidity removal and other physicochemical properties of surface water sample were measured before and after the treatment. The acacia catechu powder can remove turbidity up to 91% at an optimal dosage of about 3.0 mL/L. On the other hand, the powder of acacia catechu can also remove the total dissolved solids (TDS) by 57.3%. However, this powder did not have significant effects on its physical and chemical properties.

6.2.5 PLANTAGO OVATE AS A NATURAL COAGULANT FOR WATER TREATMENT

It is another plant coagulant that is extracted from *Plantago ovate* (blood plantain or blond psyllium) by using Ferric chloride-induced crude extract (FCE). The turbid river water was treated with the FCE that showed maximum turbidity removal at pH level of 8.0 and optimum dosage of 0.8 mg/L of DOC. The turbidity removal was due to the increase in humic acid content during the treatment. And the results clearly showed that FCE was an eco-friendly biocoagulant [8].

6.2.6   CACTUS

Recently cacti species have been used for water treatment, in addition to *Nirmali* and *M. oleifera*. The common cactus genus for water treatment is Opuntia, which is known as 'nopal' in Mexico or 'prickly pear' in North America. The cactus is also used for medical purpose and dietary food source. The other than Opuntia, *Cactus latifaria* has also been successfully used as a natural coagulant. The higher coagulation activity is due to the presence of mucilage, which is viscous and complex carbohydrate present in inner and outer parts. The galacturonic acid present in Opuntia is possibly the active ingredient acts in coagulation mechanism, and it accounts for about 50% of turbidity removal. There are not enough research reports on coagulation mechanism of galacturonic acid that exists predominantly in polymeric form to provide 'bridge for particles' for absorption. There might be presence of some functional groups along chains of poly galacturonic acid, showing chemisorption between the charged particles. The presence of hydroxyl group also infers the possible intra-molecular interactions to distort the relative linearity of chain.

6.2.7   CHITIN AS NATURAL COAGULANT FOR WATER TREATMENT

The chitin is one of the natural coagulants used for water treatment. The turbidity removal was tested by using different dosages of chitin @ 0.5, 1, 1.5, and 2 mg/L. Where the pH level is stable while treating with chitin and 93% of hardness can be removed. The optimum dosage was 1 and 1.5 mg/L for effective removal of turbidity.

6.2.8   COAGULATION ACTIVITY OF NATURAL COAGULANTS FROM SEEDS OF DIFFERENT LEGUMINOUS SPECIES

Other than tree species, some of the leguminous plants have also been used as a natural coagulant in the water treatment, which includes *Phaseolus vulgaris* (beans), *Robinia pseudoacacia* (black locust), *Ceratonia siliqua* (carob tree), and *Amorpha fruticose* (lead plant). The active component was extracted from the seeds by using distilled water. The different turbid water has been tested with this extract to find the coagulation activity. The study confirmed that these natural coagulants also have their positive coagulation activity. From all the four species studied, the seed extract from *Ceratonia siliqua* was best among these four plants. Dosage @ 20 mg/L of coagulant gave the 100% coagulation activity for clarification of water.

6.2.9 WORLDWIDE POTENTIAL USE OF NATURAL COAGULANTS

There are large numbers of plant materials that have been used for water treatment. Naturally, occurring plant coagulants are safe for human health; and these are: *Moringa oleifera, Moringa stenopetala, Vicia faba, Canavalia ensiformis, Bombax constatum,* and okra. The seed powder of *M. oleifera* was the best coagulant to replace the chemical coagulant (Alum). The natural coagulants are readily available in some parts of the world than the chemical coagulants. In most of the developing countries, the chemical coagulants are not easily available and are more expensive.

6.2.10 INTEGRATION OF CHEMICAL AND NATURAL COAGULANTS FOR WATER TREATMENT

Some studies were conducted to evaluate the combined use of natural and chemical coagulants in treating the WW, i.e., Moringa oleifera + Alum. The combined use of these chemical and natural coagulants gave higher efficiency of chemical oxygen demand (COD) removal. The effective dosages were 50 mg and 100 mg of *M. oleifera* + Alum.

6.3 SUMMARY

This chapter focuses on the WW treatment by using the indigenous plant materials that are locally and easily available at a lower cost; and have lower side effects compared with chemical coagulants, such as aluminum sulfate, Alum, and other common chemicals. The disadvantages of using chemical coagulants are the residues left after the treatment process causes several health hazards like encephalopathy, sessile dementia, Alzheimer's disease, etc.

KEYWORDS

- alum
- chemical oxygen demand
- dissolved organic carbon
- filtration
- turbidity
- wastewater

REFERENCES

1. Anastasakis, K., Kalderis, D., & Diamadopoulos, E., (2009). Flocculation behavior of mallow and okra mucilage in treating wastewater. *Desalination, 249*, 786–791.
2. Aweng, E. R., Anwar, A. I., SitiRafiqah, M. I., & Suhaimi, O., (2012). *Cassia alata* as a potential coagulant in water treatment. *Research Journal of Recent Sciences, 1*(2), 28–33.
3. Beltran-Heredia, J., Sanchez-Martın, J., & Jimenez-Giles, M., (2011). Tannin-based coagulants in the depuration of textile wastewater effluents: Elimination of anthrax quinonic dyes. *Water Air Soil Pollution, 222*(1–4), 53–64.
4. Bhatia, S., Othman, Z., & Ahmad, A. L., (2007). Pretreatment of palm oil mill effluent (POME) using *Moringa oleifera* seeds as natural coagulant. *J. Hazard. Mater., 145*, 120–126.
5. Bhuptawat, H., Folkard, G. K., & Chaudhari, S., (2007). Innovative physicochemical treatment of wastewater incorporating *Moringa oleifera* seed coagulant. *J. Hazard. Mater., 142*, 477–482.
6. Binayke, R. A., & Jadhav, M. V., (2013). Application of natural coagulants in water purification. *International Journal of Advanced Technology in Civil Engineering, 2*(1), 118–123.
7. Bolto, B., & Gregory, J., (2007). Organic polyelectrolyte's in water treatment. *Water Research, 41*(11), 2301–2324.
8. Geay, M., Marchetti, V., Clément, A., Loubinoux, B., & Gérardin, P., (2000). Decontamination of synthetic solutions containing heavy metals using chemically modified saw dusts bearing poly acrylic acid chains. *Journal of Wood Science, 46*(4), 331–333.
9. Hussain, S., Mane, V., Pradhan, V., & Farooqui, M., (2012). Efficiency of seeds of *Moringa oleifera* in estimation of water turbidity. *International Journal of Research in Pharmaceutical and Biomedical Sciences, 3*(3), 1334–1337.
10. Jahn, S. A. A., (1988). Using *Moringa* seeds as coagulants in developing countries. *Journal American Water Works Association, 80*(6), 43–50.
11. Katayon, S., Noor, M., Asma, M., Ghani, L. A. A., Thamer, A. M., Azni, I. A., Khor, B. C., & Suleyman, A. M., (2006). Effects of storage conditions of *Moringa oleifera* seeds on its performance in coagulation. *Bioresource Technology, 97*(13), 1455–1460.

12. Miller, R. G., Kopfter, F. C., Kelty, K. C., Strober, J. A., & Ulmer, N. S., (1984). The occurrence of aluminum in drinking water. *Journal of the American Water Works Association, 76,* 84–91.

13. Miller, S. M., Ezekiel, J. F., Vinka, O. C., James, A. S., & Julie, B. Z., (2008). Toward understanding the efficiency and mechanism of *Opuntia* spp. as a natural coagulant for potential application in water treatment. *Environmental Science and Technology, 42*(12), 4274–4279.

14. Ndabigengesere, A., & Narasiah, K. S., (1998). Quality of water treated by coagulation using *Moringa oleifera* seeds. *Water Research, 32,* 781–791.

15. Ogedengbe, O., (1975). The performance potential of poly electrolytes and high velocity gradients in the treatment of wastewater. *Water Research, 10,* 343–349.

16. Okuda, T., Baes, A. U., Nishijima, W., & Okada, M., (1999). Improvement of extraction method of coagulation active components from *Moringa oleifera* seed. *Water Research, 33,* 3373–3378.

17. Okuda, T., Baes, A. U., Nishijima, W., & Okada, M., (2001). Coagulation mechanism of salt solution extracted active component in *Moringa oleifera* seeds. *Water Research, 35*(15), 830–834.

18. Okuda, T., Baes, A. U., Nishijima, W., & Okada, M., (2001). Isolation and characterization of coagulant extracted from *Moringa oleifera* seed by salt solution. *Water Research, 35,* 405–410.

19. Oliveira, I. M., Cruz, V., Visconte, L. L. Y., Pacheco, É. B., Acordi, V., & Dezotti, M., (2013). Tannin treated water for use in the emulsion polymerization of SBR. *Polímeros, 23*(3), 326–330.

20. Ozacar, M., (1997). *Study on the Use of Tannins, Obtained from Oak Acorns (Valonia), as Natural Polyelectrolyte in Water Treatment* (p. 123). PhD Thesis; Sakarya University, Science Technology Institute, Sakarya.

21. Pulkkinen, E., & Mikkohen, H., (1992). Preparation and performance of tannin-based flocculants. In: Hemingway, R. W., & Laks, P. E., (eds.), *Plant Polyphenols* (pp. 953–966). Plenum Press, New York.

22. Rebhun, M., Narkis, N., & Wachs, A. M., (1969). Effect of polyelectrolytes in conjunction with bentonitic clay on contaminants removal from secondary effluents. *Water Research, 3,* 345–355.

23. Rondeau, V. J. H., & Commenges, D. D. J. F., (2001). Aluminum in drinking water and cognitive decline in elderly subjects. *The American Journal of Epidemiology, 154*(3), 288–290.

24. Sánchez-Martín, J., Beltrán-Heredia, J., & Solera-Hernández, C., (2010). Surface water and wastewater treatment using a new tannin-based coagulant. pilot plant trials. *Journal of Environmental Management, 91,* 2051–2058.

25. Sanchez-Martin, J., Ghebremichael, K., & Beltran-Heredia, J., (2010). Comparison of single-step and two-step purified coagulants from *Moringa oleifera* seed for turbidity and DOC removal. *Bioresour. Technol., 101,* 6259–6261.

26. Saulawa, S. B., Okuofu, C. A., Ismail, A., Otun, J. A., Adie, D. B., & Yakasai, I. A., (2011). Effect of deterioration on the turbidity removal potentials of *Moringa oleifera* seed extract. In: *Paper presented at the 7ᵗʰ Annual National Conference of the Society for Occupational Safety and Environmental Health* (SOSEH) (p. 8). Education Trust Fund (ETF) Conference Center, University of Abuja, Abuja FCT Nigeria.

27. Šciban, M., Klašnja, M., Antov, M., & Škrbic, B., (2009). Removal of water turbidity by natural coagulants obtained from chestnut and acorn. *Bioresource Technology*, *100*, 6639–6643.

28. Shulz, C. R., & Okun, D. A., (1984). *Surface Water Treatment for Communities in Developing Countries* (p. 300.). Practical Action Publishing; John Wiley & Sons, New York.

29. Srinivasan, R., Agarwal, M., & Mishra, A., (2002). Plantpsyllium-grafted-polyacrylonitrile: Synthesis, characterization and its use in suspended and dissolved solid removal from textile effluent. *Water Qual. Res. J. Can, 37*, 371–378.

30. Swati, M., & Govindan, V. S., (2005). Coagulation studies on natural seed extracts. *J. Indian Water Works Assoc., 37*, 145–149.

31. Tripathi, P. N., Chaudhari, M., & Bokil, S. D., (1976). Nirmali seed a naturally occurring coagulant. *Indian J. Environ. Health*, *18*, 272–281.

32. Warhurst, A. M., McConnachie, G. L., & Pollard, S. J. T., (1996). The production of activated carbon for water treatment in Malawi from the waste seed husks of *Moringa oleifera*. *Water Science and Technology*, *34*(11), 177–184.

CHAPTER 7

IMPACT OF BIOFILM ON CLOGGING OF DRIP IRRIGATION EMITTERS

NIVEDITA KHAWAS, VINOD KUMAR TRIPATHI, and ANIL KUMAR

ABSTRACT

This chapter investigates the impact of different types of water on emitter clogging in Brambe village of Mandar block in Ranchi district of Jharkhand. The characterization of water was done for various water quality parameters. The pH for wastewater (WW) was lower than pond water but for groundwater, it was moderate. The EC, Fe^{+2}, Ca^{+2}, Mg^{+2}, CO_3^{-2}, HCO_3^-, DO, BOD, and COD values were higher for WW. The fluctuations of the emitters flow rate, DRV, and CU were extremely high. DWs and EPS were higher in case of WW when compared to pond water and groundwater.

7.1 INTRODUCTION

Water is applied to crops through nature or through conventional (flood) irrigation. Drip and flood irrigation methods are generally employed by farmers. In flood irrigation, there is an extreme fluctuation in the moisture content, temperature, and aeration of the soil, resulting in plant stress. The drip irrigation method keeps the moisture content of soil relatively constant. It also ensures the appropriate air-water mixture in the root-zone. The water losses in drip irrigation due to evaporation or runoff are less in comparison with other irrigation methods. The drip irrigation method delivers the right amount of water to the plants. The ability of crops to withstand drought resulting from any breakdown in the irrigation system is reduced as the water of the wetted zone gets depleted and the surrounding region is dry.

The major issue associated with the drip irrigation system is the clogging of emitters. It disrupts the uniformity of emitter discharge for delivering an equal amount of water. Irrigation with treated/untreated wastewater (WW) is one of the conservation methods that can be effective with a drip irrigation system. However, untreated WW contains organic and inorganic chemicals that can lead to emitter clogging and poor performance of drip irrigation. Emitter clogging depends on the quality of water, types of emitters, filtration systems, and environmental conditions. Clogging can result from physical (grit), chemical (scale), and biological (bacteria, algae) agents. Clogging is a combined effect of one or more than one of these factors. Biological clogging of emitters results in bacterial growth that further build-up slime. It combines with mineral aggregates to flocculate and clog the openings of emitters. Clogging problems are common when using water with high biological activity (surface water) and when biofilms form a physical barrier or precipitates with minerals, such as iron, manganese, and sulfur that are present in water.

Biofilm formation is initiated by colonist microbial species attached randomly to the surface forming a monolayer. These exist in an organized structured habitat associated with microorganisms that are enclosed in an exopolysaccharides matrix into microcolonies. Initially, the biofilm formation process is characterized by a high range of different microorganisms and then followed by secondary colonization of bacteria that are benefited from a protective environment in the biofilm [24]. The advantage of attaching to a surface is the ability to build up a foundation to a preferred environment for bacterial growth.

The microorganisms, when have a close spatial arrangement, are advantageous because of the potential for interaction and co-metabolism. It is the degradation of the substrate (surface) in which the degradation of the secondary substrate is due to the presence of the primary substrate. The structure is considered as an immobilized enzyme system in which their habitat (milieu) and the enzyme activities are constantly changing and evolving to an approximately steady state [6].

During development of biofilm, substrate concentrations become heterogeneous; allowing formation of microfiches, (the function, or position of microorganisms within microbial community) characterized by environmental conditions and provides growth conditions that are suitable for new species. The presence of relatively large channels and pores within the matrix structure leads to expansion that might allow the entry of colonizing cells, present in water, and their establishment within the biofilm [6].

In the fertigation method, soluble fertilizers (which consist of organic compounds) are used in drip irrigation, and these act as nutrients for the biofilm inside the irrigation pipes. However, the population of microbial species may decline in the middle stages of biofilm development due to a shortage of space, nutrients, and other resources. Successive colonization processes influence biofilm structure starting from the initial colonization phase and over the time affect microbial species distribution, both on the surface and within the biofilm structure.

Biofilms include pathogenic (infectious agent) and non-pathogenic microorganisms (incapable of causing diseases) and clog the opening of emitters, which not only lead to disruption in the system but also require high-maintenance cost. The disturbance in the system will lead to insufficient water distribution for crops and can affect the crop yield.

This chapter focuses deals on the state-of-art and potential use of biofilms in drip irrigation systems and effects on emitter clogging.

7.2 RESEARCH STUDIES ON BIOFILM TO REDUCE CLOGGING OF EMITTERS

The biofilm is held together by polymers mainly composed of extracellular proteins and carbohydrates [10, 11, 22, 23]. The characteristics of biofilms attached on 7 types of emitters under reclaimed water have been systematically explored with phospholipids fatty acids (PLFAs) and other microbiological testing methods; and quantitative study among biofilms components and emitter clogging was done for the smooth operation of drip irrigation system.

The degree of clogging showed that the emitters at downstream were less clogged than the emitters at the beginning of the laterals. Increasing rates (IR) of DW of biofilms inside the emitter revealed that initially, the emitters at the beginning of the lateral were less clogged whereas emitters at downstream were highly clogged. Later on, the head part of the lateral was highly clogged than the end part. The biofilms were easier to detach and transported along water. The low flow rate and turbulence resulted in accumulation of biofilm in the end part. Growth of biofilm was faster with increase in running time and later was decreased. With cold weather, microbial activity was inhibited, and sticky substances were reduced. The adsorption capacity of biofilms was decreased, and growth rate was reduced. After moving the drip irrigation system indoor, microorganisms experienced shortage of nutrients and were then even dead. The sticky substance reduced the bond strength; and adhesion of biofilm was detached, and biofilm surfaces became rougher [25].

The bio-fouling process in different types of pressure compensating emitters was evaluated. Subsurface emitters were operated under specific thermal range (16–24°C). This involved a series of definable web structure in the matrix of premature biofilms. Majority of organic particles contributed in growth of interior biofilm. Higher flow rate due to shear force proved as an anti-clogging parameter. The best period of flushing was determined with the protein-carbohydrates ratio in extracellular polymeric substance (EPS) [17].

In Mediterranean countries, treated municipal wastewater (TWW) is being used. Fruits and crops were analyzed for fecal coliform, E. coli (Escherichia coli), fecal streptococci salmonella, and helminth egg. TWW increased the yield of the crops (eggplant, tomato) without doing considerable damage to the emitters and drip irrigation system. E. coli increased level was noticed. Level of microbial contamination was reduced by applying post-treatment health protection control measures according to the World Health Organization (WHO) [4, 28].

Moyne et al. [14] investigated the impact of drip and overhead sprinkler irrigation on the persistence of attenuated E. coli in the lettuce-phyllosphere. Rifampicin-resistant attenuated that E. coli was inoculated onto the soil beds after seeding. When E. coli was injected into 2-weeks old plants, the organisms were enriched in 1 of 120 or 0 of 240 plants almost 21 to 28 days post-inoculation, respectively. They concluded that the population size of E. coli declined rapidly and on 7th-day counts were near or below the limit of detection (10 cells per plant) for 82% or more of the samples. Drip irrigation and overhead irrigation did not consistently influence E. coli survival.

Biofilm is complex with the association of cells in it. The major component of these cells was the mixture of polysaccharides that were secreted by the cells within the biofilm structure. The similar structure was found in the processed food, in which a mixture of macromolecules was found. These macromolecules interacted with each other in many ways and a recognizable structure was formed. These structures involved cells, water, ions, and soluble low and high molecular products that were trapped, and the major component was water. The EPS (extracellular polymers) synthesized by microbial cells varied in their chemical and physical properties.

Polysaccharides interacted with molecular species that involved lectins, proteins, lipids. The EPS reflected the polymers that were synthesized by planktonic cells and demonstrated using antibodies against EPS from planktonic cells, also by comparison of the enzymic products following digestion of planktonic. The resultant effect was the variation in polysaccharides composition, which was synthesized within the biofilm. The microbial cells in biofilm depend greatly on physiological state of the biofilm. The amount of EPS that were synthesized within biofilm was dependent on availability of carbon substrate. There is a limitation in other nutrients (such as nitrogen, potassium) that will promote the growth of EPS. This slow growth of bacteria enhances the production of EPS [27].

Microbial community was analyzed by Yan et al. [24]. Biofilm structure within the emitter flow path and the biofilm biomass in emitters with different structural dimensions were measured. Emitter discharge was recorded at beginning of experiment and at an interval of 360 h. Individual emitter flow rates and its distribution was determined. Due to closely-spaced emitters, a friction loss along drip line was negligible and pressure was considered constant. Discharge variation was attributed to the degree of clogging [24].

The cohesiveness of the biofilm found by the physicochemical properties of its EPS that had led to failure of several methods out of which disinfection and flushing had led to reduction in nutrients in the drinking water. Effective cleaning procedures resulted in breaking the matrix and affected the elastic properties of bacterial biofilms. The change in the cohesive strength of drinking water biofilms (2 months old) was evaluated under increasing hydrodynamic shear stress (from 0.2 to 10 Pa) and shock chlorination (applied concentration of 10 mg of Cl_2 per liter for 60 min contact time). Biofilm erosion (loss of cells per unit surface area) and cohesiveness (the changes in the detachment due to shear stress) and cluster volumes measured using atomic force microscopy (AFM) were studied. Biofilm removal dependent on the process of erosion and coalescence of the biofilm clusters when there was rapid increase in the hydrodynamic constraint. Indeed, 56% of the biofilm cells were removed and the number of 50–300 μm^3 clusters were decreased and the number of the smaller cells were increased (<50 μm^3) and were larger (>600 μm^3) ones. The compactness of the biofilm EPS was increased with hydrodynamic stress. Shock chlorination could remove cells (75%) from the biofilm. Oxidation stress resulted in a decrease in the cohesive strength profile of the remaining biofilm in the drinking water [12].

7.3 MATERIAL AND METHODS

The study was conducted in the Brambe village of Mandar block in Ranchi district of Jharkhand state, India, and lies at latitude 23°45'N and longitude of 85°30'E. Drip irrigation system was setup on the experimental plot as shown in Figure 7.1.

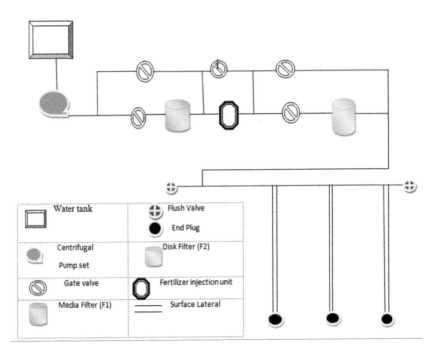

FIGURE 7.1 Schematic layout of experiment for wastewater, pond water, and groundwater (separate units were used for each type of water).

Water quality parameters of three types of water (i.e., WW, pond water, and groundwater) were analyzed. Disk filter of 25 m^3h^{-1} discharge rates and media filter of 10 m^3h^{-1} were used in drip irrigation system, and flow rate was maintained constant at working pressure of one kg/cm^2. The water quality parameters were measured in the laboratory of the Center for Applied Chemistry, Central University of Jharkhand, Brambe village. Thirty emitters were used in each lateral at an emitter spacing of 30 cm. The laterals were manufactured by polyvinyl chloride having 18 m in length.

7.3.1 EMITTER DISCHARGE

The outflow rates of all emitters were measured at preliminary stage at 1 kg/ cm^2. The outflow data used for calculation were as follows:

1. **Discharge Ratio Variation (DRV):** DRV of drip irrigation emitters was estimated by Christiansen uniformity coefficient as follows:

$$DRV = 100 \frac{\pounds_{i=1}^{n} q_i}{n\bar{q}_{new}} \tag{7.1}$$

where, the outflow is rates of drip irrigation emitter no. in the unit (Lh^{-1}); $\bar{q}_{new}$ is the rated outflow rate of emitter provided by the manufacturer (Lh^{-1}); andis the total number of emitters.

2. **Coefficient of Uniformity (CU):** CU of drip irrigation emitters was determined as follow:

$$CU = 100 \left(1 - \frac{\pounds_{i=1}^{n} |q_{i-\bar{q}}|}{n\bar{q}} \right) \tag{7.2}$$

where, the outflow rates of ith emitter (Lh^{-1}); $\bar{q}$ is the average outflow of emitters tested in each treatment (Lh^{-1}); and n is the total number of emitters.

7.3.2 BIOFILM SAMPLING AND TESTING

Biofilm samples were collected from the emitters carefully from the up-stream of the first ten emitters. Later, samples were taken from ten emitters from the mid-portion and ten emitters from the end part of the drip irrigation laterals, respectively. The samples were then weighed with high precision electronic balance (accurate to 0.0001 g), respectively. The biofilm sample was then scrapped out using distilled water; and was divided into two parts for determination of DWs and extracellular polymers (EPS) tests:

1. **Dry Weights (DWs):** After scrapping out the biofilm sample from the emitter, the sample was dried at a constant temperature of 60°C for 1 h. After 1 h, the sample was taken out from the oven and was weighed again. Thus, the DWs of biofilm attached to inner surface of emitters were the difference between the initial reading (taken before dry the sample in the oven) and final reading (taken after drying the sample into the oven).
2. **Extracellular Polymers (EPS):** The EPS of biofilm inside the emitter includes extracellular polysaccharides and extracellular

protein. The protein was estimated using Bradford's method. To determine the total protein in a mixture, dyes were used that exhibited changes in their spectral properties to bind the protein. The absorbance was measured at 595 nm wavelength in the glass cuvette against the blank reagent in spectrophotometer. The reading of blank reagent was taken by zeroing the base. The cuvette used should be thoroughly washed with distilled water or ethanol to remove the bluish stain. The protein fraction was calculated as follows:

$$\text{Protien}\ (\frac{1}{g}) = \frac{\text{Absorbance}}{\text{slope}} \times \text{dilution factor} \tag{7.3}$$

7.4 RESULTS AND DISCUSSION

7.4.1 WATER QUALITY PARAMETERS

The potential for emitter clogging problem varies with source of irrigation water. Water quality parameters (such as turbidity, pH, EC, Fe^{+2}, Ca^{+2}, Mg^{+2}, CO_3^{-2}, HCO^-, DO, BOD, and COD) were analyzed and are presented in Table 7.1.

7.4.2 FLOW RATE VARIATION

The variations in average flow rate for three types of water during the study are presented in Figure 7.2. The flow rate was initially 2 lph as per manufacturer recommendation, which was later decreased to 2.05, 1.96, and 1.81 lph after operating the system for 108 h, respectively. For 336 h, the flow rate decreased to 23.9% under groundwater, 26.02% under pond water, and 40.6% under WW. Later after 540 h, the flow rate decreased to 35.6%, 35.20%, and 45.55% for groundwater, pond water, and WW, respectively.

     Yan et al. [24] observed that the discharge rate dropped during the first 200 h, recovered slightly after the next 540 h and continued to decrease through the end of irrigation; and after 360 h of irrigation with reclaimed WW, the emitters average flow reduction ranged from 14.4 to 72.2% with an average value of 34.7%. Maximum numbers of emitters were clogged partially, and no emitter was clogged completely (Table 7.2).

TABLE 7.1 Range of Water Quality Parameters for Three Different Types of Water

Water Quality Parameter	Wastewater			Pond Water			Groundwater		
	Mean	S.D.	Range	Mean	S.D.	Range	Mean	S.D.	Range
pH	6.16	0.166	5.9–6.5	8.65	0.604	8.21–9.01	7.19	0.1067	7.06–7.32
BOD_5	400	18.25	430–400	250	17.26	237–275	205	7.25	195–212
Ca^{+2}	128	12.07	115–144	98	4.89	92–104	80	4.74	75.3–86.5
CO_3^{-2}	327	64.63	236–380	203.33	8.96	191–212	103	2.12	100.4–105.6
COD	800	32.07	757–834	510	76.73	411–598	450	12.75	440–468
DO	40	4.76	48–40	2.5	1.03	1.75–3.95	2	0.22	1.75–2.3
EC	3.29	0.341	2.93–3.75	2.06	0.23	1.78–2.34	0.66	0.09	0.58–0.79
Fe^{+2}	2.91	0.074	2.83–3.01	2.01	0.16	1.73–2.3	1.8	0.37	1.4–2.3
HCO_3^-	360	85.73	285–480	280	74.06	212–383	140	18.56	121–165.8
Mg^{+2}	23.69	2.07	21.02–26.03	16.15	0.38	15.88–17.01	11.2	0.87	10.2–12.3
Turbidity	21	1.08	22.02–19.5	10	2.16	8–13	4	1.41	3–6

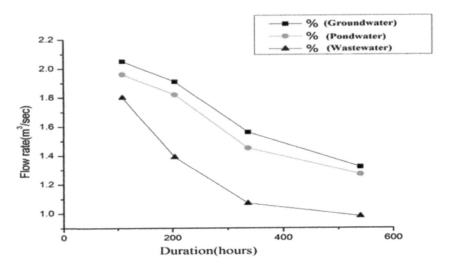

FIGURE 7.2 Variation of flow rate with increase in running time of drip irrigation system.

7.4.3 DISCHARGE RATIO VARIATION (DRV) AND UNIFORMITY COEFFICIENT (CU)

DRV and CU for three types of water showed decrease in DRV with increase in running time (Table 7.2) under drip irrigation. Initially, the discharge rate of all the emitters given by the manufacturer was 2.1 lph that was reduced to 2.05 lph (2.3% reduction for 108 h. Thereafter, the rate of decrease in discharge rate continued to increase with increase in running time. Discharge rate was further decreased from 2.05 lph to 1.32 lph (35.60% reduction). For pond water, the discharge rate decreased to 35.20% from 108 h to 540 h. For WW, flow rate was 1.39 for 204 h, which was decreased by 29.49% after operating the system for 540 h.

7.4.4 DRY WEIGHT (DW) OF BIOFILM AND THEIR EFFECTS ON CLOGGING OF EMITTERS

Average DWs of biofilms based on 30 emitters in each lateral for three types of water has been presented in Table 7.3.

   DWs of biofilms attached in the emitter showed an increasing trend with the increase in the operating hours of the drip irrigation system using three different types of water. The average DWs attached were 30.2 mg, 54.17 mg,

TABLE 7.2 Fluctuation of Discharge Ratio Variation (DRV) and Uniformity Coefficient with Drip System Running Time

Types of Water	1st Sample (108 h/18 Days)			2nd Sample (204 h/34 Days)			3rd Sample (336 h/56 Days)			4th Sample (540 h/90 Days)		
	Avg q (lph)	DRV	CU	Avg q (lph)	DRV	CU	Avg q (lph)	DRV	CU	Avg q (lph)	DRV	CU
Ground water	2.05	97.61	96.41	1.91	90.95	88.09	1.56	74.2	70.7	1.32	62.85	56.35
Pond water	1.96	93.33	92.19	1.82	85.71	83.01	1.45	69.04	65.79	1.27	60.47	54.24
Wastewater	1.8	85.71	84.66	1.39	66.19	64.11	1.07	53.5	50.98	0.98	46.66	41.85

TABLE 7.3   Dry Weights (DWs) of Biofilm Sample

Type of Water	1st Sample (in mg) (108 h/18 Days)				2nd Sample (in mg) (204 h/34 Days)			
	Head	Mid	End	Avg	Head	Mid	End	Avg
Ground water	18.7	25.1	31.3	25	23.2	26.2	41.2	30.2
Pond water	20.9	44.4	66.5	44.9	35.7	54.2	72.5	54.1
Wastewater	33.4	54.4	72	53.2	44.8	66.9	92	67.9
Type of Water	3rd Sample (in mg) (336 h/56 Days)				4th Sample (in mg) (540 h/90 Days)			
	Head	Mid	End	Avg	Head	Mid	End	Avg
Ground water	52.9	70.1	76.3	66.43	89.9	103.6	104.9	99.4
Pond water	76	125.1	164	121.7	121.5	203.5	239.8	188.3
Wastewater	116.1	143.4	213.6	157.7	196.6	214.3	335.2	248.7

and 67.92 mg using groundwater, pond water, and WW, respectively after an operation of 204 h of drip system. Further, after 540 h of operation, DWs of biofilms were increased to 99.40 mg, 188.32 mg, and 248.72 mg for groundwater, pond water, and WW, respectively. DWs for WW were higher than that of pond water and groundwater. On the other hand, the IR of attached biofilms DWs was slow initially but latter increased rapidly. After a certain interval of time, the rate of increase of attached biofilms was decreased.

Among the three kinds of water, the biofilms DWs order was $DW_{wastewater} > DW_{pond\ water} > DW_{groundwater}$. The IR of biofilms from 108 to 204 h was 20.65% for groundwater, 23.33% for pond water, and 27.53% for WW. Similarly, operating the system between 204–336 h, the IR of biofilms was 119.9% for groundwater, 124.7% for pond water, and 132.3% for WW. After 336–540 h, it was 49.6% for groundwater, 54.6% for pond water, and 57.6% for WW. For a lateral of 18 m in length, the clogging in head, middle, and end parta of the lateral were different. According to Horn et al. [9], the intensity of shear force leads to detachment of biofilms. As shear force of water was decreased in the end part of lateral, the growth of biofilm was increased.

The IR of biofilms was slow initially and then rapidly increased, because of the sticky substance secreted by the microbes in proliferation stage. The inner microbes were highly active and getting more nutrients to secrete metabolites and increased rapidly. The detachment of biofilm was observed at the end part where the shear force is reduced, and biofilm accumulation was more. When the biofilm grew to certain thickness, microorganisms required more nutrients for further growth. It was also observed that due to higher turbulence and hydraulic shear stress, the flow rate was higher in the head part. Due to this reason, it got easily detached in the head part and was transported along with water. Weaker stress implies favorable conditions for biofilm growth and finally clogging, according to Chang et al.; Nicolella et al.; and Cloete et al. [3, 5, 15]. Oliver et al. [17] studied that if the entry section was made longer with narrowly spaced more surface area comes in contact with water and create more friction and lowers the average velocity at the entrance. Therefore, lower velocity leads to lower shear stress and more biofilm growth at longer entrance.

7.4.5   EXTRACELLULAR POLYMER (EPS)

The value of protein was higher for WW compared to that for pond water and groundwater (Table 7.4).

TABLE 7.4   Concentration of Unknown Diluted Protein

Dilution Concentration	Absorbance		
	Groundwater	Pond Water	Wastewater
Undiluted (100)	1.048	1.518	1.813
1:10	1.047	1.250	1.675
1:100	1.022	1.090	1.477

The concentration of protein was higher in WW, because the concentration of DO was high that involved more microbial activity. Concentration of iron was higher, as some filamentous microbes, which oxidized Fe^{+2} to Fe^{+3}, lead to precipitation and more clogging of emitters. Protein concentration for pond water was higher than that of groundwater because there are sufficient nutrients and organic waste present due to human activities, which provided an ideal environment for microbial growth. Groundwater on the other hand was free from pathogenic bacteria but had high mineral content that varies from place to place. Therefore, it was recommended to use WW after proper water treatment.

The relationship here was close to linear. The slope of this plot was 0.0277. As the amount of standard protein sample was increased, the absorbance observed in the spectrophotometer was also increased. With the help of this standard calibration, further unknown sample reading was observed.

7.4.6 *TECHNIQUES TO OVERCOME CLOGGING*

Various techniques have been used to reduce the clogging of emitters. These treatment techniques depend on the type of issue in the system. In a drip irrigation system, filtration is the essential component to prevent emitter clogging [13, 18]. Filtration prevents inorganic particles and organic materials in water from entering the drip irrigation system. Media filters are used to prevent the passing of suspended solids, such as gravel, and sand. Particles, which are captured, are smaller than the size of openings in the mesh of the filter and are removed by both physical and chemical mechanisms [1].

Capra and Scicolone [7] noticed that tertiary treatment and chlorination were effective to reduce clogging caused by bacteria and algae. In arid and semi-arid developing countries, where extremely stringent quality water is used for irrigation, it leads to high cost for the treatment. Chemical treatment consists of injection of chemicals to maintain the pH of water to a permissible limit. Acids have been used to lower the water pH and the potential to reduce chemical precipitation [19]. The treatment with nitric acid (65%at pH 5) and sodium hypochlorite (12%) are methods [20].

Hill and Brene [8] investigated the prevention of clogging of four emitter types using WW treated with a combination of filters (a sand filter and a screen filter with openings of 100 μm in diameter) and with intermediate chlorination (free residual chlorine in the concentration of 0.4 mgl^{-1}). The results indicated that the strains of antagonistic bacterial as genus *Bacillus* spp (ERZ, OSU-142) and *Burkholdria* spp. (OSU-7) were successfully used for treatment of biological clogging of the emitters. These strains acted as anti-clogging agents for treatment of emitters in drip irrigation system [21]

7.5 SUMMARY

Research on emitter clogging in drip irrigation has always been a challenging task because of complexity, time, and patience. This chapter focuses on effects of emitter clogging using three different types of water in drip irrigation. The characteristics of three types of water were determined and then compared. Various water quality parameters for groundwater, pond water, and WW were analyzed, and these were: turbidity, pH, EC, Fe^{+2}, Ca^{+2}, Mg^{+2}, CO_3^{-2}, HCO_3^{-}, DO, BOD, and COD. For WW, pH was lower than that for pond water. The presence of iron was one of the main causes of clogging. The pond water was polluted due to the presence of concentration of heavy metals in runoff. WW with suitable water treatment can be reused.

Monitoring flow rate, DRV, and CU of the emitters is the most appropriate way to check the effectiveness of drip irrigation system. The emitters with discharge rate of 2.1 lph, which was rapidly decreased when the system was operated for 540 h. Emitters used for WW application were at higher risk for clogging. DRV and CU were higher for groundwater than WW. If the fluctuations of the emitters flow rate, DRV, and CU are extremely high, it is recommended to replace the clogged emitter or use proper treatment. DWs and EPS were higher in the case of WW compared with pond water and groundwater. In Brambe village, pond water was highly contaminated with organic (fish, snails, larvae, vegetations) and inorganic wastes due to anthropogenic activity compared with groundwater. WW was highly contaminated with microbes and heavy metals due to which it showed higher risks of clogging. With an increase in the running time of the irrigation system, there was an increase in DWs and EPS. Chemical treatment (chlorination, acidification) affected the performance of emitters, filters, or laterals. Monitoring water quality conditions, instead of implementation actual cleanup activities, can be a remedy applied to areas (pond).

KEYWORDS

- biochemical oxygen demand
- biofilm
- chemical oxygen demand
- clogging
- drip irrigation system
- emitters
- extracellular polymer
- turbidity
- water quality

REFERENCES

1. Adin, A., & Alon, G., (1986). Mechanisms and process parameters of filter screens. *Journal Irrigation Drainage Engineering*, *112*(4), 293–304.
2. Capra, A., & Scicolone, B., (2007). Recycling poor quality urban wastewater by drip irrigation systems. *Journal of Cleaner Production*, *15*, 1529–1534.

3. Chang, H. T., Rittmann, B. E., Amar, D., Heim, R., Ehlinger, O., & Lesty, Y., (1991). Biofilm detachment mechanisms in a liquid-fluidized bed. *Biotechnology and Bioengineering*, *38*(5), 499–506.

4. Cirelli, G. L., Consoli, S., Licciardello, F., Aiello, R., Giuffrida, F., & Leonardi, C., (2014). Treated municipal wastewater reuse in vegetable production. *Agricultural Water Management*, *104,* 163–170.

5. Cloete, T., Westaard, D., & Van, V. S., (2003). Dynamic response of biofilm to pipe surface and fluid velocity. *Water Science and Technology*, *47*(5), 57–59.

6. Costerton, J., (1995). Overview of microbial biofilms. *Journal of Industrial Microbiology*, *15*(3), 137–140.

7. Goodwin, J. A. S., & Forster, C. F., (1985). A further examination into the composition of activated sludge using a cation exchange resin. *Water Resources*, *30*(8), 1749–1758.

8. Hills, D. J., & Brenes, M. J., (2001). Micro irrigation of wastewater effluent using drip tape. *Applied Engineering in Agriculture*, *17*, 303–308.

9. Horn, H., Reiff, H., & Morgenroth, E., (1985). Simulation of growth and detachment in biofilm systems under defined hydrodynamic conditions. *Biotechnology and Bioengineering*, *81*(5), 607–617.

10. Lasa, I., & Penades, J. R., (2006). Bap a family of surface proteins involved in biofilm formation. *Res. Microbiol.*, *157* (2), 99–107.

11. Marshall, K. C., & Characklis, W. G., (1990). *Biofilms* (pp. 731–796.). Wiley and Sons, New York.

12. Mathieu, L., Bertrand, I., Abe, Y., Angel, E., Block, J. C., Skali-Lami, S., & Francius, G., (2014). Drinking water biofilm cohesiveness changes under chlorination or hydrodynamic stress. *Journal of Water Resources*, *55*, 175–184. doi: 10.1016/j. waters, 2014.

13. McDonald, D. R., Lau, L. S., Wu, I. P., Gee, H. K., & Young, S. C. H., (1984). *Improved Emitter and Network System Design for Wastewater Reuse in Drip Irrigation* (Vol. 163, p. 87). Water Resources Research Center, Honolulu, Hawaii.

14. Moyne, A. L., Sudarshana, B., Mysore, R., Blessington, T., Steven, T., Cahn, M., & Linda, H. J., (2011). Fate of *Escherichia coli* O157:H7 in field inoculated lettuce. *Food Microbiology*, *28*(8), 1417–1425.

15. Nicolella, C., Chiarle, S., Di Felice, R., & Rovatti, M., (1997). Mechanisms of biofilm detachment in fluidized bed reactors. *Water Science and Technology*, *36*(1), 229–235.

16. Niu, W., Liu, L., & Chen, X., (2012). Influence of fine particle size and concentration on the clogging of labyrinth emitters. *Irrigation Science*, *31*, 545–555.

17. Olivera, M. M. H., Hewaa, G. A., & Pezzanitia, D., (2014). Biofouling of subsurface type drip emitters applying reclaimed water under medium soil thermal variation. *Agricultural Water Management*, *133*, 12–23.

18. Oron, G., Shelef, G., & Turzynski, B., (1979). Trickle irrigation using treated wastewater. *Journal Irrigation Drainage Division*, *105*(IR2), 175–186.

19. Pitts, D. J., Haman, D. Z., & Smajstrla, A. G., (1990). *Causes and Prevention of Emitter Plugging in Micro Irrigation Systems* (p. 9). Gainesville, FL: University of Florida, Florida Cooperative Extension Service Bulletin 258.

20. Ribeiro, T. A., P. R., Paterniani, J. E. S., & Coletti, C., (2008). Chemical treatment to unclog dripper irrigation system due to biological problem. *Science Agriculture*, *65*(1), 1–9.

21. Sahin, U., Anapal, O., Donmez, M. F., & Sahin, F., (2005). Biological treatment of clogged emitters in a drip irrigation system. *Journal Environmental Management, 76*(4), 338–341.

22. Tripathi, V. K., Rajput, T. B. S., & Patel, N., (2014). Performance of different filter combinations with surface and subsurface drip irrigation systems for utilizing municipal wastewater. *Irrigation Science, 32*(5), 379–391. Online: doi: 10.1007/ s00271-014-0436-2.

23. Whitchurch, C. B., (2002). The clogging mechanism in the emitter. *Science*, 14–27.

24. Yan, D., Bai, Z., Rowan, M., Gu, L., Shumei, R., & Yang, P., (2009). Biofilm structure and its influence on clogging in drip irrigation emitters distributing reclaimed wastewater. *Journal of Environmental Science, 21*, 834–841.

25. Zhang, J., Zhao, W. H., Tang, Y. P., Wei, Z. Y., & Liu, B. H., (2007). Numerical investigation of the clogging mechanism in labyrinth channel of the emitter. *International Journal for Numerical Methods in Engineering, 70*(13), 1598–1612. https://onlinelibrary.wiley.com/ toc/10970207/2007/70/13 (accessed on 20 June 2020).

26. Zhou, B., Yunkai, L., Yiting, P., Yaoze, L. Z. Z., & Jiang, Y., (2013). Quantitative relationship between biofilms components and emitter clogging under reclaimed water drip irrigation. *Irrigation Science, 31*, 1251–1263.

27. Sutherland, I. W., (2001). The biofilm matrix: An immobilized but dynamic microbial environment. *Trends in Microbiology, 28*(8), 222–227.

28. Tripathi, V. K., Rajput, T. B. S., & Patel, N., (2016). Biometric properties and selected chemical concentration of cauliflower influenced by wastewater applied through surface and subsurface drip irrigation system. *Journal of Cleaner Production, 139*, 396–406. doi: 10.1016/j.jclepro.2016.08.054.

Part III:
Wastewater Use: Principles and
Field Practices

SUSTAINABLE SOIL CONSERVATION AND MANAGEMENT: PRINCIPLES, ISSUES, AND STRATEGIES

SHRIKAANT KULKARNI

ABSTRACT

Soil conservation and management in a sustained manner are of critical importance to the well-being of mankind. The diligent use in management is of vital and critical importance now than earlier to fulfill the ever-increasing production of food in tune with the demand of the rapidly increasing global population. Concerns have been raised worldwide about soil degradation and environmental pollution due to anthropological reasons like overuse of agrochemicals, and apathy towards judicious use of land resources and adoption of best environmentally benign practices for diligent conservation of soil despite extensive research, abundantly available literature on adopted soil and water conservation strategies, and management. Managing soils subjected to heavy use and replenishing already eroded/degraded soils should be the key priorities for achieving sustainable agronomic and forestry production. Managing for high-quality soils asks for managing water resources by adopting the best irrigation and drainage practices, managing nutrients, and managing nitrogen and phosphorus.

Management must precede conservation to restore and improve the eroded and degraded lands and ecosystems across the world. Thus, this chapter presents a balanced review and discussion about the consequences of soil erosion and its severity, evaluation of the soil health, management of soil fertility and its conservation, soil quality, biology, and resilience, chemical dynamics, and its measures. Special emphasis has been given to management techniques for resource-deprived regions; conservation and management over generic factors and mechanisms of erosion. It recognizes possible discrepancies of past and present research initiatives in water and soil conservation and prescribes measures for its enhancement. The issue of immense interest to both soil conservationists and policymakers is to promote a better understanding of principles on soil erosion and adopting measures of strategic importance for bringing about sustainable soil conservation and management.

8.1 INTRODUCTION

World soil resources are limited and distributed unevenly across diverse regions, fragile, and vulnerable to degradation by misuse and mismanagement of soil, and prone to extreme events related to the sudden climate change (SCC). The land area susceptible to degradation processes is estimated at 3,500 Mha thereby affecting the livelihood and well-being of the huge underprivileged population living in regions subjected to severe climate and marginal agricultural lands. Further, the resource-deficient farmers are unable to invest much in soil and water conservation techniques and in declining soil fertility [1]. Thus, a sustained use of farming practices extractive in nature by small landholders of the tropical and semitropical regions and for rapid economic returns by large-scale farmers have aggravated the problem of widespread distribution of depleted and degraded soils, often with a tailored topsoil layer because of the rapid soil erosion. The area equipped on the other hand for supplementary irrigation has expanded to about 287 Mha, and it has also severely depleted groundwater and secondary salinization.

The food production must be drastically increased to feed a huge and ever-growing world population of more than 10 billion, to fulfill the ever-growing food needs of growing population and shift in the dietary habits more animal-based than plant-based. By 2030, it is estimated that the global cereal demand for food and animal feed may rise to 2.8 billion Mg/year, which is double the production in 2005. The scope for bringing new land under cultivation is getting reduced and therefore, the productivity must be increased from the existing agricultural land-area, time, and use of energy-based inputs (i.e., fertilizer, irrigation). In accordance with the concept of "zero net land degradation," any new land erosion must be checked by the restoration of earlier degraded land [2].

8.2   CAUSES OF SOIL DEGRADATION ACROSS GLOBE

8.2.1   SOIL EROSION

Soil erosion or loss is mainly affected by water, wind, and tillage during agricultural production. In addition, landslides or gravitational erosion may be another cause particularly in a very steep sloppy area. Extremely wet soil conditions are responsible for the soil erosion [3]. Dry soil is the cause of wind erosion. Tillage erosion is cause of either steep or undulating topography notwithstanding moist soil conditions. Erosion is the cumulative effect of a force offered by water, wind, or gravity responsible for erosion, an inherent vulnerable soil, other than factors related to management or landscape. Soil's susceptibility or erodibility depends on texture (namely, silts greater vulnerable over clays and sands), mechanical strength and aggregate size (amount of organic matter), and soil moisture. Although there are no universal solutions to the erosion problems, yet right and sound management techniques can check soil erosion [4].

8.2.2   WATER EROSION

Water erosion when rainfall rates supersede soil's infiltration capacity and runoff on bare, but steep sloping land. The water-streamlets dislodge the concentrated soil and transport it to low lying areas. Runoff water during its movement down the slope removes more soil, agrochemicals, and nutrients, which make their way to water bodies like streams, lakes, and estuaries. Deterioration in the soil health in agriculture and urban watersheds reflect on further increase in runoff during heavy rainfall and leading to flooding. The lower infiltration capacity lowers both the quantum of water available to plants that sip through the soil into underground water reservoirs leading to drying up of streams during in particular droughts [5].

   Soil erosion poses a grave threat to the uncovered surface and therefore vulnerable to destruction due to raindrops and wind energy. Degraded soils promote erosion, leading to a reduction in soil health. Therefore, a cycle vicious in nature with a cascading effect on erosion, leads to further vulnerability to erosion. Soil degradation is the result of removal of organic matter with which the surface layer enriched. Erosion selectively removes the finer soil particles amenable to transportation. Highly eroded soils, therefore, become devoid of organic matter and are bereft of physicochemical and biological properties, hampering the sustainability of crops and have detrimental environmental impacts [6].

8.2.3 WIND EROSION

Wind erosion occurs when the dry and loose soil, with smooth but barren land, with the landscape does not provide for too many obstructions to wind. The wind moves and carries coarser soil particles, which will further remove some more soil particles and raise effective soil damage. The fine soil particles (silt and sand) are light in weight form suspensions. They can be carried over larger distances, cutting across continents and oceans. Wind erosion influences soil qualitatively by way of the degradation of topsoil organic content rich and can be responsible for crop damage from abrasive forces. In addition, wind erosion influences quality of air, which is a cause grave concern to the adjoining areas. The capacity of wind to damage a soil is governed by how far and how well the soil is managed, as strong agglomeration reduces its less amenability to dispersion and transportation. Moreover, many soil enriching practices like mulching as well as the cover crops use to provide the protection to the surface of soil due to water and wind erosion [7].

8.2.4 LANDSLIDES

Landslides take place when the soils get super-saturated on steep slopes with rains over a prolonged time period. The pressure of population has led to the farming of sloping hillsides. The consistent rains saturate the soil at the landscape locations, where water is received water from up sloping areas. This results in two fallouts:

- Increases the mass of the soil; and
- Decreases the cohesiveness and the ability to resist the gravitational force of the soil.

Agricultural lands, pastures on steep lands with shallow-rooted grass in mountainous regions are more vulnerable to slumping than forests as the former lack deep roots that can hold on soil lumps in cohesion. Landslides may turn into mudslides depending upon the soil types [8].

8.2.5 TILLAGE EROSION

Tillage brings about degradation of land over and above promotion of water and wind erosion by breaking down lumps of masses leading to exposure of soil in elemental form. It may bring about erosion by the direct movement of the slope down to low lying field regions. Tillage erosion is caused by gravity, which moves more soil by the plow or harrows down the slope and typically happens at greater speed as against uphill travel, thus worsening the situation furthermore. Tillage along with the contour too leads to soil slipping down the slope. A unique characteristic of tillage erosion as against other types of erosion is that it is not influenced by the drastic weather conditions and takes place progressively with every tillage exercise. Tillage erosion cascades the effect by furthering soil degradation by wind or water. Tillage erosion does not bring about off-site damage [9].

8.2.6 SOIL TILTH AND COMPACTION

Soil compacts and gets dense, when soil particles are forced together intimately. Soil compaction is affected by many factors and is visible in different forms. Compaction at surface covers surface, while plow layer compaction covers subsoil.

8.2.6.1 SURFACE COMPACTION

Surface compaction occurs in almost all agricultural soils, which are worked intensively. Degradation of soil aggregation is fallout of erosion, reduction in organic matter contents, and force applied by the weight of field equipment. The first two factors lead to cut down in the supplies of adhering binding materials and a subsequent erosion of aggregates. Surface crusting and plow layer compaction both result due to certain factors but the former occurs specifically when the surface of the soil is left exposed to crop residue and further the energy of raindrops breaks down wet lumps, segregates them apart into particles that precipitate into a thinner but denser surface crust. The blocking of the soil lowers water percolation, and the surface settles into a hard mass on drying. Crusting immediately following the planting may postpone or check seedling appearance. The crust not so severe though limits germination, can contain seepage of water. Soils with surface hardening are susceptible to more runoff and rate of erosion. Surface crusting can be reduced by covering the soil with more residues and maintaining strong soil aggregation on the surface [10].

8.2.6.2   SUBSOIL COMPACTION

Soil compaction due to heavy and tillage equipment wet soils. Primarily subsoil compaction occurs due to a combination of factors and is responsible for plow layer compaction. High soil water contents with little internal cohesion flow like a liquid on a slope due to gravitational force like sliding muds during highly wet-periods. Reduced water contents with more cohesion can be readily molded due to plasticity of the mass, which on drying becomes friable, i.e., it breaks down under pressure rather than getting molded. The plastic limit lies between plastic and friable soil, which reflects on agricultural implications. Soil wetter than the plastic limit, amenable to compaction as soil lumps, is held on in interspersed, dense mass if tilled or traveled on. On the other hand, the friable soil (i.e., the moisture content is lower than the plastic limit) shatters, and thereby compaction is resisted. Thus, the compaction potential is influenced by the timing of field operations and soil moisture [11].

8.2.7   SOIL CONSISTENCY

Soil consistency refers to how soil behaves in response to external forces. The soil consistency is greatly influenced by its texture, e.g., coarse-grained sandy soils are amenable to drainage, and they turn into friable from plastic soils rapidly. Fine-grained soils take longer drying periods to drain water enough to get friable. Scheduling of field operations may be delayed due to surplus drying time. Surface hardening and compaction of the plow layer are specifically prevalent with thoroughly tilled soils [12].

Natural aggregates break down, and organic matter decays leading to further increase in compaction. Although the last seed-bed may be perfect for planting, yet rainfall following planting may bring about surface blocking and further deposition, resulting into a compact plow layer and a surface crust that bring down the plant growth. Re-wetting of such soils makes them softer and the moisture provides a temporary relief to plants.

8.2.8 CONSEQUENCES OF COMPACTION

Soil gets denser with a substantial reduction in pores due to the intimacy of the particles on compaction. Notably, inter-particle spaces are eliminated. Segregation of particles on compaction is particularly responsible for reducing the infiltration and percolation of water capacity of fine- and medium-textured soils due to a reduction in inter-particle spaces, and capacity of exchange of air with the atmosphere. Compaction influences soils with larger particle size, but to a smaller degree. In such soils, the inter-particle spaces are reasonably large and not affected much by the aggregation to provide for a reasonably good water and air circulation [13].

Compacted soil hardens on drying, and has low water holding capacity when subjected to high suction. This can hamper the growth of roots and the activity of soil organisms. Compacted soils have greater impermeability to moisture compared to a well-structured soil as the later one has reasonably large inter-aggregate pores to allow better percolation. The impermeability of a wet, very rich soil is 300 psi usually far lower than the critical limit at which growth of root comes to a standstill for most of the crops. A high-quality soil increase in its strength on drying and may still has the capacity below the critical level for varied moisture ranges. Compacted soil has a very restricted moisture carrying capacity for better root growth to take place. Such soil has greater degree of impermeability even in the presence of high moisture due to the hardness in the soil. Compacted soil is vulnerable to rapid hardening than well-structured soil, so that it exceeds the critical level that resists root growth.

Smoothly growing roots require pores with diameters (> 0.1 mm), same as most of the root tips size. Roots should get into the pores and hold onto it before carrying on with the growth. Compacted soils that lack big pores won't permit plants to be anchored effectively. Roots in eroded soils don't grow to the subsoil below the tilled area as it is too compact and too hard for the growth to take place. Deeper root growth is typically very vital for agriculture, which is rain-fed. The compaction of subsoil reduces the soil volume, where from roots can siphon off the water, and enhance the possibility of crop productivity subjected to drought. Plant growth is affected, above a certain volume of soil, to grow the roots. A root system that can stifle mechanical hindrances, propagates a hormone-based signal to the plant stem, which then controls the both the respiration and growth in tandem with the same.

The plant appears to respond to a stimulus like water stress as a part of natural survival mechanism. It is not so easy to distinguish the consequences of compaction and drought as few of the hormones involved are same and the increase in mechanical resistance is in tune with dryness of the soil. The roots then will be short, thick, and fine with root hairs. The few thick roots with thickened tissue may come across few poor soil domains, and thereby showing abnormal patterns devoid of capacity of extracting water and nutrients [14].

8.2.9 CONSEQUENCES OF SOIL EROSION

Soil erosion amounts to onsite and offsite detrimental agronomic, ecologic, environmental, and economic problems. It influences quality of agricultural, forest, pasture, etc. Cropped soils are more vulnerable to erosion among the cropping and the latter because they are often barren with lack or inadequate residue cover. The effect of on-site erosion is the lowering of soil productivity, while the cause of the offsite one is the transport of the deposit and chemicals far from its source to natural water bodies through streams and wind depositions [15].

8.2.10 ONSITE EFFECTS OF EROSION

- Compaction of soil;
- Lowered crop productivity;
- Lowered soil fertility;
- Nutrient and organic matter reduction;
- Ridding off contaminants, and nutrients storage;
- Soils functional capacity to generate crops;
- Topsoil thickness reduction, amenable to erosion of soil structure;
- Weak emergence of seedling.

8.2.11 OFFSITE EFFECTS OF EROSION

- Affects water bodies and protective shelter belts;
- Alteration in the landscape characteristic profile;
- Alters the physicochemical, and biological processes, e.g., compact soils are susceptible to structural degradation, salinization, and checked microbial activity which are physical, chemical, and biological processes respectively than non-compacted ones;
- Financial loss;

- Raises tree mortality;
- Reduces livestock productivity;
- Stems down habitat of wildlife.

Nature of soil will decide the type of process predominant in it. For example, salinization is predominant in lands irrigated but poorly, internally drained over drained better [16].

8.2.12 DRIVING FORCES OF SOIL EROSION

Anthropological activities like deforestation, overgrazing, Excessive cultivation, mismanagement of land, cultivation on down slopes, and urbanization are instrumental in enhancing the potential of soil erosion. Moreover, soil mismanagement, topography, climate; political, socio-economic, parameters too affect soil degradation. In the developing world, soil erosion is connected to the level of poverty. Farmers with resource crunch don't have adequate means to adopt conservation practices. Marginal farmers working for subsistence must use extractive over conservation practices on small farm landholdings (0.5 to 2 ha) over the years for the food production that reduces soil erosion potential [17]. The three important factors responsible for rapid soil erosion are:

- Deforestation: 30% contribution in soil erosion;
- Overgrazing: 35% contribution in soil erosion; and
- Mismanagement of lands cultivated: 28% contribution in soil erosion.

8.3 NEED OF SOIL CONSERVATION

Soil is the one of the vital basic resources. Soil is an invaluable resource that offers us with food, fiber, and fuel. Food quality and security are necessary for the very existence of human being that is assured by soil. Importance of soil is realized only when there is a decline in innate resilience due to substantial erosion or degradation of soil and fall in food production. Although conventionally, the soil's major role has been as a carrier for the growth of plant, yet these days' issues like concerns of food security, quality of environment, climate change at the global front, and dumping ground for urban/industrial waste are associated with soil [18]. Soils across the world now are managed to:

- Clarify the air;
- Mitigate the demand for food which rising continuously;
- Purify water;
- Sequester carbon to rid-off the greenhouse gas emissions like CO_2.

Soil is looked upon as a conventional, dynamic resource but is vulnerable to rapid degradation and is misused on a large scale. Productive lands are limited (<11%) but the subsistence of over 6 billion people is dependent upon it and population is further increasing at a reasonable rate. Thus, degradation all over the limited soil resources can drastically endanger global issues like food security and environmental quality. Soil conservation is beneficial in the areas like agronomics, environment, and economics. The onsite and offsite approximated costs due to erosion prevention for restoration of nutrients lost, cleaning of water bodies, conveyances, and preventing degradation (Table 8.1).

There has been a need to maintain the multi-functionality of soil, which demands diligent use of it for fulfilling the present and future needs. The degree to which the prudence is professed and practiced reflects upon the sustainability in land use, food security, the air quality and water reservoirs, and the very survival of mankind. Soil conservation has been conventionally thought of only in terms of crop production, unlike present-day soil conservation that is assessed based on benefits accrued in the form increase in crop productivity, checking water contamination, and decline in levels of greenhouse gases (GHG) responsible for overheating the atmosphere [19].

TABLE 8.1   Estimated Cost Due to Erosion

Region	USA	Global	USA (Water Erosion)	USA (Wind Erosion)
	Billion (US$)			
Estimated cost	38	400	12–42	11–32

8.4 MISMANAGEMENT OF CULTIVATED LANDS: CAUSE OF SOIL EROSION

- Intensive cultivation enhances water runoff and soil degradation, which takes off the site the nutrients and pesticides, reducing both soil and water quality.
- Cultivation-switch over.

- Extension of cropping to slopes, marginal and shallow landholdings.
- Indiscriminate use of agrochemicals.
- Intensive agricultural practices like plowing.
- Irrigation with poor feed water quality.
- Lack of vegetation over soils responsible for degradation.

Eroded soils take far more time from 5 to 40 years to regain restoration of the soil in full. Due to huge population pressure and scanty arable land, farmers are using downslope hilly areas or mountainous, marginal, or eroded lands for crop cultivation.

8.5 GLOBAL SCENARIO OF SOIL EROSION

Soil erosion will not reach to threatening level soon in the developed world unlike developing ones who are facing population pressure, scanty land resources, and resource-poor farmers, etc. Soil erosion is a serious concern since the advent of agriculture to the mankind. However, the problem has aggravated in terms of the degree of severity and magnitude during the 20th century because of population pressure and poor management of cultivable lands in Africa and South Asia. The annual rate of erosion in these regions varies from 30 and 40 Mg/ha, due to cropping in rows on marginal lands, with slopes and hilly regions. Soil erosion leads to acute and severe malnutrition and poverty in the developing world, wherein farming community can ill-afford to adopt erosion-combating technology measures.

The gravity of the problem of soil erosion is region-centric. The major threatening regions of soil erosion today are sub-Saharan Africa, Africa, Asia, China-Loess Plateau, the Andean region, the Caribbean (e.g., Haiti), and lower Himalayas. The amount of soil erosion in large degree results from deforestation, excessive grazing, and soil mismanagement. Globally, soil erosion is a problem underway. More emphasis is laid on other agricultural-related issues than soil degradation and its implications. It is unfortunate that soil erosion issues, though severe, are not seriously looked into at large.

Global estimation shows that about 1,960 Mha of land is vulnerable to soil erosion ($\approx$15%) of the land area worldwide, ($\approx$50%) is badly degraded, with the majority of this land is not cultivated. In a few countries, about ($\approx$50%) of the agricultural lands of prime importance are severely degraded. The incumbent land area under cultivation is close to the abandoned land area from the advent of agriculture. Approximately 75×10^9 Mg of soil is being discarded across the world, which is equivalent to about 400.00 billion

US$/year with nutrients, soil, and water losses valued at US$70/person/year. Approximately 6×109 Mg per year of the land area is abandoned only in China and India [20, 21]. The management of soil erosion can be through the adoption of measures as shown in Figure 8.1.

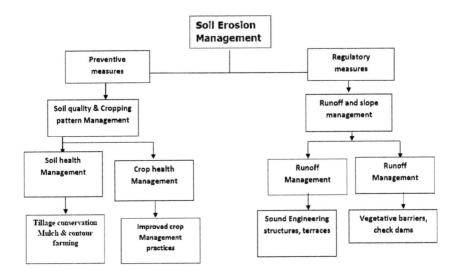

FIGURE 8.1   Soil erosion management.

8.5.1   DRYLANDS

Drylands are prone to erosion due to wind because of the limited cover of crop residue and adverse climatic conditions. The total area of dryland susceptible to erosion is close to 3.6 billion ha that is about 60% of the total across the world. About 9 to 11 Mha of drylands/year are getting infertile. The rate of soil degradation in drylands is increasing gradually in developing countries. Further, about 30% of the people residing in drylands are characterized by poor crop productivity and livestock.

8.5.2   SOIL CONSERVATION PROGRAM

Soil conservation is a multidisciplinary subject encompassing many issues than the perceived till today. It is a field woven intrinsically with numerous other subjects belonging to natural sciences (e.g., geology, agronomy, soil science, hydrology, climatology, agricultural engineering) and social sciences. Therefore, unraveling the mechanism of soil erosion demands drawing the knowledge in the form of fundamental principles of subjects like agronomy, climatology forestry, hydrology, engineering structures, etc. Soil conservation embodies runoff control and soil erosion, management, and techniques like irrigation, use of amendments, fertilizers, and drainage leading to preserving and/or enhancing the soil productivity. A successful soil conservation program asks for active participation of landowners, farming community, experts of economics, social sciences, think-tank involved in policymaking decisions, and the public at large. An impetus in the form of knowledge drawn from new disciplines is in demand for further developing technology in soil conservation, which is worth adopting across the board [22].

8.6 LATEST SCENARIO ON SOIL AND WATER CONSERVATION FRONT

The understanding of causes, factors, and mechanisms underlying soil erosion and the relevant soil properties has led to a considerable advancement in developing effective practices for conserving soil in the 20th century. The thorough knowledge of the factors responsible for estimating the amount of risks associated with soil erosion has resulted in the evolution and adoption of practices to control soil erosion worldwide. Despite advancements in technology, the extent of soil erosion is of higher-order yet. Better land conservation and management practices and policies have resulted in a decline in rates of water and wind erosion in the USA. The incidents like the Great Depression and the Dust Bowl (1930) have led to the generation of interest and promotion of research in evolving and implementing soil conservation techniques. The earlier policies focused on the significance to conserving the soil and thereby the onsite effects like crop productivity. However, in the aftermath of 1980's, policies of soil conservation have emphasized upon onsite and offsite impacting soil degradation.

A host of USDA initiatives is in existence that brings about a decline in soil erosion and bettering the quality of water and wildlife habitat. The Food Security Act in 1985 developed the conservation reserve program (CRP) that supports owners of land and farming community for their prudence. The CRP offers technical and financial inputs to producers to put into practice the approved conservation practices on cropland, which is vulnerable to high degradability. Adoption of non-tillage farming (a practice in which crops are allowed to grow with no soil turning) and conservation tillage too have played a role of their own in the soil erosion reduction to a certain degree. Although these initiatives have contributed much in nurturing better soil management, yet there is a long way to go to harness the full potential. Water pollution due to sediment and chemicals remains a major cause of concern.

There has been a significant development in the soil and water conservation in the developed world than in the third-world countries wherein food security is a major issue. More sound and concrete practices of soil conservation are sought to check soil erosion keeping in view the integrated agro-, politico-socio-economic approaches. Soil erosion would pose a grave threat to the sustenance of agriculture and environment unless farming systems adopt economically feasible, viable, and eco-friendly soil conservation practices. Figurative data obtained on soil erosion in developing world are largely short and discrete particularly in areas vulnerable to soil erosion. Therefore, some people see soil erosion crisis as exaggerated while others take it as serious and threatening to the consistency in the agricultural production. Assessment of consequences of erosion is not done properly at times due to either non-existent or inadequacy in the credibility of data on the soil erosion rates.

8.7   STRATEGIES FOR SOIL CONSERVATION

The soil and water conservation problems should be resolved by using consistent efforts of soil prudence and advance technology. It requires a diligent and comprehensive approach to address problems that are confronted not only by farmers but also the society at large. Aholistic program in water and soil conservation and in connection with politico-socio-economic constraints of each area is to be undertaken [23]. Some of the strategic initiatives to be taken up are:

- Adoption of conservation approaches which retain the soil on surface and check both the onsite and offsite fallouts of erosion of soil, minimize soil, transport, and check rate of water runoff.
- Development of effective economic and conservation approaches to recover eroded soils and improve upon the productivity of agricultural lands.
- Drastically degraded soils restoration by having vegetation covers, and eroded and marginal lands by enrichment of soil with organic matter buffers and establishing a sound nutrient base.
- Enrichment of the permissible limits of erosion for all kinds of soils and regions.

- Establishment of a trans-disciplinary and cooperative approach in areas with resource-scare farmers for improved soil and water conservation.
- Evolving conservation approaches which cut down both the contamination of water bodies and GHG emissions. Treatment of water contaminated not only involves lot much of cost based on varied nature of contaminants, their concentration level but also reduces the quality standards of water beyond correction. Thus, water loss due to runoff and contaminants should be checked by enhancing the percolation and soil water retentivity.
- Evolving technologies to eradicate loopholes in the tillage systems (e.g., low or no tillage) like low crop productivity, overuse of agrochemicals, organic, and nutrient stratification of the soil surface, with planting and soil warming exercises interfered by residue mulch.
- Erowing emphasis on educational and research initiatives accompanied by the sound technology transfer to the owners of land and end consumers by strengthening network connectivity.
- Identification and evolving at site-centric conservation techniques depending upon local and region-specific bio-physical, politico-socio-cultural factors. There is no universal practice which is applicable across all situations.
- Implementation of soil conservation approaches like increasing vegetation cover, strip cropping, using crop residue stumps, planting of N-fixing trees, growing green manure crops, retaining riparian buffers, bordering fields, using agroforestry, and other measures of biological origin accompanied by mechanical means like terraces.

Moreover, the following are some of the site-centric and basic initiatives to be taken up:

- Collecting and synthesizing data on the consequences of erosion on crop productivity. Reliable data is necessary for credible estimates of the amount of soil erosion pre and post-implementation of regulatory practices in order to create awareness among policymakers, farming community, and the public in general about the negative fallouts of erosion as far as maintenance and strengthening of issues like food security and environmental quality are concerned.

- Obtaining data from the farm with reference to erosion (wind, water, and tillage) and its reflection on soil productivity in particular from developing nations.
- Solving site-selective issues regarding conservation water and soil research.
- Undertaking fundamental research pertaining to mechanisms (precipitation and transport) underlying soil erosion transport and precipitation in relation to issues like soil dynamics, climate change, and greenhouse gas emissions. Better exploration of mechanisms of soil erosion would reflect upon evolving at tandem techniques for controlling and improving upon a pool of soil, air, and water resources [24].

8.7.1 POLICY IMPERATIVES

Funding for carrying out research in water and soil conservation is depleting. Both Government and other private organizations throughout the world should give more emphasis on the support for research to evolve at innovations in technological and strategic developments. Conservation policies should be aimed at meeting social costs and bettering the farmers economically. The policy initiatives required a special attention in this regard are:

- Enhancing accessibility to educational initiatives to create awareness about the benefits that can accrue by way of soil conservation stewardship.
- Enhancing soil conservation initiatives with the help of technical and financial support.
- Evolving at regulatory norms for the conversion of eroded soils into those covered with vegetation permanently.
- Identifying key areas for R & D of innovative technologies on priority.
- Increase in financial support for a better implementation of conservation programs.
- Meeting training needs and providing financial support for good soil conservation practices (soil prudence).
- Aiding for information exchange among farmers, institutions, and other stakeholders and end-users.

- Providing necessary support for active and collaborative research, socially-oriented activities, and programs for training for resource-stricken farmers in the developing world [25].

8.7.2   SPECIFIC STRATEGIES FOR ADDRESSING THE GLOBAL ISSUES

The upsurge in food production has been confined only to developed countries since World War II unlike developing countries (Africa and South Asia), wherein there has been a concern of food security. In fact, the crop production in developing world should increase by two-folds till 2050 to maintain food security. Although there is mass-scale production of food worldwide, yet the produced food does not reach to the needy ones [26]. Moreover, the effect of soil erosion on crop productivity is yet to be ascertained fully, which demands the following measures:

- Assessment of the impact of soil erosion on food yields.
- Balanced analysis from the economic point of view of impacts of soil erosion due to factors like soil order, type of crop, etc., at national, and continental levels.
- Evolving at mathematical models for the purpose of assessment of correlation between soil erosion and crop production for a host of crops, soils, climates.
- Quantitation of soil erosion rates against the quantum of crop production to know how far the soil erosion is voluminous.
- Use of Geographic information system-based models developed especially for evaluating crop productivity and soil erosion patterns across a wide spectrum of geographical regions.

Soil restoration and enhancement in crop productivity in the developing world may be achieved through the following measures:

- Boosting retention of crop residue on land and employing fertigation techniques.
- Checking and bringing about the reversal in the patterns of soil erosion as well as reclaiming degraded infertile soils by employing more efficient conservation practices.
- Developing and upgrading rainwater harvesting for irrigation in water-scarce regions.

- Enhancing availability of fertilizers to the farmers who are resource-deficient.
- Finding out innovative varieties of crops adaptable to soils degraded and resistant to both biotic as well as abiotic agents.
- Giving an impetus to the use of farming (not tilled) with Nitrogen-fixing (cover) and rotational crops.
- Soil erosion and landscapes management by using chemical fertilizers rationally and more use of organic fertilizers as a part of integrated nutrient management system.

8.7.3   RESEARCH ON BIOLOGICAL AND AGRONOMIC INITIATIVES FOR CONTROLLING SOIL EROSION

- Evaluating the efficiency of tillage systems based cover crops, nature of soil, and climate by obtaining practical data from farms which is site-specific across different soils and ecosystems.
- Exhibiting the advantages of vegetation cover on farms for marginal farmers namely in developing world.
- Optimizing the rates of use of soil conditioners like poly-acrylamide and zeolites for maintaining varied soil and climatic conditions.
- Promotion of biological initiatives for checking soil erosion, adopting cover crops and improving Carbon sequestration over a host of ecosystems.
- Using bio- and information technology and nanotechnology to resolve the longstanding issues related to soil erosion and deposition of sediments.

8.7.4   SOIL FERTILITY: CONSERVATION AND MANAGEMENT

Soils should be studied with application of definite approaches because of their highly complicated nature. The purpose of soil science should be aimed at growing the plants according to agronomic, horticultural, and silvicultural studies. It is found from just a casual study of the soil that the crop productivity depends upon the interplay of many heterogeneous factors. Hence, for the better know-how of the heterogeneity in the host of factors, soil science as a discipline is classified into different subject areas, like: physics, chemistry, mineralogy, microbiology, fertility, genesis, morphology, technology, engineering, and conservation of soil. However, there is no definite boundary between them and are not of much importance in isolation without integrating with other fields as they are intricately woven with one another. They are all studied with the only purpose of maintaining and/or increasing the land productivity. Further, it is a fact that soil not rich in all the necessary plant nutrients is like a desert and cannot have adequate crop production if water is scarce. Similarly, a dense plow soil layer stifles root penetration into the subsoil and thereby the nutrients get inaccessible to crops. High crop productivity is governed by efficient farming and optimum plant growth, which in turn is influenced by soil fertility and productivity [27].

Soil fertility refers to the quality or property that helps a soil to offer the adequate nutrients in the sufficient quantities, with right balance, to give an impetus to the growth potential of particular crops, or sequence of crops given a proper system of management. A fertile soil is productive. Yet, it does not mean that a fertile land is always productive. The soil, the atmosphere and the oceans constitute the biosphere, is in the form of a thin blanket around the earth called biosphere in which living things exists. The soil is the most complex yet easily destroyed of all these constituents.

In the developing nations, wherein a major chunk of the world's population now lives, the soil also provides other than food most of the fuel as firewood and large amount of fiber essential to make clothes, ropes, etc. The soil is thus; although the world's most precious, entity looked down upon as a mere dirt. If famine and malnutrition are to be combated, then it is only the soil, which must be valued most. However, we are losing the ground due to rampant soil degradation of one form or another, forcing us to bring much of new land into production every year. It has been estimated that soil erosion has so far brought about loss of about 2,000 Mha of productive soil for crop production. The better a soil is farmed, the healthier and more productive it becomes. However, adopting wrong techniques and the wrong crops are responsible for disastrous effects. For example, decline in production is fallout of increase in erosion, which has a cascading effect in furthermore lowering the crop production and even more erosion. Cropland becomes a wasteland. One of the important factors responsible for soil erosion is poverty as the poor do not have any other alternative than exploitation of the soil. Soil erosion problem is closely related to the problems of rural development and is a threat in the reckoning confronting humanity, though we can get over it with adequate and proper but timely measures [28].

Apart from soil erosion, the other causes of land degradation (turning of once productive fertile crop land into a waste one) are:

- Chemical poisoning;
- Mining land;
- Salt accumulation/salinization;
- Thus, arid areas can be irrigated, and waterlogged once drained;
- Continuous cropping for about 5–6 years on tropical soils, without any concrete measures to restore the soil fertility, may drop down the yield to a minimum.

8.8 MANAGING WATER: IRRIGATION AND DRAINAGE

Deficiency and excess of water are vital factors, which limit the crop yield. Greater than 50% of the worldwide supply of food is estimated to be dependent on one or other kind of water management system. In fact, first civilizations like Mesopotamia-literal meaning the "land between the rivers, Tigris, and Euphrates" emerged when farmers started controlling water, leading to more consistent yields and food supplies. High productivity in the areas, which are well-drained and irrigated, led to the development of branding in trade [29].

Moreover, innovative water management practices made the communities to get aligned, work in harmony on schemes like irrigation and drainage, and set up norms for allocations of water. However, failure on the front of water management was also the cause for erosion of civilizations. Mesopotamia due to salinization of lands and ditches accumulating sediment deposits led to reduction in soil fertility and thereby failing to sustain many of the large civilizations. Today, majority of the agricultural lands rely on one or the other kind of water management scheme. In US, on an average crop productivity of farms under irrigation is more by 118% and 30% for wheat and corn crops respectively than the dryland. On the global front, 18% of the cultivable land is under irrigation, which amount to 40% of the total food produced across the world. A major chunk of land in the western part of U.S. and other dry areas of the world would not have produced much without bringing land under irrigation. Further, in U.S. (California) horticulture crops heavily rely on irrigation. Even in wet regions, majority of cash crops are cultivated by irrigation of lands during dry seasons to ensure crop quality and consistent supplies of agricultural produce. It is attributed to the less resilience of the soils to natural calamities like droughts because of intensive use [30].

8.8.1   IRRIGATION

There are various kinds of irrigation systems, based on source and method of application of water, and system size. The three most important sources of water are: surface, groundwater, and recycled one. Irrigation systems on small farms to bigger ones are subjected to regulations of governmental entities. Water irrigation methods cover a host of conventional ones (flood or furrow), which depend on flow of water under gravitational force and nonconventional (sprinkler, drip irrigation) pumped waters [31].

8.8.1.1   SMALL-SCALE IRRIGATION SYSTEMS

These involve pumping water directly from either streams or ponds and are supplemental irrigation systems in humid regions, where small quantity of water may be required for better productivity. Such systems, confined to and managed by a solitary farm, have little impact on the environment [32].

8.8.1.2   LARGE-SCALE IRRIGATION SYSTEMS

These have evolved worldwide with strong governmental support. For example, the Central Valley Project in California is an example of a huge return against the amount of investment. The Imperial Irrigation project in Southern California (1940) was undertaken to bring about substantive economic change as a major contributor to the food or fiber production at the national and international levels. However, such systems frequently have ill-effects in the form of displacement of people and the flood of highly yielding cropland or very vital wetlands [33].

8.8.2   SURFACE WATER SOURCES

Conventionally, the major sources of water for irrigation supplies are streams, rivers, ponds, and lakes. In the past, diversion river waters were diverted to store in ponds. For example, Anazasi (U.S.) and Nabateans (Jordan) are small-scale systems, which were fed by diversions of small stream waters [34].

8.8.2.1 GROUND WATER

Good aquifers are comparatively cheap sources of water for irrigation. It does not require any substantive investment in the construction of infra-structural facilities like dams and canals by governmental institutions. It does not affect either regional hydrology or ecosystems. However, pumping water from deep water bodies demands a reasonable amount of energy. Overhead sprinklers with center-pivots are often preferred, and individual systems with capacity of irrigating land of 120–500 ha, specially pumped from a localized well. A sound source of groundwater and low salt levels are vital for the success of such systems. For example, the majority of the western U.S. Great Plains (Dust Bowl area) employs irrigation using the huge (174,000 miles^2) Ogallala aquifer that is easily available due to a shallow water reservoir. It is, however, a questionable and unsustainable practice if pumping and recharging in wells are not at comparable rates and wells are deep requiring more energy for pumping [35, 36].

8.8.2.2 RECYCLED WASTEWATER

In recent years, alternative sources of irrigation water have been promoted by Governments due to scarcity of it. Recycled wastewater is a good alternative for irrigation as it does not demand the same quality of water as for drinking. It is used in the areas, where:

- Groundwater sources are scarce or must be conveyed over larger area;
- Highly populated areas generate large amount of near the irrigation projects;
- Many irrigation projects in U.S. are collaborating with municipal authorities to provide safe recycled, notwithstanding a few doubts are raised about their effects in the long run.

The countries like Israel and Australia have been using recycled with their systems to meet the irrigation needs, given their large scale advancement in agriculture and scarcity of water [37].

8.8.3 IRRIGATION METHODS

Furrow (or flood) irrigation refers to traditional method widespread across the globe. The principle of this method is the simple filling of a farm for a given time span, permeating the water to percolate, water is passed through the furrows, sips down, and moves into the ridges laterally under gravitational force which require for almost leveled-farm. These are so far the very inexpensive in terms of installation and running, but devoid of exact and even application rates. These systems are vulnerable to salinization, as they can readily lower water tables. It is also employed for rice cropping, wherein water is ponded in dikes [38].

8.8.3.1   SPRINKLER IRRIGATION SYSTEMS

The principle used is application of water by using sprinkler heads subjected to pressure, which demands conduits or pipes and pumps. Normally this kind of systems need:

- Overhead sprinklers (traveling);
- Sprinklers (stationary) on risers.

These systems are characterized by precision in water application rates, more efficiency in water use than flood irrigation. The disadvantage is the high initial investment and high energy cost. Spraying gun (traveling type) has high efficiency. Localized irrigation with small-sized pumps and micro-sprinklers with low-diameter "spaghetti tubing," are relatively inexpensive [39].

8.8.3.2   DRIP (TRICKLE OR MICRO) IRRIGATION SYSTEMS

These systems make use of flexible ("spaghetti tubing") PE tubing, used preferentially in crops with beds employing a line source having emitters with equal spacing or point-source emitters, which apply water directly to plants. The advantages of this system are:

- Easy installation;
- Greater degree of control;
- Judicious utilization of water;
- Less energy-intensive;
- Relatively inexpensive;
- Requires low pressure for operation.

For example: In scale down systems like market gardens, pressure may be employed by way of a gravitational force based hydraulic head with the help of a water reservoir mounted on the platform or rooftop. Subsurface systems of this type are now also put into use, which require the positioning of the tubing and emitters in close vicinity of the roots to allow water to move laterally from the drip conveyance line into the soil [40, 41]. For more details, the reader can browse through two book series (by Dr. Megh R Goyal, Senior Editor-in-Chief) by Apple Academic Press Inc., namely: http://appleacademicpress.com/Research-Advances-in-Sustainable-Micro-Irrigation; http://appleacademicpress.com/Innovations-and-Challenges-in-Micro-Irrigation.

8.8.3.3   MANUAL IRRIGATION

It refers to using cans, buckets, and hoses in gardens, inverted bottles, etc., for watering that does not conform to the requirements of large-scale agriculture. It can be used in gardens and agriculture on small landholdings preferably in underdeveloped nations [42].

8.8.3.4   FERTIGATION

It is an efficient technique by which fertilizer can be applied to plants by using sprinkler- or trickle irrigation (pressurized systems). The fertilizer is added to irrigation water to dilute to lower doses (low concentration) of fertilizer, which are very easily taken up by the crop (spoon-feeding) as it grows [43].

8.8.4   ENVIRONMENTAL THREATS AND MANAGEMENT PRACTICES

The practice of irrigation offers innumerable benefits, however, it is a cause of great concern also. The major threat posed in terms of damage to soil productivity particularly in dry areas is the deposition of mineral salts, e.g., sodium (salinization). It causes more difficulty in water absorption. For example, on accumulation of sodium, lumps of soil break down make soils denser, and difficult to work. Over the years, many irrigated regions have become infertile due to piling of salts and it has reached a threatening proportion in many regions [44].

8.8.4.1   SALINIZATION

It is the irrigation water getting evaporated, leaving behind salts. It is prominently observed in flood irrigation that makes it to use more water to increase the water table already salty. On reaching the water table close to the surface, soil water is transported to the surface by capillary action, where it is subjected to evaporation leaving behind salts. If not properly managed, it leaves lands infertile within a short time span. It can also happen with drip systems, particularly in dry climate as leaching of salts is prevented through natural precipitation. In humid regions, accumulation of salts is generally not a concern, but irrigation in excess is a concern about losses in the nutrient contents and agrochemicals leaching to deeper layers. High rates of application and quantities may enhance leaching of nitrates and pesticides causing pollution of groundwater [45].

8.8.4.2   IRRIGATION MANAGEMENT

Sustenance in irrigation management and prevention of piling of mineral salts requires concrete planning, suitable equipment, and continuous observation. The very first step is to build capacity in the soil so that right quantity of water is consumed by the crop. Soils, which are organic matter-deficient and sodium-rich, have reduced capacities of infiltration attributed to surface sealing (blocking) and crusting (surface hardening) due to lowering in stability of aggregates. Healthy soils are better than soils, which are compact and deprived of organic matter contents in terms of water available to plant. The soils bereft of organic matter have poor root health, density, and subsoil crusting containing volume for rooting. It is mentioned that 1% reduction in organic matter results into soil water holding capacity of 16,500 gallons [46].

8.8.4.3   REDUCING TILLAGE

Organic fertilizers, responsible for preventing compaction and taking rotational crops can enhance water storage capacity. Experimental results have shown that reduced tillage and use of crop-rotation crops can increase water capacity available to plant by about 34%. Techniques like zone tillage can enhance rooting depth and moreover bring about increase in organic matter and storage capacity of water in the long run [47].

8.8.4.4   INCREASING SURFACE COVER

Cover crops enhance organic matter content in the soil and thereby offer mulching, but care must be taken with cover crops, as during their growth they consume large amount of water which would be required for salts to leach or to make it available to the cash crop. This can be achieved by keeping a watch on the soil, the plant, or weather parameters by using the need-based water. Soil sensing with tensiometers or capacitance probes can assess soil humidity [48].

8.8.4.5   DRAINAGE

Soils, which are by nature drained and are aerated inadequately, in general possess greater degree of organic matter content. Improper drainage renders them not suitable for growing many of the crops except some plants like rice and cranberries, which consume water a lot. However, soils drained synthetically are high yielding, as the greater degree of organic matter content makes available the necessary nutrients. For many years, swamps have been changed into high yielding agricultural lands by way of making ponds and canals, accompanied by sound pumping systems to drain out the water from soggy or water-logged areas.

Drainage has benefits such as: deep land volume, sufficient aeration for growing normal plants. If crops like grasses for pastures or hay tolerating shallow rooting conditions are grown then the water table may however be retained either near the surface or drainage lines which can be laid apart from one another, thereby containing installation and maintenance costs substantially low. Crops, like corn and soybeans demand an aerated zone deep enough and subsurface drain lines that should be installed about 3–4 feet deep at spacing of 20–80 feet, depending on soil properties. Drainage saves time for farm exercises and reduces potential of damage by compaction. Farmers in humid areas have few dry days to undertake spring and fall farm operations. Insufficient drainage hampers farm operations to be carried out before the arrival of rainfall. However, with adequate drainage, farm operations can begin within days after the onset of rain. Inadequately drained soils help accrue benefits on agronomical and environmental because these check both compaction and harm to soil structure. It further addresses high nitrogen losses through denitrification. As a thumb rule, croplands that are regularly saturated should be subjected to drainage or reverted to natural vegetation or greener pastures [49, 50].

8.9 MANAGING NUTRIENTS

Plants require 18 elements and nitrogen, phosphorus, and potassium is inadequate in soils most of the times. However, magnesium, sulfur, zinc, boron, and manganese may also be deficient, but not across the board. Southeastern states or areas with heavy rainfall (Pacific Northwest of USA) having minerals weathered largely may have shortfall of sulfur, magnesium, and few of the other micronutrients. Calcareous (high pH) soils mainly in dry areas may have iron, zinc, copper, and manganese deficiencies. An emphasis has been laid on N and P management over many years. Although all these nutrients are vital nutrients in regard to soil productivity, yet they are responsible for causing environmental threats [51] such as:

- Excessive use of chemical-based fertilizers;
- Misusing sewage sludge containing biosolids, and composts;
- More livestock on comparatively lower land area has led to pollution of water aquifers in many parts of the U.S. due to overuse of both N and P which have their own potential environmental consequences [51];
- Soil and crop mismanagement.

Too much of N availability in the early season lead to growth of weeds, having low nutrient buffers. It is detrimental for crop growth if nutrient contents are not available in right and balanced amounts, and at right time. Excessive N results into far more growth of foliage or fruits won't bear as against other nutrients. Plants subjected to fluctuations in nutrients have abnormal growth. For example, with sudden variations of N levels in plants are unable to generate enough natural chemicals, which drag the useful insects feeding on leaves or fruits. Lower K-levels are responsible for aggravating rot (stalk) of corn, while pod rot of peanuts is related to greater amount of K within the fruiting zone. When soil loses N and P and leaches into groundwater or reaches surface water through run-offs, societies in entirety may have to face deteriorated water quality [52].

8.9.1 ORGANIC MATTER AND NUTRIENT AVAILABILITY

The better strategic initiative to manage nutrients is to enhance the organic contents in soils quantitatively with special reference to nitrogen and phosphorus. Organic matter and any fresh residues used can supply N to soil.

Phosphorus and sulfur are derived from organic contents too. Presence of organic matter in soil offers a tendency to better attach with cations like K^+, Ca^{++} and Mg^{++} ions and subsequently form metal complexes (chelates) naturally, which can further help in retaining and easily absorbing mineral salts (micronutrients: Z_n, C_u, and M_n) amenable to plants. Moreover, the better soil health and the products of decomposition of organic matter gives an impetus to growth of healthy root system that can provide better capacity to absorb nutrients from sprawling soil surface area [53].

8.9.2   IMPROVING NUTRIENT CYCLE ON THE FARM

Plants should use nutrients in cycles to return as either crop residue or manure with greater efficiency on the field as the economic and environmental priorities. Reduction in nutrient flows over a longer distance and improved methods of nutrient cycling should be the major objectives to support crops.

8.9.3   REDUCE UNINTENDED LOSS

Promotion of improved water infiltration, root condition, organic content, and physical properties should be maintained. Organic content may be preserved and further strengthened by adding various sources of organic matter other than employing the techniques for checking losses by tillage and other methods for conservation purpose. Excess water for irrigation than required leads to nutrient losses both due to runoff and leaching. However, in dry regions application of surplus water can be justified to bring about the leaching of accumulated salts below the root zone [54, 55].

8.9.4   ENHANCING NUTRIENT UPTAKE EFFICIENCY

Careful application of inorganic fertilizers and organic amendments, in conjunction with irrigation practices, will enhance nutrient uptake, sound placement, and synchronizing better efficiency of nutrients. At times, change in cropping times and a pattern too generates an enhanced compatibility between the crop needs and time at which nutrients are made available to crops [56].

8.9.5 TAPPING LOCALIZED NUTRIENT SOURCES

Organic materials are made available locally in the form waste from markets, food waste, weeds, etc. Some of these materials may or may not add nutritional value. However, extracting useful nutrients from agricultural point of view from the "organic waste streams" is of great significance to develop eco-benign® nutrient flows. Nutrient cycling in a true sense is what it matters most when people buy locally sold foods. Farms, which are supported by communities, in return generate waste to the same farm so as to compost it there itself, to complete a natural cycle [57].

8.9.6 REDUCE LOSS OF NUTRIENTS

The excellent option to check nutrient loss/acre and to use legumes intermittently is to mix it with ruminant animal product in a farm. Bringing up animals with bought feed enriches a farm with nutrient concentrations [58].

8.9.7 ENHANCE ANIMAL POPULATION

It may be achieved by either buying land or getting it on rent to produce more amounts of animal feeds, sound application of manures or by regulating animal numbers.

8.9.8 DEVELOP LOCAL PARTNERSHIPS

Organic matter management involves both management of nutrients and crop rotations. This is particularly useful when a farmer is raising more number of animals, greater degree of feed and a farm in the neighborhood demands nutrients and insufficient land area. There would be a win-win situation for both farms, provided they cooperate on the front of these practices. Organic farmers are in the search of cost-effective animal waste derived from manures and composts. The landscape industry too makes use of reasonable quantity of composts. The exchange of compost at local or regional level can benefit nutrient deficient soils [59].

8.9.9   ISSUES IN THE UTILIZATION OF FERTILIZERS AND AMENDMENTS

- How much is required quantitatively?
- What are the methods of application to be adopted for the fertilizer or amendment?
- What are the sources to be utilized?
- What is the appropriate time for the application of either fertilizer or amendment?

8.9.10   NUTRIENT SOURCES

Many fertilizers and amendments are used usually in agricultural fields. Chemical fertilizers namely urea, triple superphosphate, and potassium chloride have the following advantages and disadvantages.

Advantages:

- Convenience in storing and usage;
- Predictable in behavior and the abundance of the nutrients in soils are very well known;
- Simple to mix to fulfill the requirements of nutrients of specific fields reflecting upon predictable effects;
- The nutrient application time, pace, and uniformity are easy to monitor while using commercially available and chemical-based fertilizers.

Disadvantages:

- Nitrogenous materials like urea, ammonia, and ammonium that are used commonly are amenable to form acids;
- The manufacturing of nitrogen-based fertilizers consumes a substantial amount of energy (25–30% against that required for bringing up a corn crop) [60];
- Their application in moist regions demands lime to be added with a greater frequency.

8.10 SELECTION OF COMMERCIAL FERTILIZERS

Integrated nutrient management program should have organic fertilizers to have a healthy soil, although some farms require commercial fertilizers as supplements for achieving better productivity to fulfill ever-growing demands of the population over and these sound practices like growing cover crops, crop rotations, reduced tillage, and integration of plant and animal agriculture, etc., are employed in practice on farms. Blends like 10-20-20 or 20-10-10 of N: P_2O_5: K_2O or other ones which are commonly used. Following are the situations in which the cheapest source is not preferred:

- Anhydrous ammonia is although cheapest N form has got constraints in injecting it in soil with stones and vulnerable to losses in too wet soil and therefore in such circumstances other sources may be preferred.
- Di-ammonium phosphate (DAP) is a better option when both N and P are required because it is compatible with concentrated superphosphate in terms of cost and N and P contents.
- Potash (KCl) is not preferred although cheapest one. Potassium magnesium sulfate would be a good option when magnesium and lime is not required in the farm [61, 62].

8.11 METHODS AND TIMING OF FERTILIZER APPLICATION

The method of application would decide the time at which fertilizer is applied. Broadcast-application is best to enhance the nutrient level in bulk in soil and it provides even distribution of fertilizer over the entire field, which then is embedded by using tillage. It is used to strengthen P and K in the wake of their deficiency. Broadcasting in conjunction with distribution of fertilizer is normally undertaken either during fall or spring ahead of tillage. Topdressing (broadcasting) is nothing but is used usually to apply N to crops (e.g., wheat) that cover soil surface in entirety. Amendments (e.g., lime, and gypsum) applied in greater amounts are also broadcast before introducing it into the soil [63].

8.12   TILLAGE OPERATIONS VERSUS FERTILITY MANAGEMENT

- Moldboard, and chisel plow, disk harrow;
- Zone and ridge till.

It is very much a possibility to employ fertilizers and amendments. In the wake of roots being active in soil and if there is no-tillage then it is difficult to blend the nutrients with the soil and to evenly strengthen the level of soil fertility. Huge amounts of ammonia can be wasted as it volatilizes, e.g., urea as a fertilizer may either remain on the surface of soil or as a run-off during rainy season [63, 64]. Tillage and crop residue management trend is shown in Figure 8.2.

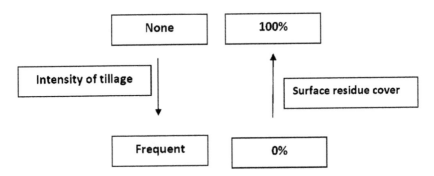

FIGURE 8.2   Trends in tillage and crop residue cover management.

8.13   NITROGEN AND PHOSPHORUS MANAGEMENT

Nitrogen and phosphorus both are very vital for the plant growth in reasonable amounts. However, more N can be detrimental to environment in the following ways:

- Nitrate is hazardous to infants and young animals too as it hamper the oxygen carrying capacity of blood.
- Nitrates are responsible for stimulating the growth of algae and aquatic plants similar to agricultural plants.
- Nitrates further the growth of pathogenic microbes (e.g., water in the Mexico Gulf or the Chesapeake Bay).
- Soil nitrate gets leached to groundwater and contaminates it.

- Surplus N as runoff, and nitrous oxide (N_2O) loss, which is a potent greenhouse gas. Loss of Nitrogen content is an economic drag on farmers if not managed well.

Denitrification a microbial process wherein the bacteria in soil break down nitrates into nitrous oxide (N_2O) and N_2. In case of P, the major problems are:

- Decomposition of plant debris causes deterioration, leading to death of aquatic life.
- Loss of even residual concentration of P has substantial effect on the water quality because P is instrumental in containing the growth of freshwater aquatic weeds and algae.
- Losses of phosphorus in farms are normally low.
- Losses to freshwater reservoirs [65].
- When present to a greater degree in say lake by anthropological activities (agriculture, domestic septic tanks, runoff), raise excessive growth of plant debris called eutrophication subsequently becomes a problem for fishing, swimming, and boating.

Improved soil structure and proper vegetation cover crops can contain loss of N and P by checking percolation, denitrification, as well as runoff. Moreover, production, transport, and use of N fertilizers are al energy consuming exercises against P fertilizers. Therefore, more emphasis on N-fixation biologically and its cycling in an efficient manner will help to check the dwindling of non-renewable resource and will be economical [66]. The management strategies that are employed due to variation in the behavior of N and P are:

- Apply fertilizer in different lots if they are vulnerable to losses as a potential problem for N due to leaching or denitrification.
- Applying nutrients only when leaching or runoff problems are at minimum.
- Check tillage.
- Checking losses and improving uptake of nutrients.
- Conducting manure and compost tests to assess their impact on contributions of nutrients.
- Making available nutrients from all possible sources.
- Nutrient contributions on the decomposing of crop residues with reference to N only.
- Nutrient sources altogether are taken into consideration.

- Strike a balance to import and export of farm on meeting the crop needs.
- Testing soils to characterize them for assessing present nutrients quantitatively.
- Use of nutrient sources with greater efficiency.
- Using locally available fertilizers as and when possible.
- Using perennial forage crops on rotation.
- Using vegetation cover crops.

8.13.1   DETERMINING NUTRIENT AVAILABILITY

Sound N and P management involve availability of nutrients to plants from the soil, such as: organic matter, manure, compost, crop rotation, and cover crop. Fertilizer plays only supplementary role. Organic farmers fulfill all their demands by way of these organic sources, due to high price of organic fertilizers. Determination of fertilizer N-requirement is not so easy, as soil tests in themselves are unable to solve all issues. The N-available to plant (namely nitrate) can vary rapidly, as N is lost by leaching or denitrification [67].

8.13.1.1   FIXED AND ADAPTIVE METHODS

Estimation of crop N-Needs demands many approaches, which can be classified as fixed and adaptive to weather methods. Fixed approach assumes that the N-requirements remain fixed during the seasons but may vary because of earlier crop. This method is very handy in planning and work well in dry conditions, but not so precise in moist conditions.

8.13.1.2   THE MASS BALANCE APPROACH

Fixed-approach is usually employed to determine N-contents in fertilizer. The assessment is made based on the yield and the corresponding N-uptake minus non-fertilizer N-content in soil from the mineralized organic matter, earlier crops, and organic amendments. Although more N is required for better productivity, the weather pattern that gives higher yields implies:

- Root systems that are large and healthy with better capacity to uptake more N;
- The weather pattern is responsible for stimulating nitrate levels to higher order in the soil.

The maximum return to N (MRTN) approach has been adopted by many frontier U.S. corn-producing states, which is one more fixed approach discarding mass-balance approach altogether.

8.13.1.3   THE ADAPTIVE APPROACH

It attempts to consider weather during growing seasons, soil nature, and management effects and is required to be estimated or modeled:

- The pre-plant nitrate test (PPNT) quantifies either nitrates independently or with ammonium in the soil (0–2 ft) in the early season in dry climate for fertilizer N at the time of planting.
- The pre-side dress nitrate test (PSNT) or late spring nitrate test (LSNT) is aimed at quantifying nitrate contents at the surface of the soil (0–12 in) and provides for adaptive side-dress or top-dress of N.

Environmental information systems and simulation models are other methods in the successful use for N management, specifically for wheat and corn. This is an adaptive approach, which makes use of advanced environmental databases, such as, radar-based high-resolution precipitation estimates that give necessary technical inputs for developing computer models. The simulation of N-mineralization and losses is done to assess N-contents in soil and supplementary N-needs through fertilizers [68].

8.14   ASSESSMENT AT THE END OF THE SEASON

To assess fertility of the soil, farmers at times plant crops as strips in field with variation in N rates to compare productivity as the season concludes. The nitrate test (lower stalk) is applied to approximate the corn N-rates. None of these methods is either fixed or adaptive. Adaptive management takes into consideration the experiments by farmer under local conditions [69].

8.15   MANAGEMENT PLANNING FOR N AND P

The behavior of N and P is in different soils. However, their management is done by similar approaches. The consideration is given to following important points while planning for N and P management strategies:

- Degrading sods;
- Enhance nutrients in manures;
- Organic residues.

Evaluation of the available N and P on farm sources beforehand is a prerequisite before employing supplements from outside. At times, fertility of the soil due to availability of on-farm sources is more than expected to meet the crop needs. The soil test should reflect on all the nutrients added through manure, if added before sampling of soil [70].

8.15.1   RELYING ON LEGUMES

Legume cover crops with high yields (like hairy vetch and crimson clover and forage crop (grass, clover) in rotation) can supply enough N required by the next row crops.

8.15.2   REDUCING N AND P LOSSES

Apply efficiently N and P fertilizers as and when required by most plants. If N cycling is done effectively, then the losses of nutrients are reduced. Although demand for N fertilizer requirement is high in the early period, long-term effect of no-tillage enhances organic matter over conventional tillage, which is attributed to larger N mineralization, thereby farmer is benefited economically [71].

8.15.3   WORKING TOWARDS BALANCING

N and P are lost from soils due to runoff, leaching of nitrate and sometimes P, denitrification, and ammonia volatilization from urea and manures applied in the surface. Despite taking all the necessary steps to reduce avoidable losses, still N and P loss may take place to a certain extent.

8.15.4 USING ORGANIC SOURCES OF P AND K

Depending on the N needs of soils, manures, and organic amendments are used at the required rates. However, this may add normally more P and K over and above the requirement of crop. It is observed on soil characterization that P and K concentrations are on the rise when nutrients for N are used for years together to soil. It is advisable to use reduced level of P through different P-sources. It matters most as how best the manure is distributed across the fields [72].

8.15.5 MANAGING HIGH P-SOILS

Higher P concentration levels in soils are attributed to either applications of P in excess as fertilizers or through more amount of manure. This may be a cause of concern for livestock farms having small land, wherein feed is made available from outside at different scales. However, it is of paramount importance to check the risk of harm to environment due to high levels of P-soils that generate runoff on steep slopes, fine texture, weak structure, or poor drainage [73]. Higher-P soils can be dealt with following practices:

- Check animal P intake to the abysmally low levels.
- Check runoff and erosion to rock bottom levels. P creates a problem only in surface waters on getting into it.
- Reduce or eliminate using excess of P by either having more land or apply it over more land area.
- Regular monitoring of P levels in soils.

8.16 SOIL MANAGEMENT PRACTICES FOR RESOURCE DEPRIVED REGIONS

Recovery of eroded soils and management of present high yielding soils are key priorities especially for small- and marginal farm owners with <2 ha. Insecurity of food from the reduction in crop productivity is attributed to overexploitation and dwindling of nutrients in deprived regions like Africa. Since bringing in more land area under cultivation is a major constraint in most of the regions because of lack or inadequate of cultivable agricultural lands, the only remedy to food security is to improve the crop yield/ha [74].

Major challenges involve reduction in soil degradation within tolerable or prescribed limits and to enhance the soil fertility well-managed by poor farmers. Soil management tools at the disposal of small farmers may not be same and influenced by the factors like the ecosystem, topography, and climate [75]. Different measures must be adopted to manage soil erosion and bettering the soil fertility like:

8.16.1   TERRACES

These structures bring about interception, retention, slowdown, and diversion of runoff to safer vents, checking soil degradation and loss of nutrients. Small farmers can use different types of terraces like conservation benches, orchard, intermittent, and continuous terraces, ditches on the hillside, etc. [76].

8.16.2   RAINFED PONDS

Harvesting of rainwater is a strategic initiative to reduce soil degradation, water storage for crops, and enhance crop productivity on farms which are sloping; Construction of ponds is a remedial measure to harvest runoff and rainwater during rainy seasons. During the dry seasons, this water can be used both for irrigation and growing of crops [77].

8.16.3   INORGANIC FERTILIZERS

Using inorganic fertilizers at right time and prescribed rate enhances efficiency and efficacy of their use leading to increase in crop productivity. Proper utilization and management of fertilizers with efficiency help to check environmental degradation. Fertilizers should be cost-effective and available on time. Using about 9 kg/ha of N in India may raise crop productivity by 50% provided it is used at the required time and location [78, 79]. The performance of fertilizers is related to:

- Average rainfall;
- Crop types;
- Soil types and texture;
- Tillage management.

8.16.4   MANURING

It is a well-known traditional technique to enhance soil fertility. Manures in developing countries are mostly poor both qualitatively and quantitatively, which is governed by the animal type and quality of forage. The manure efficiency can be improved by keeping the manure in areas like holes or furrows at which plants are vulnerable to grow rather than broadcasting (scattering) across the field. N (60%) and P (10%) may be wasted due to poor management and broadcasting of manure. Adoption of such approach improves the efficiency of manure by reducing the wastage of manure and byproducts by way of degradation, volatilization, and leaching. Guidelines pertaining to manure management must be framed to check rates of application and intimate contact between seeds and manure [80, 81].

8.16.5   GRAIN LEGUMES

These are introduced into the conventional cropping patterns to strengthen soil fertility and cycling of N. Green manures can form because of either rotation or intercropping of grass and tree-legumes with row crops. Legumes contribute reasonably to the fixation of nitrogen in soil biologically and thereby reduce use of inorganic fertilizers. Legumes growing is an eco-friendly and inexpensive option to recover soil fertility. Row crop rotations with either grasses or tree-legumes have its own economic benefits and therefore proper practices to set up and manage must be introduced [82].

8.16.6   AGROFORESTRY

Soil humidity and wind erosion may be contained by tree plantations in and around croplands. Trees can also fix N biologically. For example, Sesbania, Tephrosia, Gliricidia, Leucaena, Calliandra, Senna, and Flemingia, etc., are employed for improving soil fertility in the African continent. Agroforestry technology in conjunction with grain legumes must be used more aggressively and on a large-scale for better results [83].

8.17   INTEGRATED NUTRIENT MANAGEMENT APPROACHES

Organic and inorganic fertilizers must be used together as a better alternative than using either of them in isolation (e.g., manure, compost). The combination approach reduces overuse of inorganic over organic fertilizers. An integrated approach for nutrient management refers to using application methods, timing, quantity, and fertilizer type together with grain legumes and agroforestry initiatives [84].

8.17.1 TILLAGE MANAGEMENT

Mulch and stripped ones are preferred tillage systems to restore degraded soils [85].

8.17.2 RESIDUE MANAGEMENT

Protective cover with residue after rain helps to contain erosion. Inadequate crop residue cover and its competitive use are responsible for poor soil fertility [86]. Getting enough crop residue cover and use of different grass and legume plants can be useful to have good amount of residue cover [87, 88].

8.17.3 CONSERVATION BUFFERS

Grass, wind barriers, fencing to fields, and riparian buffers are some of the preventive measures that can be adopted for protecting soil from erosion [89]. Measures like combining grass barriers with food crops in parallel rows can contain sediment eradication during its off-site transport to the farther low lying areas [90].

8.17.4 CROPPING PATTERNS

Crop rotation, multiplicity in cropping, stripped, contour, cover cropping, contour farming are various options to preserve soil fertility with water [91]. Dense canopy and more biomass generating crops are instrumental in intercepting raindrops and protect the soil from getting eroded. A well-organized way of arranging strip crops on the contour, and cultivating variety of crops/ year can check soil degradation [92].

8.18 SUMMARY

Sustainable soil and water conservation has become imperative across the world and in developing and underdeveloped countries following substantial degree of soil erosion or degradation, that has taken place over the time. The problem has aggravated because of resource-poor farming community in this part of the world. Available management techniques are terracing, rainfed ponds, use of right combination of inorganic fertilizers and organic amendments, manuring, grain legumes, agroforestry, integrated nutrient management, tillage management, residue management, cropping patterns to conserve soil and restore it, where it has degraded due to poor soil and water management practices, overuse of chemical-based fertilizers, poor water management practices, etc. The problem is more threatening and imminent in tropical and semitropical regions, which need to take sustainable action on war footing to improve the soil productivity to feed growing population. All strategies and management practices must be devoted for the said cause.

KEYWORDS

- agrochemicals
- agronomics
- anthropological
- chemical dynamics
- soil conservation
- strategic measures
- water conservation

REFERENCES

1. Abdul-Baki, A. A., Teasdale, J. R., & Korcak, R. F., (1997). Nitrogen requirements of fresh-market tomatoes on hairy vetch and black polyethylene mulch. *Hort. Science, 32*(2), 217–221.
2. Ahn, J. K., & Chang, I. M., (2000). Allelopathic potential of rice hulls on germination and seedling growth of barnyard grass. *Agronomy Journal, 92*, 1162–1167.

3. Akemo, M. C., Bennett, M. A., & Regnier, E. E., (2000). Tomato growth in spring sown cover crops. *Hort. Science, 35,* 843–848.

4. Alabouvette, C., Lemanceau, P., & Steinberg, C., (1996). Biological control of *Fusarium* wilts: Opportunities for developing a commercial product. In: Hall, R., (ed.), *Principles and Practices of Managing Soil Borne Plant Pathogens* (pp. 192–212) St. Paul, MN: The American Phytopathological Society.

5. Alexander, R. B., Smith, R. A., & Schwarz, G. E., (2000). Effect of stream channel size on the delivery of nitrogen to the Gulf of Mexico.*Nature, 403,* 758–761.

6. Alexander, R. B., Smith, R. A., & Schwarz, G. E., (2008). Differences in phosphorus and nitrogen delivery to the Gulf of Mexico from the Mississippi River Basin. *Environ. Sci. Technol., 42*(3), 822–830.

7. All, J. N., & Musick, G. J., (1986). Management of vertebrate and invertebrate pests. In: *No-Tillage and Surface Tillage Agriculture* (pp. 347–388). New York: John Wiley and Sons.

8. Aulakh, M. S., Khera, T. S., & Doran, J. W., (2000). Yields and nitrogen dynamics in a rice-wheat system using green manure and inorganic fertilizer. *Soil Science of America Journal, 64,* 1867–1876.

9. Ayanlaja, S. A., Owa, S. O., Adigun, M. O., Senjobi, B. A., & Olaleye, A. O., (2001). Leachate from earthworm castings break seed dormancy and preferentially promotes radicle growth in jute. *Hort. Science, 36*(1), 143–144.

10. Aylsworth, J. D., (1996). No-till transplanter improves yields, cuts costs. *American Vegetable Grower, 18,* 20–25.

11. Babcock, J. M., & Bing, J. W., (2001). Genetically enhanced Cry1F corn: Broad spectrum Lepidoptera resistance. *Down to Earth, 56,* 10–15.

12. Bastida, F., & Moreno, J. L., (2006). Microbiological degradation index of soils in a semi-arid climate. *Soil Biol. Biochem., 38,* 3463–3478.

13. Bastida, F., Zsolany, Z., Hernandez, T., & Garcia, D. C., (2008). Past, present, and future soil quality indices: A biological perspective. *Geoderma, 147,* 159–171.

14. Benito, E., Santiago, J. L., & De Blas, E., (2003). Deforestation of water-repellent soils in Galicia (NW Spain): Effects on surface runoff and erosion under simulated rainfall. *Earth Surf Processes Landforms, 28,* 145–155.

15. Bennett, H. H., (1947). *Elements of Soil Conservation* (p. 406). McGraw-Hill, New York.

16. Bertol, I., Engel, F. L., Mafra, A. L., Bertol, O. J., & Ritter, S. R., (2007). Phosphorus, potassium, and organic carbon concentrations in runoff water and sediments under different tillage systems during soybean growth. *Soil Till Res., 94,* 142–150.

17. Bertol, I., Engel, F. L., Mafra, A. L., Bertol, O. J., & Ritter, S. R., (2007). Phosphorus, potassium, and organic carbon concentrations in runoff water and sediments under different tillage systems during soybean growth. *Soil Till Res., 94,* 142–150.

18. Bielders, C. L., Michels, K., & Rajot, J. L., (2000). On-farm evaluation of ridging and residue management practices to reduce wind erosion in Niger. *Soil Sci. Soc. Am. J., 64,* 1776–1785.

19. Blanco-Canqui, H., Lal, R., Post, W. M., Izaurralde, R. C., & Owens, L. B., (2006). Soil structural parameters and organic carbon in no-till corn with variable Stover retention rates. *Soil Sci., 171,* 468–482.

20. Du-Preez, C. C., van, C. W., Huyssteen, P., & Mnkeni, N. S., (2011). Land use and soil organic matter in South Africa 1: A review on spatial variability and the influence of a rangeland stock production. *S. Afr. J. Sci.*, *107*(5/6), 1–8.

21. Du-Preez, C. C., van, C. W., Huyssteen, P., & Mnkeni, N. S., (2011). Land use and soil organic matter in South Africa 2: A review on the influence of arable crop production. *S. Afr. J. Sci.*, *107*(5/6), 1–8.

22. Du, Z., Shufu, L., Kejiang, L., & Tusheng, R., (2009). Soil organic carbon and physical quality as influenced by long-term application of residue and mineral fertilizer in the North China Plain. *Aust. J. Soil Res.*, *47*, 585–591.

23. Dukes, M. D., Zotarelli, L., Liu, G. D., & Simonne, E. H., (2013). Principles and practices of irrigation management for vegetables. In: *Fundamentals of Irrigation and On-Farm Water Management* (pp. 17–27.). Florida: Springer.

24. Dungait, J. A. J., Hopkins, D. W., Gregory, A. S., & Whitmore, A. P., (2012). Soil organic matter turnover is governed by accessibility not recalcitrance. *Global Clim. Bio.*, *18*, 1781–1796.

25. Eickhout, B., Bouwman, A. F., & Van, Z. H., (2006). The role of nitrogen in world food production and environmental sustainability. *Agric. Ecosyst. Environ.*, *116*, 4–14.

26. Eswaran, H., Lal, R., & Reich, P. F., (2001). Land degradation: An overview. In: Bridges, E. M., Hannam, I. D., Oldeman, L. R., Pening, D. V. F. W. T., Scherr, S. J., & Sompatpanit, S., (eds.), *Responses to Land Degradation, Proc. 2ⁿᵈ International Conference on Land Degradation and Desertification*. Khon Kaen, Thailand; Oxford Press, New Delhi: India. Online: https://www.nrcs.usda.gov/wps/portal/nrcs/detail/soils/use/?cid=nrcs142p2_054028 (accessed on 20 June 2020).

27. FAO, (2002). *Comprehensive Africa Agriculture Development Programme*. http://www.fao.org/3/y6831e/y6831e-03.htm (accessed on 20 June 2020).

28. FAO, (1996). *Our Land Our Future* (p. 52). Food and Agriculture Organization (FAO) and United Nations Environment Programme, Rome. http://www.fao.org/3/a-bc145e.pdf (accessed on 20 June 2020).

29. Follett, R. F., (2001). Soil management concepts and carbon sequestration in cropland soils. *Soil Till Res.*, *61*, 77–92.

30. Frazao, L. A., Piccolo, M. C., Feigl, B. J., Cerri, C. C., & Cerri, C. E. P., (2010). Inorganic nitrogen, microbial biomass, and microbial activity of a sandy Brazilian Cerrado soil under different land uses. *Agric. Ecosyst. Environ.*, *135*, 161–167.

31. Ghorbani, B., & Amini, M., (2011). Assessment of sprinkler irrigation systems operation at Chaharmahal and Bakhtiari provinces, Iran. In: *Proceedings of ICID 21ˢᵗ International Congress on Irrigation and Drainage* (pp. 15–23)Tehran.

32. Hamdy, A., Ragab, R., & Scarascia-Mugnozza, E., (2003). Coping with water scarcity: Water saving and increasing. *Irrigation and Drainage*, *52*, 3–20.

33. Hannam, F. A. O., (2002). *Reducing Poverty and Hunger: The Critical Role of Financing for Food, Agriculture, and Rural Development* (p. 8.). Paper presented at International Conference on Financing for Development, Monterrey, Mexico.

34. Hatfield, J. L., Sauer, T. J., & Prueger, J. H., (2001). Managing soils for greater water use efficiency: A review. *Agron. J.*, *93*, 271–280.

35. Hazarika, S., Parkinson, R., Bol, R., Dixon, L., Russel, P. J., & Donovan, S., (2009). Effect of tillage system and straw management on organic matter dynamics. *Agronomy for Sustainable Development*, *29*, 525–533.

36. Hudson, B. D., (1994). Soil organic matter and available water capacity. *J. Soil Water Conserv., 49*, 189–194.

37. Indoria, A. K., Sharma, K. L., & Reddy, K. S., (2016). Role of soil physical properties in soil health management and crop productivity in rain fed systems, part II: Management technologies and crop productivity. *Curr. Sci., 110*, 320–328.

38. Iqbal, M., Harold, M., Van, E., Anwar-Ul-Hassan., & Schindelbeck, R. R., (2014). Soil health indicators as affected by long-term application of farm manure and cropping patterns under semi-arid climates. *International Journal of Agriculture and Biology, 12*, 242–250.

39. Jorden, T. E., Correll, D. L., & Weller, D. E., (1997). Relating nutrient discharges from watersheds to land use and stream flow variability. *Water Resour. Res., 33*, 2579–2590.

40. Kaiser, J., (2004). Wounding earth's fragile skin. *Science, 304*, 1616–1618.

41. Katterer, T., Bolinder, M. A., Andren, O., Kirchmann, H., & Menichetti, L., (2011). Roots contribute more to refractory soil organic matter than above-ground crop residues, as revealed by a long-term field experiment. *Agriculture, Ecosystems and Environment, 141*(1/2), 184–192.

42. Keeney, D. R., & DeLuca, T. H., (1993). Des Moines River nitrate in relation to watershed agricultural practices: 1945 versus 1980s. *J. Environ Qual., 22*, 267–272.

43. Kenny, J. F., (2005). *Estimated use of Water in the United States* (pp. 23–24). Circular 1344; Washington D.C.: U.S. Department of the Interior: U.S. Geological Survey.

44. King, K. W., Williams, M. R., & Macrae, M. L., (2015). Phosphorus transport in agricultural subsurface drainage: A review. *Journal of Environmental Quality, 44*, 467–485.

45. Lal, R., (2007). Promoting technology adoption in sub-Saharan Africa, South Asia. *Crops, Soils, Agronomy News, 52*, 10–13.

46. Lal, R., (2003). Soil erosion and the global carbon budget. *Environ International, 29*, 437–450.

47. Lal, R., (1998). Soil erosion impact on agronomic productivity and environment quality. *Crit. Rev. Plant Sci., 17*, 319–464.

48. Lal, R., (2009). Ten tenets of sustainable soil management. *J. Soil Water Conservation, 64*, 20A–21A.

49. Lal, R., Sobecki, T. M., & Iivari, T., (2004). *Soil Degradation in the United States: Extent, Severity, and Trends* (p. 224). Boca Raton, FL: CRC Press.

50. Larney, F. J., Bullock, M. S., & Mcginn, S. M., (1995). Quantifying wind erosion on summer fallow in southern Alberta. *J. Soil Water Conserv., 50*, 91–94.

51. Li, X. Y., Liu, L. Y., & Wang, J. H., (2004). Wind tunnel simulation of Aeolian sandy soil erodibility under human disturbance. *Geomorphol., 59*, 3–11.

52. Loveland, P., & Webb, J., (2003). Is there a critical level of organic matter in the agricultural soils of temperate regions?: A review. *Soil Till Res., 70*, 1–18.

53. Mafongoya, P. L., Bationo, A., & Kihara, J., (2006). Appropriate technologies to replenish soil fertility in southern Africa. *Nutrient Cycling Agroecosyst., 76*, 137–151.

54. Mahajop, M. M. Y., (2017). *Effect of Some Organic Insecticides on the Yield and Quality of Field-Grown Tomatoes (Solanum lycopersicum L.)* (p. 138). MSc Thesis; University of Khartoum, Sudan.

55. Martens, D. A., Emmerich, W., McLain, J. E. T., & Johnsen, T. N., (2005). Atmospheric carbon mitigation potential of agricultural management in the southwestern USA. *Soil Till Res., 83*, 95–119.

56. Mason, J. W., Wegner, G. D., Quinn, G. I., & Lange, E. L., (1990). Nutrient loss via groundwater discharge from small watersheds in western and south-central Wisconsin. *J. Soil Water Conserv., 45*, 327–331.

57. McConkey, B. G., Ullrich, D. J., & Dyck, F. B., (1997). Slope position and sub soiling effects on soil water and spring wheat yield. *Can J. Soil Sci., 77*, 83–90.

58. Michelena, R. O., & Irurtia, C. B., (1995). Susceptibility of soil to wind erosion in La Pampa province, Argentina. *Arid Soil Res. Rehab., 9*, 227–234.

59. Moebius-Clune, B. N., Moebius-Clune, D. J., & Gugino, B. K., (2016). *Comprehensive Assessment of Soil Health* (p. 218). The Cornell Framework Manual, Edition 3.1.; Geneva, NY: Cornell University.

60. Morales, A. C., & Mongcopa, C. J., (2008). *Best Practices in Irrigation and Drainage learning from Successful Projects* (p. 98). Asia Development Bank.

61. Morton, L. W., & Brown, S. S., (2011). *Pathways for Getting to Better Water Quality: The Citizen Effect* (p. 262). Springer Science Business, New York.

62. Mwase, W., Sefasi, A., Njoloma, J., Nyoka, B. I., Manduwa, D., & Nyaika, J., (2015). Factors affecting adoption of agro forestry and evergreen agriculture in southern Africa. *Environment and Natural Resources Research, 5*(2), 148–155.

63. NRC (National Research Council), (2010). *Toward Sustainable Agricultural Systems in the 21st Century.* The National Academies Press. Washington, D.C.

64. O'Neal, M. R., Nearing, M. A., & Vining, R. C., (2005). Climate change impacts on soil erosion in Midwest United States with changes in crop management. *Catena, 61*, 165–184.

65. Ogbodo, E. N., & Nnabude, P. A., (2012). Effect of tillage and crop residue on soil chemical properties and rice yields on an acid ultisol at Abakaliki Southeastern Nigeria. *Nigeria Journal of Soil Science, 22*(1), 73–85.

66. Ojiem, J. O., Vanlauwe, B., De Ridder, N., & Giller, K. E., (2007). Niche-based assessment of contributions of legumes to the nitrogen economy of Western Kenya smallholder farms. *Plant Soil, 292*, 119–135.

67. Oldeman, L. R., (1994). The global extent of land degradation. In: Greenland, D. J., & Szabolcs, I., (eds.), *Land Resilience and Sustainable Land Use* (pp. 99–118).CAB International, Wallingford: UK.

68. Olson, B. M., Bremer, E., McKenzie, R. H., & Bennett, D. R., (2010). Phosphorus accumulation and leaching in two irrigated soils with incremental rates of cattle manure. *Can. J. Soil Sci., 90*, 355–362.

69. Olson, K. R., Reed, M. L., & Morton, W., (2011). Multifunctional Mississippi river leveed bottomlands and settling basins: Sny Island Levee Drainage District. *J. Soil Water Conserv., 66*(4), 90A–96A.

70. Ozlu, E., (2016). *Long-Term Impacts of Annual Cattle Manure and Fertilizer on Soil Quality Under Corn-Soybean Rotation in Eastern South Dakota* (p. 98). MSc Thesis; South Dakota State University.

71. Pagliai, M., Vignozzi, N., & Pellegrini, S., (2004). Soil structure and the effect of management practices. *Soil Till Res., 79*, 131–143.

72. Pimentel, D., (2000). Land use: U.S. soil erosion rates-myth and reality. *Science, 289*, 248–250.

73. Pimentel, D., (2006). Soil erosion: A food and environmental threat. *Environ Develop Sust., 8*, 119–137.

74. Pimentel, D., Harvey, C., & Resosudarmo, P., (1995). Environmental and economic costs of soil erosion and conservation benefits. *Science, 267,* 1117–1123.

75. Pimentel, D., & Lal, R., (2007). Biofuels and the environment. *Science, 317,* 897–906.

76. Ramamoorthy, B., & Velayutham, M., (2011). The "law of optimum" and soil test-based fertilizer use for targeted yield of crops and soil fertility management for sustainable agriculture. *Madras Agric. J., 98*(10–12), 295–307.

77. Reeves, D. W., (1997). The role of soil organic matter in maintaining soil quality in continuous cropping systems. *Soil Till Res., 43,* 131–167.

78. Ruan, H. X., Ahuja, L. R., Green, T. R., & Benjamin, J. G., (2001). Residue cover and surface sealing effects on infiltration: Numerical simulations for field applications. *Soil Sci. Soc. Am. J., 65,* 853–861.

79. Ryals, R., Kaiser, M., Torn, M., Berhe, A., & Silver, W., (2014). Impacts of organic matter amendments on carbon and nitrogen dynamics in grassland soils. *Soil Biology and Biochemistry, 68,* 52–61.

80. Sainju, U. M., Singh, B. P., & Whitehead, W. F., (2002). Long-term effects of tillage cover crops, and nitrogen fertilization on organic carbon and nitrogen concentrations in sandy loam soils in Georgia, USA. *Soil Till Res., 63,* 167–179.

81. Schoenau, J. J., & Davis, J. G., (2006). Optimizing soil and plant responses to land applied manure nutrients in the great plains of North America. *Can. J. Soil. Sci., 86,* 587–595.

82. Services, R. D., (2007). *Best Management Practices for Water Use and Irrigation in the Tasmanian Dairy Industry* (p. 125). Tasmania, Australia.

83. Shipitalo, M. J., & Owens, L. B., (2006). Tillage system, application rate, and extreme event effects on herbicide losses in surface runoff. *J. Environ Qual., 35*(6), 2186–2194.

84. Smith, C. M., Peterson, J. M., & Leatherman, J. C., (2007). Attitudes of great plains producers about best management practices, conservation programs and water quality. *J. Soil Water Conserv., 62*(5), 97A–105A.

85. SWCS (Soil and Water Conservation Society), (2003). Conservation implications of climate change: Soil erosion and runoff from cropland. In: *A Report from the SWCS.* Ankeny, Iowa.

86. Trigalet, S., Van, O. K., Roisin, C., & Van, W. B., (2014). Carbon associated with clay and fine silt as an indicator for SOC decadal evolution under different residue management practices. *Agriculture, Ecosystems and Environment, 196,* 1–9.

87. Troeh, F. R., Hobbs, J. A., & Donahue, R. L., (2004). *Soil and Water Conservation for Productivity and Environmental Protection* (4th edn., p. 672) Prentice Hall, New Jersey.

88. Troeh, F. R., Hobbs, J. A., & Donahue, R. L., (1999). *Soil and Water Conservation* (p. 348). Prentice-Hall, New Jersey.

89. UN (United Nations), (2015). *The Millennium Development Goals Report.* UNEP, New York. (Accessed on 20 June 2020).

90. USDA-NRCS (2020). Soil erosion; *National Resources Inventory (NRI).* http://www.nrcs.usda.gov/technical/land/nri03/SoilErosion-mrb.pdf (accessed on 20 June 2020).

91. Wang, E., Xin, C., & Williams, J. R., (2006). Predicting soil erosion for alternative land uses. *J. Environ. Qual., 35,* 459–467.

92. Zhao, H. L., Cui, J. Y., Zhou, R. L., Zhang, T. H., Zhao, X. Y., & Drake, S., (2007). Soil properties, crop productivity and irrigation effects on five croplands of Inner Mongolia. *Soil Till Res., 93,* 346–355.

CHAPTER 9

EX-SITU AND IN-SITU CROP RESIDUE MANAGEMENT TECHNOLOGIES IN TROPICAL COUNTRIES

PEBBETI CHANDANA, Y. LAVANYA, and K. KIRAN KUMAR REDDY

ABSTRACT

Crop sublimates/residues are plant parts that are unaccounted for economic purposes in the agricultural field after crop harvesting. Farmers usually burn the crop residues, which is a cost-effective option compared to any other means, however, it creates enormous environmental issues. The various uses of residues should be explored, such as ruminant feed, fodder, animal bedding material, mushroom cultivation, etc. The Government of India (GOI) has attempted to restrict the burning of these residues by providing subsidies for mechanization and campaigns conducted to convert crop residue into energy. Composting, biochar production, and mechanization are few effective valorization techniques while preserving the nutrients present in the crop residue in the soil. Smart mechanization aids in increasing production, productivity, and profitability of a crop within the specified time and is a key factor for crop residue management.

9.1 INTRODUCTION

The copious scope accessible for agricultural activities increases the production of agro-byproducts and this has led to an exacerbation in waste generation. Agricultural waste is not considered under the municipal solid waste (MSW). The MSW is mainly regulated by public entities, whereas agricultural waste is typically handled by farmers. This has caused increased intricacy in the ditching of agricultural waste [4].

The concept of crop residue management resulted from the major cropping system of rice-wheat in India. Crop residues are plant parts that are unaccounted for economic purposes in the field after harvesting. Crop residues are the nutrient baskets for plants and wellspring of organic carbon for soil microorganisms. Hence, recycling or processing is a must for the effective utilization of nutrients. Retention of residues in the field itself is the management option by reducing soil erosion, runoff, evaporation, and land preparation costs and in other way, it is one of the major impediments since it harbors pests. However, increased generation of crop residues every year shows that retention of residues or conservation agriculture may not be successful for the crops.

According to the National Policy for Management of Crop Residues (NPCMR) [27], India generates about 500 Mt annually of crop residues. Crop residue generation is highest in Uttar Pradesh (60 Mt), followed by Punjab (51 Mt) and Maharashtra (46 Mt). Among different crops, cereals generate maximum residues (352 Mt), followed by fibers (66 Mt), oilseeds (29 Mt), pulses (13 Mt), and sugarcane (12 Mt). Harvesting of rice using combine harvesters has become a common practice in India, but the major handicap in this process is that the residues are left behind the machinery in a narrow swath in the field, which intervenes in the immediate sowing of wheat within 10–20 days. This constraint has been helped by happy seeder, even then, when the custom hiring prices of machinery are thousands of Indian rupees versus one rupee of a box of matches, farmers find it as a cost-effective method; although they are well-aware about the environmental pollution and health problems. It also reduces the activity of soil microbes.

This chapter focuses on: (1) crop residue management used for multifarious purposes like electricity generation, biogas generation, composting, etc., (2) and smart mechanization for *in-situ* crop sublimate management in tropical countries.

9.2 PAY OFF FOR RESIDUE BURNING

Ravindra et al. [31] concluded that 488 Mt of total crop residue was produced during 2017 in India, and about 24% of it was burnt in agricultural fields. Atmospheric emissions from crop residue burning were estimated using Intergovernmental Panel on Climate Change (IPCC) methodology, which resulted in 824 Gg of particulate matter, 58 Gg of elemental carbon, and 239 Gg of organic carbon. Also, the 211 Tg of CO_2 equivalent greenhouse gases (GHG: CO_2, CH_4, N_2O) were effused to the atmosphere. There is a potential to generate 120 TWh of electricity when the crop residues are used as alternate energy sources. This comes to 10% of the total energy production in India. Substantial crop residue burning in Delhi made it as the most polluted city in the World in the first week of November 2016, enforcing the Government of India (GOI) to announce Delhi air pollution as an emergency [www.theguardian.com/World/India]. It is estimated that one ton of rice residue burning liberates 13 kg particulate matter, 60 kg carbon monoxide (CO), 1460 kg carbon dioxide (CO_2), 3.5 kg nitrous oxide, 0.2 kg sulfur dioxide (SO_2) [2].

Niveta et al. [26] observed that burning of residue lead to enormous loss of plant nutrients particularly organic carbon. The entire amount of organic carbon, 80–90% nitrogen (N), 25% phosphorus (P), 20% potassium (K) and 50% sulfur present in crop residues are lost during burning in the form of gaseous and particulate matters, resulting in atmospheric pollution and global warming and also causes adverse effect on soil health. Soil temperature rises upon burning of residues and causes depletion of the bacterial and fungal population. Biomass smoke possesses carcinogenic substances, which lead to lung cancer or airborne diseases. Microbes usually regenerate after a few days of burning but continuous ablaze results in demolishing of microbial population permanently.

9.3 REASONS FOR ON-FARM BURNING OF CROP RESIDUES

Farmers and policymakers are well-aware of the adverse consequences of on-farm burning of crop residues. However, farmers are compelled to burn the residues, because of increased mechanization, particularly the use of combine harvesters, declining number of livestock, the long period required for composting, and unavailability of alternative economically viable solutions.

9.4 NOVEL SOLUTIONS FOR BURNING PROBLEMS

Historically cereal residues are mainly used as cattle feed. Rice residues are used in boilers for parboiling rice or as domestic fuel. Residues of groundnut are burnt as fuel in lime kilns and brick kilns. Cotton, chili, pulses, oilseed, and other crop residues are mainly used as fuel for household needs. Other than these, the crop residues can also be used in several ways as described in this section.

9.4.1 FODDER

The use of crop residue for livestock as fodder is an age-old practice but declining number of livestock in the recent past is the major drawback that causes decline in the crop residues as fodder. Lack of timely availability of fodder and ablaze of crop sublimates coincides in tropical countries, following marked increase in the price of fodder. Wheat and maize straws are commonly used as fodder for cattle. The use of rice straw as livestock feed is meager as it contains high silica especially in leaves compared to stem ensuing in low digestibility and nutritive values for livestock, therefore the crop should be cut close to the ground [25]. Also, crop residues need to be fortified for enriching the nutrients and providing nutrient-rich feed to livestock. For a complete nutritional diet of animals, enriching the residues with urea, calcium hydroxide, and molasses and supplementing with legume (sun-hemp, horse gram, cowpea, and gram) and green fodders straws are essential. Agro-industrial wastes rich in proteins can be used to enhance the nutritive value and utilization of rice straw as livestock feed [42].

Grinding usually reduces digestibility, but the little that is digested is well utilized by the ruminant. This is profitable to small farmers only when small machines are used for grinding or chopping the straw. Chemical treatment of rice straw is more economical to farmers compared to machinery and is not harmful. Alkaline agents like sodium hydroxide (NaOH), ammonia (NH_3), urea, chlorine, and lime have been widely accepted for application on farms. These alkaline agents get absorbed into the cell-wall and chemically break down the ester bonds between lignin, hemicelluloses, and cellulose, and structural fibers become swollen physically, which enable the rumen micro-organisms to attack more easily the structural carbohydrates, improving degradability and palatability of the rice straw. [33]. Sarnklong et al. [32] observed that unconventional alkali treatment made from filtrate of a 10% rice hulls ash solution enriched with urea and minerals increased volatile fatty acid production, ammonia nitrogen, and rumen microbial protein synthesis.

9.4.2 COMPOSTING

Composting is usually practiced in backyards in India from their own crop residue and it can be returned easily to agricultural lands. Vermicompost is similar to the nutrient content of animal manure and food waste. Compost plays a pivotal role in sustaining soil fertility and thereby achieving sustainable agricultural productivity, since it is a rich source of organic matter.

Also, it can save application of fertilizers and pesticides. Increased yields, resistance to external factors (such as drought, disease, and toxicity) are beneficial effects of compost amended soils [17]. Composting is mediated by micro-organisms like bacteria, fungi, actinomycetes, algae, and protozoa, which are naturally present in residues or added artificially to facilitate decomposition.

Vermicompost application can enhance soil NPK content and highest nutrient uptake by mustard was observed in the residual effect for second cropping [28]. The first cropping of mustard yielded highest with the application of spent mushroom waste (V1) and coconut husk vermicompost (V2) @ 10–15 tons ha^{-1}. Increased productivity was noticed in the second cropping, whereas in the third and fourth sequential cropping it was reduced in the vermicompost V1 and V2, whereas sugarcane trash vermicompost (V3) application improved the productivity of the third and fourth cropping too.

Rice residues from one hectare land on composting yields 3 tons of manure, which can be fortified with P using indigenous source of low grade rock phosphate to make it value-added compost with 1.5% N, 2.3% P, and 2.5% K [3]. The compost prepared from various crop residues using earthworms and its use in agricultural fields yields good quality compost and is a green technology, generating local employment, and economic benefits [11, 22]. Rural youth must be trained scientifically for quality compost generation.

9.4.3 BEDDING FOR ANIMALS

Crop residues for animal bedding are generally used for compost preparation. Each kilogram of straw laid in the animal shed absorbs about 2–3 kg of urine, which enriches it with N.

9.4.4 GENERATION OF BIOMASS ENERGY

Energy consumption in India was augmented by 7.9% in 2018 and is expected to be 18% by 2035. India might be the second largest contributor to the increase in global energy demand by 2035 [7]. Growing energy demands of India on one pan and limited domestic oil or gas reserves on other pan of a balance calls for the country to bring it into equilibrium using alternative sources of energy. Biomass is the competent source of energy and can be as substitute for fossil fuels, solar, and wind power since it can be stored for future use, inexpensive, energy-efficient, and environment-friendly.

However, transportation cost and infrastructural settings (harvest machinery, modes of collection, etc.), are some of the factors that curb the usage of crop residues for energy generation. Here comes the role of cooperatives or government in providing incentives for the purchase of required machinery.

In India, around 500 biomass power and cogeneration plants have been installed by the Ministry of New and Renewable Energy, GOI. These biomass power plants generate 13% of the total renewable power supply. Adoption of advanced technology with higher energy efficiency and ensuring availability of crop residue throughout the year enhances renewable energy. However, there are many challenges in utilizing crop residue and also there is a need for financial assistance to promote the growth in biomass power sector. Biomass power accounted for only 1.60% of total estimated potential renewable power in 2018 in India [7]. Annual bioenergy potential from the surplus residue is around 17% of energy consumption of India [13]. Ravindra et al. [31] showed that 120 TWh of electricity can be generated from crop residues for biomass power plants, which is 10% of the total energy production of India.

Punjab, one of the most residue-burning states of India, has only eight biomass energy plants and six more are in the pipeline. The 0.12 million tons (MTs) of residues per year are required for a 12-MW rice residue power plant; hence it needs a large dumping yard. Also, these biomass energy plants produce large amount of bio-ash and there is a question for its management in a profitable way. Usually, it is discarded in landfills or depressions created by brick kilns [24]. Gasification, biomethanation, and ethanol generation are effective processes for energy production from residues possessing lignocellulosic biomass [35].

9.4.4.1 GASIFICATION

Production of gas due to partial combustion of crop residues by thermochemical process is called gasification. Removal of impurities for purification of gas is the major problem in biomass gasification for power generation. The crop residues can be used in the gasifiers for "producer gas" generation. Producer gas is fed into the engines coupled with alternators for electricity generation. One ton of biomass can produce 300 kWh of electricity [14].

The Central Institute of Agricultural Engineering (CIAE) in Bhopal has developed a power plant running on 'producer gas' generated from biomass. Most of the furnaces in Punjab use 25–30% of rice residue mixed with

70–75% of other biomass and the present utilization of rice straw is only 0.5 million tons annually. Higher silica content in rice straw causes clinker formation in the boilers, which is the major drawback in the practical utility of this technology. The efficiency of 75% has been achieved by circulating fluidized bed gasifier and downdraft gasifier in China [47].

9.4.4.2   BIOGAS GENERATION AND BIO-METHANATION

Crop residues at the village level can be best utilized by adopting the concept of biogas production. This can meet the cooking fuel needs along with manure management of crop residue. Biogas slurry is the by-product of this process, which is rich in nutrients and can be used as manure. The local youth must be trained and engaged in its production, which provides employment opportunities and will generate a local market to further enhance the production. Community biogas plants (such as KVIC (Khadi and Village Industries Commission) design) must be encouraged. The government should support fascinated people to build biogas plants and provide training. Further, the conversion of biogas to biomethane (known as biomethanation) paves the way for commercialization. The benefit for biomethanation plant users is the organic fertilizer obtained from biogas residue [13, 40].

9.4.4.3   BIO-FUEL GENERATION

Crop residue can be used to produce biofuels and other useful products like bio-oil, an important action to reduce dependence on fossil fuel. Conversion of lignocellulosic biomass into alcohol is of immense importance as ethanol can either be blended with gasoline as a fuel extender and octane-enhancing agent or used as a fuel in internal combustion engines. One ton of dry matter of various crop residues generates about 382 to 471 liters of ethanol [37]. Few limitations in this process are: costly hydrolytic cellulase enzyme, high energy-requiring operating conditions, and unavailability of a natural robust commercial organism to ferment pentose and hexose sugars simultaneously either as single species or in a combination of other species.

Bio-oil can be produced from crop residues by the process of fast pyrolysis, which is a thermal disintegration process wherein the temperature of biomass is raised to 400–500°C within a few seconds. About 75% of dry weight (DW) of biomass is converted into condensable vapors. A dark brown viscous liquid, called bio-oil, is formed when the condensate gets cooled quickly within a couple of seconds [14].

9.4.5 BIOCHAR PRODUCTION

Biochar production from crop residues is a potential measure for controlling GHG emissions and is an excellent residue management strategy. Biochar is a carbonized porous material obtained when biomass undergoes pyrolysis in oxygen evacuated conditions, which helps in sequestering carbon, mitigating GHG, improving soil health and water retention capacity [15, 34, 46].

Mohammadi et al. [21] concluded that biochar production from locally available crop residues reduced 38–49% of carbon footprints from rice production. The properties that decide the quality of biochar are: type of feedstock, heating rate, and pyrolysis temperature and residence time [18]. Biochar has various applications, such as the wastewater (WW) treatment, construction industry, food industry, cosmetic industry, metallurgy, and many other chemical applications.

9.4.6 MUSHROOM CULTURE

Use of crop residue for mushroom cultivation transforms inedible crop residues into edible food with high moisture content, protein, and an amino acid composition equal to that of milk or meat [12]. Rice and wheat straws are suitable stratum for the usually cultivated edible mushrooms, like, *Agaricus bisporus*, and *Volvariella volvacea*. Thongklang and Luangharn [44] reported that sorghum mixed with corn cobs had the highest growth rate (16.83 mm/day), followed by sorghum mixed with rice husks (11.07 mm/day); and among the four substrates, the best substrate to cultivate *Pleurotus ostreatus* is sawdust + rice husks with an average wet weight harvest of 277.50 ± 79.74 g in a 40-day production cycle.

9.4.7 RAW MATERIAL FOR INDUSTRY

Agro-industry waste can also be utilized to produce nano-silica, an industrial product, which serves as a raw material for manufacturing nanomedicines, cosmetics, solar cells, etc. [23]. The prospect of converting crop residue into commercial products intensifies the farmer's income. Crop residues are composed of cellulose, hemicellulose, and lignin; therefore, they can be used to prepare biodegradable and compostable materials, such as disposable leaf and straw plates/bowls, cardboard, and packaging materials instead of plastic. A new area for micro- and small enterprise development has been evolved for commercialization of crop residues [48].

9.4.8 SURFACE SOIL MULCH

Crop residues as surface soil mulch serve as a better option for soil and moisture conservation by avoiding evaporation. Singh et al. [39] reported that the surface retention of crop residues as mulch is known to have various uses: conserve soil water, moderates thermal regimes, curbs the weed germination and improve soil health by the building of soil microbial populations results in increasing soil organic carbon, which ultimately helps in improving crop yields and saving of water for irrigation. Mulching with rice straw in wheat increased wheat grain yield reduced crop water-use by 3–11% and improved water use efficiency (WUE) by 25% compared with no mulch. Mulching produced 40% higher root length densities compared to no-mulch in lower layers (>0.15 m), probably due to greater retention of soil moisture in deeper layers [5].

9.4.9 OPTIMUM TIME OF SOWING AND PROMOTION OF HYBRID SEEDS

Delay in sowing of the first crop due to various reasons reduces the time gap for plowing/sowing of the next crop, which encourages the farmers for burning instead of residue management. Hybrids possessing with shorter growing period helps to increase the sowing timeframe, which aids in the management of residues.

9.4.10 CROP DIVERSIFICATION

Continuous cropping with rice and wheat can be deviated at times with alternate crops having higher nutrient value and should be used as fodder and manure. An alternative crop rotation system with selective crops could effectively reduce the GHG emissions from paddy fields and increase the monetary benefits [1].

9.4.11   NITROGEN ENRICHMENT

Tengfei et al. [43] reported that nitrogen enrichment with 180 and 270 kg of N ha^{-1} regulates straw decomposition compared to 0 and 90 kg N ha^{-1}, in the early stage of decomposition due to slow release rate of straw N components because of inorganic N addition and that accelerated the microbial enzyme activity for their own metabolism and further straw decomposition. Although the N fertilizer rate had a minute effect on the rate of straw decomposition, yet it was a salient factor resulting in the variation of hydrolytic enzyme activities.

9.4.12   FALLOW PERIOD DECOMPOSITION

Intensive double or triple rice cropping systems for years with short aerobic periods results in a change in the qualitative composition of soil organic matter towards more phenolic compounds, accumulation of reduced substances, and acidification of the rice rhizosphere due to greater root-induced Fe^{2+} oxidation. In this scenario, the time of incorporation of organic materials is more important than the amount.

According to Dobermann and Witt [6], allowing dry-fallow period between rice crops and early incorporation of the rice residues immediately after the harvest results in aerobic decomposition of crop residues leading to increased N availability; reduced methane emissions; re-oxidation of reduced substances following P availability; large shifts in microbial communities; reduced weed growth and irrigation water saving by reducing the land soaking period for rice. Accumulation of phytotoxic substances (e.g., phenolic acid and acetic acid) is observed under anaerobic decomposition of crop residues and in aerobic soils due to rapid metabolization by microorganisms.

9.4.13   MICROBIAL DECOMPOSITION

Agricultural waste management using microbes is an eco-friendly option because they degrade the complex substances either by aerobic or anaerobic processes to simpler ones present in the biomass that can be recycled [8, 9]. These processes including microbes promote plant growth through the production of growth accelerating metabolites, reduce the soil toxicity, and provide plant nutrients through sequestration from soil [10].

Anaerobic digestion is an upgrading technique since it converts various types of organic waste, crop residues, and manure into highly energetic biogas. Biomethanation involves anaerobic digestion of biomass wherein microbial conversion occurs in an aqueous environment without any pretreatment. Anaerobic digestion involves hydrolytic bacteria in the first phase and production of acetic and formic acid in the second stage which is reduced to carbon dioxide and methane by acetotrophic, methylotrophic, and hydrogenotrophic bacteria in the third stage [19].

White rot fungi are effective lignin degraders, which are used to improve the nutritive value of fodder for animal nutrition [16]. Bacteria grows quickly on freshly added plant residues, hence dominates in the initial phases of decomposition. Fungi decompose more recalcitrant material and grow more slowly than bacteria, thus thrive in the later stage of straw decomposition [30].

Timothy et al. [45] indicated that *Ceriporiopsis subvermispora* and bacterium *Cellulomonas* sp., selected for their cellulolytic capabilities and *Azospirillum brasilense* for its nitrogen-fixing bacterium sprayed over the residue and mixed in soil supplemented with 0.3% molasses, showed a significant difference in the visual decomposition of residue, bacterial, and fungal populations, soil pH, nitrogen, and available phosphorus when compared to the control and the treatment in which the consortium was sprayed over the residue but not mixed.

9.4.14 CONSERVATION AGRICULTURE

The resource conservation technologies with modernization in sublimate management curbs straw burning, enhances organic carbon in soil, increases resource use efficiency, and also reduces greenhouse gas emissions [29]. Zero-tillage, permanent crop cover with recycling of crop residues and crop rotation are principles of conservation agriculture. Sowing of a crop in the presence of residues is possible with zero-till seed-cum-fertilizer drill/planters such as happy Seeder and rotary-disc drill, wherein direct drilling of seeds occur. These machines aid in crop residue management by conserving moisture and nutrients as well as controlling weeds in addition to moderating soil temperature.

There are several challenges in using crop residues in conservation agriculture right from sowing to pest infestation. Weed control in conservation agriculture requires herbicide application, thus hindering the healthy soil and environment. Nutrient management also becomes cumbersome because of higher crop residues, reduced alternatives for application of manures and loss of N since it is entirely applied as basal and later top dressing is diffi-cult and specialized equipments are necessary, which leads to higher costs. Further restraining factors in incorporation of crop residues in conservation agriculture by farmers are lack of technical expertise in machinery, percep-tion of lesser productivity and institutional constraints. Also, farmers prefer clean tilled fields versus untilled scruffy fields.

9.5 SMART MECHANIZATION

The rapid decline in farmworkers has led to an increasing of the workload of farmer without consideration of the increased risk to their health and safety as they age. Smart mechanization is one of the potential to resolve labor scarcity and increase agricultural productivity of all other inputs used in production. This can also help attract youth to agriculture for the establish-ment of profitable business enterprises.

9.5.1 ROLE OF SMART FARM MECHANIZATION TO CURB STUBBLE BURNING

In North India, paddy is sown in late *kharif* season and in a wider area, which results in less time for farmers to plant wheat crops in winter. Particularly, Punjab and Haryana may be headed for smoky winter in the absence of adequate machinery to help farmers in the region to clear their fields by burning of paddy stubble before the planting of wheat. It leads to increase in the particulate matter level soaring to a hazardous degree in the air every year. Agricultural machines could be one of the solutions, but it has some critical issues that are listed by the Ministry of Agriculture, GOI [20].

Using agricultural machines like combine harvester, farmers chose conservation tillage technologies like minimum and zero-till after removal or burn left-over paddy straw which starts from early-October to plant wheat crop in rice-wheat cropping sequence within a short period of timer. However, there is obstacle in using zero-till seed drill, direct drilling of wheat into combine-harvested paddy fields due to: straw acquisition in seed drill furrow openers; and under heavy residue conditions there is need for frequent lifting of the implement that results in irregular seed depth. The availability of suitable machinery was a major constraint to direct drilling of next crop in previous crop stubbles. Therefore, promotion of agricultural machines for *in-situ* management of crop residue machineries could be one of the solutions for reduction in crop stubble burning viz.

9.5.1.1   SUPER STRAW MANAGEMENT SYSTEM (SMS) TO BE FITTED WITH HARVESTER

The Super SMS is an additional straw management system (SMS) that could be fitted to self-propelled combine harvesters. It cuts the straw into 10–15 cm pieces and scatters it around behind the tail of the combine whereas normal combines leave 15–30 cm long stubbles. This straw works as mulch to conserve soil moisture for better establishment of wheat crop in rice-wheat cropping system [41].

9.5.1.2   HAPPY SEEDER

It is mainly used for direct drilling of new crops without any burning of previous crop residue. It is a tractor-mounted device comprised of a rotor unit attached at front of the seed drill that cuts and drops the straw in between the rows. This chopped straw acts as mulch by deposition over the sown area [36]. This zero-till technology is environmental friendly to check air pollution. The operational cost of happy seeder sowing is 50–60% lower than conventional method with conservation of soil moisture [38].

9.5.1.2.1   Turbo Happy Seeder: Salient Features

1. Increases fuel use efficiency and reduces power requirement.
2. Prevent clogging/choking of machine under heavy straw load.
3. Improve work rate of the happy seeder and traction of ground wheel.
4. Reduce vibration.

9.5.1.2.2   Concurrent Use of Super Straw Management System Fitted Combine Harvester and Turbo Happy Seeder: Implications

1. 2–4% increase in wheat yield compared to conventional method of wheat crop sown.

2. Reduces the cost of labor, fuels, etc., resulting effective/reasonable cost of production.
3. Using turbo happy seeder (3–4 years continuously) results increase in 10–15% nutrient use efficiency compared to conventional.
4. Continuous recycling of residue act as mulch results reduces in evaporation losses and amount of irrigation requirement, by saving up to 10 ha-cm of water (1.0 million liters) leads to more crop per drop of water.
5. Reduces weed growth and crop lodging under unfavorable condition.
6. Enhance soil health properties by improving soil organic matter over time.
7. Improves human (on-farm and off-farm) health and environment by reduction in air pollution.
8. Prevent loss of nutrients in soil: 1 ton of stubble burning leads to depletion of 5.5 kg of N, 2.3 kg of P, 25 kg K, and more than 1 kg of sulfur, besides organic carbon [24].

9.5.1.3   BALER: RESIDUE CLEARING OPERATION

1. It is used to harvest residues (grasses and legumes), so that windrow can clear residue.
2. It compresses a cut and raked recyclable residues into compact bales with bailing pressure.

Depending on various sizes and shape with automatic wire tying mechanism, the baler is classified as:

1. **Round:** Most commonly used.
2. **Rectangular:** Used in large-scale feedlot production.
3. **Square:** Less commonly used.
4. **Industrial:** Used in recyclable industries (plastics, paper).

9.5.1.3.1   Advantages

1. Crop residues are turned in to bale make it easy to handle, transport, and store.
2. Saves the environment from greenhouse gas emissions, curb stubble burning.
3. Baled crop residue is used for biofuel and animal feeding.

4. Creates an alternative business for farmers to sell bales to power plants.

9.5.1.4 HAY RAKE

1. It is almost similar to baler;
2. It is used to cut hay straw into windrows for later collection and made into bailing by a baler;
3. Raking is done whenever hay moisture content should be less than 35–45%;
4. It reduces the workload of the baler machine for further bailing process.

9.5.1.5 MULCHING MACHINE

1. It consists of knives which are attached on the roller, rotates vertically. Cutting height is adjusted by the back wheel of that machine.
2. It is used for cut, grind, and clear vegetation/crop residues and spread uniformly on the field.
3. It acts as bio mulch results to enhance organic matter in the soil after decomposition.
4. This implement saves both time and money which is important for farmers.

9.5.1.6 ROTAVATOR

Used for land preparation and incorporation of preceding crop stubbles in the field.

9.5.1.7 ZERO TILL SEED DRILL

Land preparation along the sown area along with directly sowing of succeeding crop in preceding crop stubble

9.5.1.8 PADDY STRAW CHOPPER

Chopping of paddy stubbles into fine pieces and mix with soil.

9.6 POLICIES TO PROMOTE BEST WASTE MANAGEMENT PRACTICES FOR SUSTAINABLE AGRICULTURE

Government agencies should order measures to curb crop residue burning and regulate crop residue management. Numerous attempts were made by GOI to promote best practices of alternative sustainable agricultural waste management given by National Policy for Management of Crop Residue [14, 27]:

1. Educate the farmers about long-run depletion of plant nutrients in soil when they burn straw (ICAR, National Soil Research Institute).
2. Government can offer better minimum support price to encourage farmers to switch back high water-use paddy in a traditional wheat growing region to low water use, nutrient dense coarse grains like millets, etc.
3. Provide direct payment to farmers to deposit recyclable residues at collection centers. So that, many start-up companies are trying to develop.
4. Burning of crop residue isn't unique in India, government could advise farmers on the speed and direction of wind so that restricted burning would at least not form the sort toxic clouds hang over nearby urban areas.
5. Purchasing of in situ crop residue management machineries, government should provide financial support to farmers.
6. Provide subsidy to different groups viz., private entrepreneurs, self-help group, registered, and cooperative societies of farmers for the establishment of custom hiring centers (farm machinery bank) of *in situ* crop residue management which make benefit to small and medium farmers.
7. Provide financial assistance to organize on and off-field trainings and demonstrations among the farmers in terms of technology adoption and improvement in skills to promote the usage of agricultural machineries.
8. Financial support to different institutions like state government, Krishi Vigyan Kendra, Indian Council of Agricultural Research, public sector enterprise for the activities to be executed towards information, education, and communication.
9. Introducing carbon credit scheme attempts for carbon sequestration and mitigate the growth in concentration off GHG to benefit the farmers who follow conservation agriculture.

10. Some of the laws in India that are in operation on residue burning are:

   - Section 144 under civil procedure code-ban burning in India;
   - Air prevention and control of pollution act (1981);
   - Environment protection act (1986);
   - National tribunal act (1995);
   - National environment appellate authority act (1997).

9.7 AVOID SECTOR RELATED ESTEEM: FOCUS ON GAMUT ESTEEM

A major portion of greenhouse gas emission is from the agriculture sector. However, is this completely an agricultural sector? Based on discussions in this chapter, crop residue burning is one of adverse impacts on climate change. But from the economic viewpoint of the farmer, the price of machine hiring is 000's of rupees for *in situ* crop residue management versus one rupee of a box of matches. It is easy to make a decision even if they are aware of the negative impact on the environment. The main reason from the farmer point of view for burning resourceful recyclable residue is due to social and economic issues or both. Even though crop residue burning links up different sectors viz., agriculture, energy, economy, environment, and education, yet the government efforts are mainly on agriculture and energy. This type of sectorial thinking is a very slow process. A concept like a *gamut* thinking promotes a higher level of integration in managing environmental resources. Mushroom production, bioethanol, compost, and biochar, etc., are good examples to combat crop residue burning in terms of network thinking [4].

9.8 SUMMARY

Stubble burning at large-scale results into problems for humans, agriculture, and air poisonous due to the emission of harmful gases. The 34% and 22% from rice and wheat fields are left in the field itself. Instead of ablaze of residues, it can be used in diversified ways like roofing in rural areas, biofuel, cattle feed, mushroom, and industrial production, etc., or by employing mechanization for *in situ* crop residue management as a surface mulch. With the looming complication of burning residues, there is a need to develop law and policy measures by the government to address pollution from agriculture. To ensure effective execution, proactive measures should be monitored at regular intervals to stop the burning of stubble.

KEYWORDS

- biochar
- bioenergy
- conservation agriculture
- crop residue
- municipal solid waste
- smart mechanization

REFERENCES

1. Arunrat, N., Wang, C., & Pumijumnong, N., (2016). Alternative cropping systems for greenhouse gases mitigation in rice field: Case study in Phichit province of Thailand. *J. Clean. Prod, 133*, 657–671.

2. Bakker, R., Elbersen, W., Poppens, R., & Lesschen, J. P., (2013). *Rice Straw and Wheat Straw: Potential Feedstock's for the Bio-Based Economy* (p. 32). Netherlands Programs Food and Bio-based Research. NL Energy and Climate Change Commission.

3. Behera, B., (2018). *Recycling of Crop Residues for Improved Soil Nutrient Status and Farm Income* (p. 72). PhD Dissertation; Department of Agronomy; College of Agriculture; Orissa University of Agriculture and Technology, Bhubaneswar.

4. Bhuvaneshwari, S., Hettiarachchi, H., & Meegoda, J. N., (2019). Crop residue burning in India: Policy challenges and potential solutions. *Int. J. Environ. Res. Public Health, 16*, 832–840.

5. Chakraborty, D., Garg, R. N., Tomar, R. K., Singh, R., Sharma, S. K., Singh, R. K., Trivedi, S. M., et al., (2010). Synthetic and organic mulching and nitrogen effect on winter wheat (*Triticum aestivum* L.) in a semi-arid environment. *Agric. Water Management, 97*, 738–748.

6. Dobermann, A., & Witt, C., (2000). The potential impact of crop intensification on carbon and nitrogen cycling in intensive rice systems. In: Kirk, G. J. D., & Olk, D. C., (eds.), *Carbon and Nitrogen Dynamics in Flooded Soils* (pp. 1–25). Int. Rice Res. Inst., Los Baños, Philippines.

7. Ministry of Statistics and Programme implementation, Government of India (GOI), (2019). *Energy Statistics* (p. 123). Online: http://www.mospi.nic.in/sites/default/files/publication_reports/Energy_Statistics_2017r.pdf.pdf (accessed on 20 June 2020).

8. Franchi, E., Agazzi, G., Rolli, E., Borin, S., Marasco, R., Chiaberge, S., & Barbafieri, M., (2016). Exploiting hydrocarbon-degrader indigenous bacteria for bioremediation and phytoremediation of a multi-contaminated soil. *Chem. Eng. Technol., 39*, 1676–1684.

9. Garg, S., (2017). Bioremediation of agricultural, municipal, and industrial wastes: Chapter 15. In: Bhakta, J., (ed.), *Handbook Res. Inventive Bioremediation Tech.* (pp. 341–363). IGI Global, Hershey-PA-US.

10. Gkorezis, P., Daghio, M., Franzetti, A., Van, H. J. D., Sillen, W., & Vangronsveld, J., (2016). The interaction between plants and bacteria in the remediation of petroleum hydrocarbons: An environmental perspective. *Front Microbiol., 7,* 1836–1840.

11. Gupta, R., & Garg, V. K., (2011). Potential and possibilities of vermi-composting in sustainable solid waste management: A review. *International Journal of Environment and Waste Management, 7*(3), 210–234.

12. Harikrishna, P., (2013). *Utilization of Maize Stalks for Mushroom Cultivation and Compost Making, Department of Agricultural Microbiology and Bioenergy* (p. 210). PhD Dissertation; Acharya, N. G. Ranga Agricultural University; College of Agriculture, Rajendranagar, Hyderabad.

13. Hiloidhari, M., Das, D., & Baruah, D. C., (2014). Bioenergy potential from crop residue biomass in India. *Renew. Sustain. Energy Rev., 32,* 504–512.

14. Pathak, H., Jain, N., & Bhatia, A., (2012). *Crop Residues Management with Conservation Agriculture: Potential, Constraints and Policy Needs* (p. 32). Venus Printers and Publishers, Indian Agricultural Research Institute (IARI)-New Delhi.

15. IBI (International Biochar Initiative), (2012). *Standardized Product Definition and Product Testing Guidelines for Biochar that is Used in Soil* (p. 42).

16. Kamla, M., Tokkas, J., Anand, R. C., & Kumari, N., (2015). Pretreated rice straw as an improved fodder for ruminants: An overview. *Journal of Applied and Natural Science, 7*(1), 514–520.

17. Lei, Z., Chen, J., Zhang, Z., & Sugiura, N., (2010). Methane production from rice straw with acclimated anaerobic sludge: Effect of phosphate supplementation. *J. Bioresour. Technol., 101,* 4343–4348.

18. Mahtab, A., & Rajapakshaa, A. U., (2014). Biochar as a sorbent for contaminant management in soil and water: A review. *Chemosphere, 99,* 19–33.

19. Meegoda, J. N., Li, B., Patel, K., & Wang, L. B., (2018). There view of the processes, parameters and optimization of anaerobic digestion. *Int. J. Environ. Res. Public Health, 15,* 2224–2230.

20. Ministry of Agriculture, GOI, (2016). *Take Steps to Promote Use of Equipment's for Crop Residue Management in a Big Way: Shri Radha Mohan Singh; State Government Should Create Massive Awareness on Crop Stubble Management: Shri Singh.* New Delhi: Press Information Bureau Government of India, Ministry of Agriculture.

21. Mohammadi, A., Cowie, A., & Anh, M. T. L., (2016). Biochar use for climate-change mitigation in rice cropping systems. *J. Clean. Prod., 116,* 61–70.

22. Mor, S., Kaur, K., & Khaiwal, R., (2016). SWOT analysis of waste management practices in Chandigarh, India and prospects for sustainable cities. *Journal of Environmental Biology, 37*(3), 327–333.

23. Mor, S., Manchanda, C. K., Kansal, S. K., & Ravindra, K., (2017). Nano-silica extraction from processed agricultural residue using green technology. *J. Clean. Prod., 143,* 1284–1290.

24. NAAS (National Academy of Agricultural Sciences), (2017). *Innovative Viable Solution to Rice Residue Burning in Rice-Wheat Cropping System Through Concurrent Use of Super Straw Management System-Fitted Combines and Turbo Happy Seeder* (p. 16). Policy Brief No. 2, National Academy of Agricultural Sciences, New Delhi.

25. Na, Y. J., Lee, I. H., Park, S. S., & Lee, S. R., (2014). Effects of combination of rice straw with alfalfa pellet on milk productivity and chewing activity in lactating dairy cows. *Asian Australasian J. Anim. Sci.*, *27*, 960–964.

26. Niveta, J., Bhatia, A., & Pathak, H., (2014). Emission of air pollution from crop residue burning in India. Center for Environment Science and climate-resilient agriculture: Indian Agricultural Research Institute, New Delhi. *Aerosol and Air Quality Research*, *14*, 422–430.

27. NPMCR, (2019). Online: http://agricoop.nic.in/sites/default/files/NPMCR_1.pdf (accessed on 20 June 2020).

28. Nurhidayati, N., & Machfudz, M., (2018). Direct and residual effect of various vermicompost on soil nutrient and nutrient uptake dynamics and productivity of four mustard Pak-Coi (*Brassica rapa* L.) sequences in organic farming system. *International Journal of Recycling of Organic Waste in Agriculture*, *7*, 173–181.

29. Pathak, H., Saharawat, Y. S., Gathala, M., & Ladha, J. K., (2011). Impact of resource-conserving technologies in the rice-wheat system. *Greenhouse Gas Science and Technology*, *1*, 261–277.

30. Poll, C., Marhan, S., Ingwersen, J., & Kandeler, E., (2008). Dynamics of litter carbon turnover and microbial abundance in a rye detritus sphere. *Soil Biol. Biochem.*, *40*, 1306–1321.

31. Ravindra, K., Singh, T., & Mor, S., (2018). Emissions of air pollutants from primary crop residue burning in India and their mitigation strategies for cleaner emissions. *Journal of Cleaner Production*, *7*. Online: doi: https://doi.org/10.1016/j.jclepro.2018.10.031.

32. Sarnklong, C., Cone, J. W., Pellikaan, W., & Hendriks, W. H., (2010). Utilization of rice straw and different treatments to improve its feed value for ruminants: A review. *Asian-Australian Journal of Animal Science*, *23*, 680–692.

33. Selim, A. S. M., Pan, J., Takano, T., & Suzuki, T., (2004). Effect of ammonia treatment on physical strength of rice straw, distribution of straw particles and particle-associated bacteria in sheep rumen. *Animal Feed Science and Technology*, *115*, 117–128.

34. Shackley, S., Carter, S., Knowles, T., & Middelink, E., (2012). Sustainable gasification-biochar systems? A case-study of rice-husk gasification in Cambodia, Part 1: Context, chemical properties, environmental and health and safety issues. *Energy Policy*, *42*, 49–58.

35. Shafie, S. M., (2016). Review on paddy residue based power generation: Energy, environment and economic perspective. *Renew. Sustain. Energy Rev.*, *59*, 1089–1100.

36. Sidhu, H. S., Singh, M., & Singh, Y., (2015). Development and evaluation of the turbo happy seeder for sowing wheat into heavy rice residues in NW India. *Field Crops Res.*, *184*, 201–202.

37. Singh, B., (2018). Crop residue management through options. *International Journal of Agriculture, Environment and Biotechnology*, *11*(3), 427–432.

38. Singh, R. P., Dhaliwal, H. S., Humphreys, E., Sidhu, H. S., Singh, M., & Singh, Y., (2008). *Economic Assessment of the Happy Seeder for Rice-Wheat Systems in Punjab, India* (p. 6). Presented at AARES 52nd Annual Conference: Canberra, ACT, Australia.

39. Singh, Y., Singh, M., & Sidhu, H. S., (2010). *Options for Effective Utilization of Crop Residues* (p. 32). Research bulletin; Directorate of Research, Punjab Agricultural University, Ludhiana.

40. Sun, J., Peng, H., Chen, J., & Wang, X., (2016). Estimation of CO_2 emission via agricultural crop residue open field burning in China from 1996 to 2013. *J. Clean. Prod.*, *112*, 2625–2631.

41. Surya, A., (2019). *Haryana Not to Allow Paddy Harvesting Without Straw Management System.* Online: http://timesofindia.indiatimes.com/city/Chandigarh/Haryananottoallowpaddyharvestingwithoutstrawmanagementsystem/articleshowprint/65518569.cms (accessed on 20 June 2020).

42. Suwandyastuti, S. N. O., & Bata, M., (2010). Improvement of rice straw for ruminant feed through conventional alkali treatment and supplementation of various protein sources. *Journal of Animal Production, 12*, 82–85.

43. Tengfei, G., & Zhang, Q., (2018). Nitrogen enrichment regulates straw decomposition and its associated microbial community in a double-rice cropping system. *Scientific Reports, 8*, 1847–1853.

44. Thongklang, N., & Luangharn, T., (2016). Testing agricultural wastes for the production of *Pleurotus ostreatus*. *Mycosphere, 7*(6), 766–772.

45. Timothy, P. B., (2002). Accelerated decomposition of sugarcane crop residue using a fungal-bacterial consortium. *International Biodeterioration and Biodegradation, 50*, 41–46.

46. Verheijen, F., Jeffery, S., & Bastos, A. C., (2010). *Biochar Application to Soils: Critical Scientific Review of Effects on Soil Properties, Processes and Functions* (p. 149). European Commission, Rome, Italy.

47. Zeng, X., Ma, Y., & Ma, L., (2007). Utilization of straw in biomass energy in China. *Renewable and Sustainable Energy Reviews, 11*(5), 976–987.

48. Zhang, H., Hu, J., Qi, Y., Li, C., & Chen, J., (2017). Emission characterization, environmental impact, and control measure of PM 2.5 emitted from agricultural crop residue burning in China. *J. Clean. Prod., 149*, 629–635.

CHAPTER 10

ANALYSIS OF GROUNDWATER LEVEL: GROUNDWATER MODELING USING GIS IN KOLKATA

SUSHOBHAN MAJUMDAR

ABSTRACT

According to the latest reports from the Central Groundwater Board (CGWB), nearly 56% of the tube-wells have a low level of groundwater and it declined sharply in 2012. In India, the main reason for groundwater depletion is groundwater pumping by tube or bore wells. Among many cities in West Bengal, Kolkata suffers from problems of land subsidence and lack of drinking water. The major objective of this study was to scrutinize the level of groundwater in Kolkata city and to identify related issues. This study revealed that unscientific exploitation of groundwater increases the gap in the lower level surfaces under the city. Therefore, Kolkata city has experienced a bowl-like feature of groundwater profile, which indicates Kolkata city will face major land subsidence in the future.

10.1 INTRODUCTION

Groundwater resources (like water, mineral, etc.), play a major role by providing agricultural social security [5]. It is also the main source of irrigation water because of the easy availability of small initial investment [4]. The revolution regarding use for sustainable purposes has evolved mainly in countries in South Asia (like India, Sri Lanka, etc.), the Middle East, the USA, and in selected regions of Africa [3].

The demand for groundwater is high in areas of South Asia and Africa. The countries in South Asia and Africa are major users of groundwater of about 2010 km³ per year [2]. Because of the huge exploitation of water in most of the cases, the water table has declined thus experiencing vertical compression of subsurface materials [1].

Today, nearly 2 billion persons in the world are dependent upon groundwater, which is a renewable resource that can be recharged only by rainwater. Apart from this natural resource, there are also other human factors. The pressure of the population on groundwater is very high in urban areas than in the peri-urban or rural areas. Because of the huge exploitation and unscientific use of groundwater, the level of groundwater has declined significantly in many urban areas of developed countries. In India, it is projected that the irrigation sector will consume nearly 74% in 2025 and 72% in 2050 of the total groundwater, respectively [6]. In most of the rural areas in India, groundwater is also a major source for domestic and drinking purpose.

According to the Central Groundwater Board (CGWB), nearly 56% of tube-wells have a low level of groundwater that has been declining sharply since 2003, due to groundwater pumping.

Kolkata is one of the cities in West Bengal, which suffers from issues of land subsidence and lack of drinking water because of the lowering of the groundwater level since 90's. Also, the areas near Behala (Southern part of Kolkata Municipal Corporation (KMC)) have been suffering from the lack or shortage of drinking water. Therefore, the inhabitants in such areas have to buy drinking water daily.

This chapter focuses on the analysis of the level of groundwater in Kolkata city. The study used groundwater Modeling based on geographical information systems (GIS).

10.2 METHOD AND MATERIALS

Kolkata is an old historic city with a total area of 187.33 km² in 2012 for KMC. After 2012, the area of KMC has increased to 205 km² due to the addition of two more wards, with a total population of nearly 4.5 million. Kolkata city is situated on the deltas of the Hooghly River, which is located on the lower alluvium plain of this river. It is bounded by the Hooghly River on the Western side, northern side by several municipalities, eastern side by several rural blocks of North 24 Parganas district, and southern side by Old Ganga and *Tolly Nullah* (Figure 10.1). The soil in the KMC is alluvial in nature, which varies from silty clay to clay.

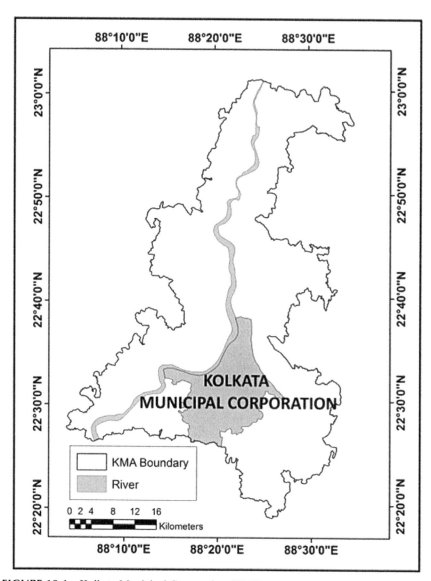

FIGURE 10.1   Kolkata Municipal Corporation (KMC).

The annual rainfall of KMC is 1647 mm. Major seasons in Kolkata are short winter, short spring, hot, and long rainy summer, which is followed by prolong monsoon. Range of temperature in winter season ranges from 10°C to 22.5°C, compared from 29.25°C to 40°C in summer months. The climate of KMC is humid subtropical.

To find out the mapping related to groundwater level in Kolkata city, data was obtained from the office of KMC, Groundwater Board of West Bengal State (GWWB), and National Groundwater Board (NGWB: New Delhi, India), etc. To scrutinize various issues related to the groundwater in Kolkata city, field survey was conducted among the various wards of Kolkata city. For the mapping purpose of profile of groundwater level in Kolkata city, various cartographical techniques were used. In Kolkata, the level of groundwater is measured mainly in two months in: June (for pre-monsoon season period) and October (for post-monsoon period).

10.3 ANALYSIS, RESULTS, AND DISCUSSION

After the partition and Independence of India, refugees from East Pakistan or Bangladesh came to Kolkata and settled down here causing huge increase in population Kolkata city. Huge pressure of population creates high demand on existing land resources including groundwater resources. Therefore, it has adverse impact on local land eco-systems. Due to the unscientific use of groundwater resources, groundwater is depleting rapidly.

Figure 10.2 shows the network stations in different areas of KMC. Careful review of these network stations indicates that the number of network stations in the eastern part of KMC is significantly lower than the other regions of KMC.

Kolkata city is located on the moribund deltaic plains of the *Bhagirathi* River. This land is in the mature delta of this river. The elevation of this city ranges from 3.5 to 6 m above mean sea level (MSL). Various low-lying marshes and wetlands are major characteristics of this region. The general gradient of the land is mainly towards the southern direction from the city core because of the Bay of Bengal. The young levee, deltaic plain, older alluvium layer on both sides of the old Ganga river are major physical characteristics of this area, which is mainly covered with younger alluvium ranging from general silty to clayey soils.

KMC is located on the 760 m of thick alluvial soils, which cover a huge layer of semi-consolidated to consolidated materials beneath the surface layer. Different features like pulses of sedimentation, marine regression, and transgression and the upliftment from the Cretaceous to Pleistocene times are major identical geographical features of this area. The borehole data of the Kolkata region shows that the subsurface layer at 300 m depth from the ground surface is from the Quaternary time (Figure 10.3).

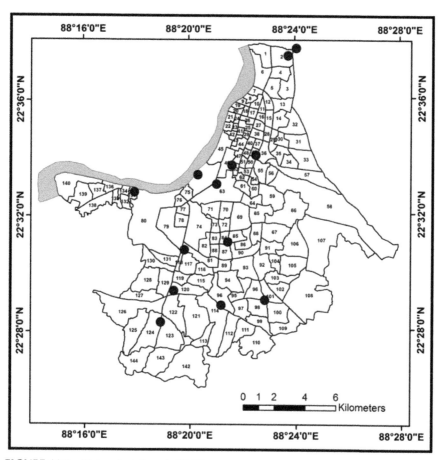

FIGURE 10.2   Network stations in KMC.

However, there are no clear-cut demarcations of the sediments between the tertiary sediments and quaternary sediments due to a lack of reliable scientific data. The alluvium in this area mainly comprises of sand, silt, and clay.

Figure 10.4 shows the location of the subsurface aquifers in the different parts of KMC. Out of these, 12 subsurface aquifers areas were selected in this research study. Most of the stations are in the southern region of the KMC.

The lithological data from various reports of KMC indicated two thick layers of clay beds in KMC. The depth of those layers is within 400 MBGL (meters below ground level) (Figure 10.5).

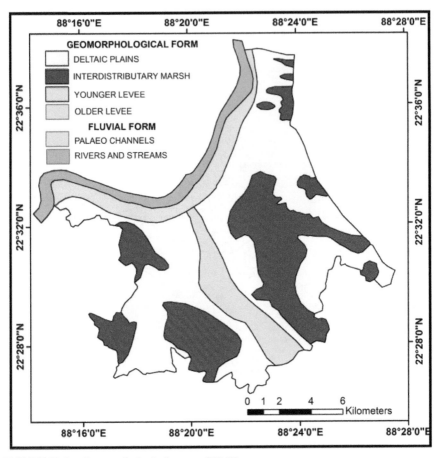

FIGURE 10.3 Geomorphological setup of KMC.

The depth of the subsurface layers consisting of basal clay is not the same throughout the region in KMC (Figure 10.4). Generally, it ranges from the 300 to 450 MBGL with land slope mainly southward. The uppermost layer of the bed ranges from 15 to 60 m; and is located above the subsurface alluvial layer of Kolkata city, which varies from place to place. Uppermost layer and subsurface layer are dark grey in color with plastic-like materials ranging from fine silt to coarse sand. The topmost layer of the clayey-bed consists of peaty materials that has been clearly found at 10 MBGL depths or where land subsidence has taken place. Different types of sand can be found between the two clayey layers in the KMC area and these layers are located in the aquifer system of the consecutive beds (Figure 10.6).

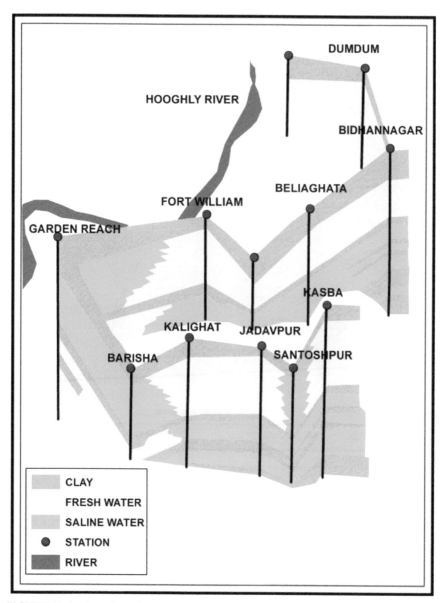

FIGURE 10.4   Location of 12 aquifers in the KMC Area.

The layer of the sand and silt layer is very thin in areas near Bollygunje, Tollygunj, Tiljola, Dhakuria, Kosba, Santoshpur, Garia, Behala, Barisha, and Thakurpukur. Most of the places in this area are marshy or swampy.

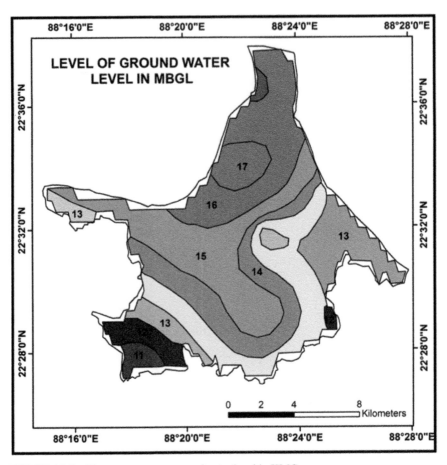

FIGURE 10.5   The pre-monsoon groundwater level in KMC.

The thickness of clayey beds in this area differs from place to place. In the riverbank of Hugli River, the lower layer is clay, but the uppermost layer contains sand particles ranging from fine to coarse grain.

Water in most of the parts of the in KMC area is brackish and groundwater occurs under confined to semi-unconfined situations. However, the water is relatively of good quality near Fort William, central part on the river bank, near Kalighat in the southern part of KMC, Kashipur in the northern part of Kolkata. Water in these areas is fresh in nature.

On the bank of the Hugli River, the ranges of aquifer are thin, which is located within 12 MBGL depths beneath the ground surface. The water is unconfined within the 17 m below ground level near Bollygunje, Tollygunj,

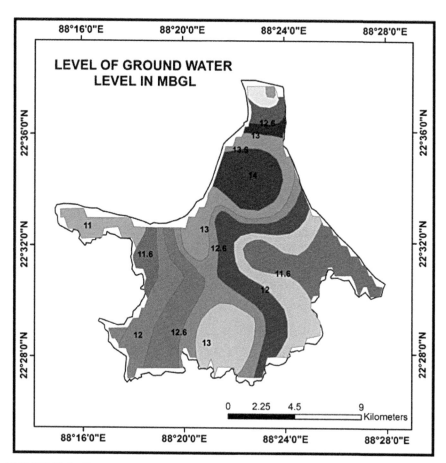

FIGURE 10.6 Post-monsoon groundwater Level in KMC.

Tiljola, Dhakuria, Kosba, Santoshpur, Garia, Behala, Barisha, and Thakur-pukur. In the pre-monsoon period, the groundwater level in KMC ranges from 11 to 17 MBGL compared from 11 to 15 MBGL in the post-monsoon period (Figures 10.5 and 10.6).

Subsurface water in Kolkata city area is mainly of two types, namely, Bicarbonate, and chloride types that are again subdivided into several categories based on the availability of cation-concentration (such as calcium-magnesium bicarbonate, sodium bicarbonate, calcium magnesium chloride, sodium chloride, etc.). Water in the south-central and southern part of KMC is bicarbonate in nature compared to chloride type in the eastern part of KMC.

Due to the presence of major groundwater trough in the Central part of KMC, gravity head recharge tube-wells have been used. In the KMC area, following suggestive measures can be considered:

- The borehole needs to be electrically logged to ascertain the location of freshwater zone because of varying nature of freshwater zone in KMC area. The layering of brackish or saline water aquifers are not in a similar pattern. Sometimes the zone of brackish water is joined with the saline water bodies. However, to identify the area, only the areas of freshwater were considered.
- The topmost layer of the groundwater recharge area is properly sealed with cement as it discourages the infiltration of the layer. For the preservation of groundwater, the water consisting of brackish water and saline water must be separated from each other with the freshwater by proper cement-scaling. For securitizing the water level beneath the ground, proper monitoring should be done. A combination of conservation measures (i.e., from saline water to fresh water) is very helpful.

10.3.1 GROUNDWATER-RELATED ISSUES AND PROBLEMS

Except for the western part of KMC, the freshwater aquifers are located within 50 to 170 MBGL depths. In the western part of KMC near the areas of Garden Reach, Behala area, and the northern part of KMC (i.e., Dum Dum area), the aquifer depth ranges from 170 to 210 MBGL, respectively. In the southern part of the KMC (near Santoshpur area), the depth of aquifer is within the 300 MBGL. Piezometric surface indicates that the depth of the layer is high in the central area of the KMC and few areas near Sealdah, Fort William, etc., due to excessive withdrawal or unscientific upliftment of groundwater.

The long-term trend of groundwater of two time periods (namely: pre-monsoon and post-monsoon) indicates that groundwater is depleting very rapidly. The trend of groundwater in KMC during 1958 to 2003 shows that the level of groundwater has been falling from 7 to 11 m within this period. Unscientific exploitation of subsurface water increases the gap in the lower level surfaces in the city. Therefore, groundwater profile follows a bowl-like feature in Kolkata city, indicating a major land subsidence in future. Figure 10.7 indicates the region for artificial recharge using rainwater.

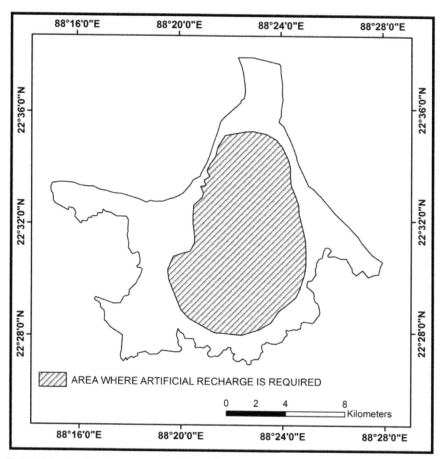

FIGURE 10.7   Land areas required for artificial recharge in KMC.

Groundwater in the west of KMC (such as B.B.D. Bag area, Dalhousie, Park Street, Garia) is of bicarbonate type compared to chloride type in the eastern part of KMC.

The subsurface water in the western and south-central parts of the city and southern part of the KMC (Ballygunge, Taltala area, Kosba area, and Santoshpur) is in the North-Northwest to south-southeast direction.

Concentration of chloride is low in KMC area. Groundwater in the northern and eastern parts of the KMC area contains Calcium, Magnesium, Chloride type; and concentration of chloride is high (280 to 620 mg/liters). In the eastern part of KMC (Tangra, Toposia, Tiljala), the aquifer depth is below 20 MBGL, because of the presence of leather industries and industrial effluents from tanneries that cause serious environmental hazards due to excessive water pollution in the bheries. Some places in KMC are arsenic-prone. Therefore, KMC has fixed a permissible limit for the upliftment of water.

10.4 SUMMARY

Due to the huge upliftment of groundwater, its profile in the central part of Kolkata city has now changed into a bowl-like feature. The field visits indicate the good scope of rainwater harvesting in the KMC area. The analysis indicated 237,000 m³ utilization of rainwater. Due to the lowering of groundwater in the KMC area, groundwater has been totally depleted. To eradicate this problem, the municipal authority has linked all areas with the piped water supply. Regular monitoring of groundwater both in the pre-monsoon and post-monsoon periods was initiated to observe the trend of the piezometric surface, because unscientific withdrawal of groundwater by tube-wells in the arsenic prone areas. Proper monitoring of the tube-wells will prevent or reduce the pollution of groundwater. The maps from this study will be helpful in the future study to take suggestive measures for the development and management of water supply in Kolkata city.

KEYWORDS

- geographical information system
- groundwater
- land subsidence
- mean sea level
- sustainable development
- unscientific exploitation

REFERENCES

1. Bouwer, H., (1977). Land subsidence and cracking due to groundwater depletion. *Groundwater, 15*(5), 358–364.
2. Mukherji, A., & Shah, T., (2005). Groundwater socio-ecology and governance: A review of institution and policies in selected countries. *Hydrogeology Journal, 13*, 328–345.

3. Scott, C., & Shah, T., (2004). Groundwater overdraft reduction through agricultural energy policy: Insight from India and Mexico. *International Journal of Water Resources Development, 20*, 149–164.

4. Shah, T., Singh, O. P., & Mukherjee, A., (2006). Some aspects of South Asia's groundwater irrigation economy: Analysis from a survey in India, Pakistan, Nepal Terai and Bangladesh. *Hydrogeology Journal, 14*, 286–309.

5. Shankar, P. V. S., Kulkarni, H., & Krishan, S., (2011). India's groundwater challenge and the way forward. *Economic and Political Weekly, 46,* 37–45.

6. Vyas, J. N., (2001). Water and energy for development in Gujarat with special focus on Sardar Sarovar project. *International Journal of Water Resources Development, 17,* 37–45.

CHAPTER 11

DISPOSAL OF EFFLUENTS FROM PULP AND PAPER INDUSTRIES THROUGH IRRIGATION

AJAY BHARTI and PANKAJ K. PANDEY

ABSTRACT

The effluents from pulp and paper industries can be utilized for lignin products, protein products, ethanol, and wood molasses. After utilizing the effluent in the form of any of these products, the pollution load in the effluent would be decreased and could be used for irrigation. Soil type and crop requirement should be considered, and the effluent should be treated if required. In assessing effluent quality criteria, it is imperative that the nature of soil and groundwater should be considered. It is not conceivable to cover every single nearby circumstance while planning water quality criteria and the approach ought to be to display rules that anxiety the administration expected to utilize the water of a specific quality. The correct decision must be made at the planning stage and while assessing specific conditions.

11.1 INTRODUCTION

Paper is one of the essential commodities in the modern age. Per capita consumption of paper is considered as a yardstick of the development of the country. The estimated manufactured paper production in India was estimated at 8.0 million tons. This generates an annual turnover of Rs. 150 billion (Rs. 60.00 = 1.00 US$). The estimated paper utilization in India was about 12 million metric tons (MT) annually [5]. Paper and paperboard production and consumption during 2009 to 2013 are shown in Tables 11.1 and 11.2, respectively. This indicates that per capita paper consumption in India is very low. The water use varies from 250 to 400 m^3 per MT of paper produced of which 90% is discharged as effluents [2, 7].

TABLE 11.1 Paper and Paperboard Production (1000 MT)

Country	Year (1000 MT)				
	2009	2010	2011	2012	2013
World	3,70,626	3,94,562	4,00,571	3,99,117	3,97,611
Asia	1,58,490	1,70,125	1,77,102	1,80,808	1,80,038
Austria	4606	5009	4901	5004	4837
Bangladesh	58	58	58	58	58
Canada	12,823	12,755	12,057	10,756	11,133
China	90,192	96,545	1,03,226	1,06,569	1,05,150
Europe	1,00,986	1,06,273	1,07,452	1,05,354	1,04,543
Finland	10,602	11,758	11,329	10,592	10,694
India	*7789*	*10,111*	*10,172*	*10,247*	*10,247*
Japan	26,268	27,364	26,609	25,957	26,093
Korea Rep	9726	11,022	11,368	11,330	11,801
Pakistan	1079	1079	1079	1079	1079
Sweden	10,932	11,410	11,298	11,417	10,782
UK	4293	4300	4342	4292	4600
USA	71,355	77,689	76,431	74,492	74,228

TABLE 11.2 Paper and Paperboard Consumption (1000 MT: MT = metric ton = 1000 kg)

Country	Consumption, 1000 MT					MT per 1000 Capita
	2009	2010	2011	2012	2013	2013
World	3,68,567	3,93,772	4,00,877	3,97,929	3,95,224	55
Asia	1,64,744	1,78,766	1,87,366	1,90,233	1,87,632	44
Austria	1951	2300	2281	2276	2247	264
Bangladesh	459	437	481	483	532	3
Canada	5945	5938	5449	5494	5536	157
China	90,303	96,901	1,02,846	1,05,724	1,03,129	73
Europe	90,967	94,603	96,201	93,395	92,159	124
Finland	1366	1386	1355	1273	1178	217
India	*8960*	*11,611*	*12,074*	*12,123*	*11,983*	*10*
Japan	27,142	27,789	27,890	27,473	27,109	213
Korea Rep	7547	9085	9240	8948	9433	191
Pakistan	1520	1480	1504	1448	1456	8
Sweden	1932	2215	1700	2328	1528	160

TABLE 11.2   *(Continued)*

Country	Consumption, 1000 MT					MT per 1000 Capita
	2009	2010	2011	2012	2013	2013
UK	10,416	10,614	10,247	9251	9385	148
USA	72,498	77,328	74,272	71,601	71,880	225

Hence, a total of 2.31×10^9 m³ to 3.69×10^9 m³ of water is discharged from different pulp and paper industry in India. This huge amount of water should not be wasted in an agricultural-based country like India. This water can irrigate 4.62×10^5 to 7.38×10^5 ha of land to a depth of 50 cm. The cost of treatment of effluent is most likely to be of the same order of magnitude as the production cost itself [7]. Therefore, the effluent should be treated only to such an extent to satisfy the soil type and crop requirement.

This chapter focuses on the disposal of effluents from pulp and paper industries through irrigation.

11.2   UTILIZATION OF EFFLUENT FROM PULP AND PAPER INDUSTRY

Wood is one of the most important sources of carbohydrate material that is potentially available on a worldwide basis [12]. Hence, the industry using forest products is the largest producer of carbohydrate material, and the waste liquors from the wood pulping industry will, if not evaporated and burnt, impose a heavy load upon the environment. The waste from pulp and paper industry can be utilized as raw material for different by-products like lignin-products, protein-products ethanol, and wood-molasses, etc. The effluent from pulp and paper industries can be regarded as a valuable raw material because it contains dissolved organic material representing about 50% of the wood.

Different by-products (like lignin sulfonates (LS), protein-products, ethanol, and wood-molasses) may be produced from chemical pulp mills without a recovery system, reducing the BOD_5 load considerably.

11.2.1   LIGNIN PRODUCTS

In chemical pulping, the fibers are liberated by degradation and dissolution of lignin by chemical reactions. About 50% of the wood is dissolved in the spent liquor. About 50% of the total solids in the spent liquor are lignin, which is the main polymeric material. Lignin is built up of phenyl-propane units which are linked together by different chemical bonds forming a three-dimensional network. In spent liquor, the molecular weight range of lignin is comprehensive, about 5000–50,000 [11]. Lignin is mainly utilized in binders and dispersants. It is also used in emulsifying, wetting, sequestering, and precipitating agents. Lignin products can even replace phenol-based and amino-based resins, glues, and thermo-plastics and foam plastics. The lignin-based adhesive is being used in plywood manufacture.

11.2.2 PROCESSES FOR THE PRODUCTION OF LIGNIN SULFONATES (LS)

The formation of lignin sulfonate may be achieved by the concentration of alcohol (0.50–0.55 of solids) formed during the sulfite pulping process. They are then dealt with as molasses or shower-dried. Ammonium calcium, magnesium, or sodium sulfonates can be delivered. However, the ammonium LS is likely the most imperative, being great wellsprings of both rough protein and vitality [8]. The synthesis of ammonium LS (Table 11.3), relies upon the types of wood and the pulping procedure, is as per the following:

TABLE 11.3 Composition and the Dry Mass of Ammonium LS

Ammonium LS Composition	Approximate Dry Mass (%)
Acetic acid	8
Inorganic	8
Lignin	45
Non-protein nitrogen	18
Other organics	9
Sugars	18

Source: FAO Animal Production and Health Division, No. 4; http://www.fao.org/agriculture/animal-production-and-health/en/ (accessed 30 June of 2019).

The sugar and volatile organic acids (VOA) in the spent sulfite liquor are fermented with yeast. The yeast is separated, and the spent liquor is evaporated, and spray dried to form lignin powder (Figure 11.1). The product contains about 75% sodium lignosulfonates, 5% carbohydrate materials, and 20% inorganics [1]. The annual lignin production in the pulp and paper industry globally is below 2%, however, the annual production of lignin sulfate (LS) as reported around 2 million MT [13].

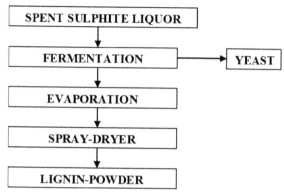

FIGURE 11.1   Production of lignin sulfonates.

11.2.3   PROTEIN PRODUCTS

Protein can be produced by aerobic fermentation of dissolved carbohydrate material in spent sulfite liquor by the Pekilo process. The spent sulfite liquor has been utilized for a long time as a substrate to produce yeast [9]. Plants for producing protein are commercially available. Most of the produced yeast is used as animal feed. The protein produced is mechanically dewatered to dryness of 30–40% and then dried to 90%. The protein content of Pekilo yeast is about 55–60% [6]. The world production of yeast from sulfite spent liquor has been estimated to be about 1 million tons per year [4].

11.2.4   ETHANOL

Ethanol can be prepared by anaerobic decomposition of the sugars of the spent sulfite liquor. The sugar quantity arriving with the spent liquor at the alcohol plant depends on the cooking conditions and is usually about 20–30% of the total solids content. About 70–80% of the sugars in the spent sulfite, liquor is fermentable by Saccharomyces yeast. The stoichiometric yield of alcohol from the hexoses according to the reaction is 51% (by weight).

$$C_6H_{12}O_6 \rightarrow 2C_2H_5OH + 2CO_2 \qquad (11.1)$$

The ethanol plants are commercially available. The production of ethanol from spent sulfite liquor has been estimated to be 1,00,000 tons per year in the Western world.

11.2.5   WOOD MOLASSES

Wood molasses is a term used to depict a gathering of items delivered by coordinate hydrolysis of timber, by concentrating sulfite alcohol, or from the waste effluents of molecule and fiber-board preparing [8]. The suspended solids from the white water from the pulp mills are filtered out, and the filtrate is evaporated and spray-dried to form wood molasses (Figure 11.2). The wood molasses contains about 97% organic material, mostly carbohydrates, and is widely used as animal food.

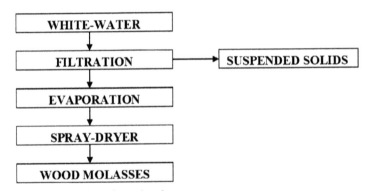

FIGURE 11.2   Production of wood molasses.

11.3   USE OF EFFLUENT FROM PULP AND PAPER INDUSTRY FOR IRRIGATION

Traditionally, irrigation has been used to improve crop yield, but in more recent years there has been a growing interest in irrigation as a method of wastewater (WW) disposal. The soil and the vegetation are used as a treatment system either to prevent WW from reaching the recipient or to remove organic pollutants by soil filtration and microbiological decomposition before entry into the recipient.

The land application systems are irrigation, infiltration-percolation, and overland flow or spray run-off system. Irrigation is the most common land application method. The effluent is discharged, by spraying or surface spreading, onto land to support plant growth. The WW is dissipated to plant uptake, to the air by evapotranspiration (ET), and to groundwater by percolation. The loading rate is 30–50 m³/ha-day for agricultural land and up to 150 m³/ha-day for grassland [10]. In the infiltration-percolation system, the effluent is applied to land principally by the same technique as in the irrigation system, but at a higher hydraulic loading rate. A smaller surface is thus required for the same effluent volume. Systems with hydraulic loading rates up to 400 m³/ha-day [10] are typically considered to be the infiltration-percolation type. The crop normally has a lower value and is chosen primarily for its water tolerance or tolerance to effluent constituents.

The quality requirements for irrigation water depend on site characteristics, loading rates, soil, crops, etc. The following parameters must always be considered before starting irrigation with WW: suspended solids (especially fibers), salinity, organic substances, nutrients, inorganic ions, sodium adsorption ratio, temperature, and bacteriological quality. The permissible limit of trace elements in irrigation water [2] is depicted in Table 11.4.

TABLE 11.4   Tolerable Limits ($mg\ l^{-1}$) for Elements in Reclaimed Water for Irrigation

Element	Long-Term Application ($mg\ l^{-1}$)	Short-Term Application ($mg\ l^{-1}$)
Al: Aluminum	5.00	20.00
As: Arsenic	0.10	2.00
B: Boron	0.75	2.00
Be: Beryllium	0.10	0.50
Cd: Cadmium	0.01	0.05
Ch: Chromium	0.10	1.00
Co: Cobalt	0.05	5.00
Cu: Copper	0.20	5.00
F: Fluoride	1.00	15.00
Fe: Iron	5.00	20.00
Li: Lithium	2.50	2.50
Mg: Manganese	0.20	10.00
Mo: Molybdenum	0.01	0.05
Ni: Nickel	0.20	2.00
Pb: Lead	5.00	10.00
Se: Selenium	0.02	0.02

TABLE 11.4 *(Continued)*

Element	Long-Term Application (mg l⁻¹)	Short-Term Application (mg l⁻¹)
V: Vanadium	0.10	1.00
Zn: Zinc	2.00	10.00

Source: Pratt [14].

Water Quality for Agriculture. Chapter 5; FAO Irrigation and Drainage Report 29; 1994; online; public document. http://www.fao.org/3/T0234E/T0234E06.htm.

Local climatic conditions will affect the design specification of the system including storage requirement and loading rates. Generally, the system operates satisfactorily in dry climates and dry weather, and freezing conditions will limit the irrigation period or reduce the capacity. Pollutants in the irrigation water are reduced by adsorption and by microbiological decomposition in the soil. Normal removal efficiencies in irrigation with WW are given in Table 11.5. The removal efficiency is somewhat lower for high load infiltration-percolation systems. Typical reductions of BOD, suspended solids, and nitrogen are 85–99%, 60–95%, and 0–50%, respectively [10].

TABLE 11.5 Removal Efficiency of Different Pollutant Parameters for Irrigation and High-Load Infiltration-Percolation

Parameter	Irrigation (% Removal)	High-Load Infiltration-Percolation (% Removal)
BOD	95–99	85–99
Nitrogen	70–90	0–50
Phosphorus	80–99	—
Suspended solids	99	60–95
Trace metals	95–99	—
Virus and bacteria	99	—

11.4 SUMMARY

Traditionally, irrigation has been used to improve crop yield, but in presently there has been a growing interest in irrigation as a method of WW disposal. The soil and the vegetation are used as a treatment system either to prevent WW from the reaching the recipient or to remove organic pollutants by soil filtration and microbiological decomposition before entry into the recipient. The water consumption in producing one tonne of paper varies from 250 to 400 m³. About 90% of this consumed water is discharged as effluent. Hence, an enormous amount of water is being discharged from the different pulp and paper industry in India. This water should not be wasted in an agricultural country like India. It can irrigate several thousand hectares of land to sufficient depth. The standard for discharge of environmental pollutants indicates that the extent of pollutants to be discharged onto the ground is far higher than that of the disposal into the water bodies. As the cost of treatment of effluent is most likely to be of the same order of magnitude as the production cost itself, the effluents need to be treated only up to such extent to satisfy the crop requirement and soil type. In this chapter, an attempt has been made for pulp and paper industries effluent disposal through irrigation.

KEYWORDS

- effluent disposal
- lignin
- paper consumption
- pollutants
- volatile organic acids
- wastewater disposal

REFERENCES

1. Belgacem, M. N., & Gandini, A., (2008). *Monomers, Polymers and Composites from Renewable Resources* (1ˢᵗ edn., p. 560). Oxford: Elsevier.
2. Charles, A. J., (2009). *Groundwater Economics* (pp. 173–229). Boca Raton, FL: CRC Press, Inc.
3. FAO, (2015). *Forest Products Yearbook, 2009–2013* [Internet]. Rome (Italy); Available from: http://www.fao.org/forestry/statistics/80570/en/ (accessed on 20 June 2020).
4. Ghosh, U. K., (1999). *Personal Communication.* Roorkee, India: Institute of Pulp and Paper Technology; Saharanpur Campus, IIT.
5. Johnson, T., Johnson, B., Mukherjee, K., & Hall, A., (2011). India: An emerging giant in the pulp and paper industry (online). In: *65ᵗʰAppita Annual Conference and Exhibition, Rotorua New Zealand: Conference Technical Papers* (pp. 135–139). Carlton, AU: Appita Inc. (Accessed on 22 June 2020).

6. Marx, J. L., (1989). *A Revolution in Biotechnology* (p. 227). Cambridge, UK: Cambridge University Press.
7. Pandey, G. N., & Carney, G. C., (1989). *Environmental Engineering* (p. 436). New Delhi: Tata McGraw-Hill Publishing Company Limited.
8. Pigden, W. J., (1977). Nutritional and economic aspects of utilizing wood-processing by-products. *FAO Animal Production and Health Division, 4,* 211–226.
9. Romantschuk, H., (1974). Feeding Cattle at the pulp mill. *Unasylva, 1974,* 15–17.
10. UNEP (United Nations Environment Programme), (1996). *Wastewater Treatment* (Vol. 2, p. 131). Industry and environment manual series: Environment management in pulp and paper industry, Rome: United Nations Publication; Technical Report 34.
11. Vishtal, A. G., & Kraslawski, A., (2011). Challenges in industrial applications of technical lignins. *Bioresources, 6*(3), 3547–3568.
12. Wiley, A. J., Harris, J. F., Saeman, J. F., & Locke, E. G., (1955). Wood industries as a source of carbohydrates. *Indus. and Eng. Chem., 47*(7), 1397–1405.
13. Will, R., & Yokose, K., (2005). *Chemical Economics Handbook.* Product review: Lignosulfonates. Chemical Industry Newsletter; London; https://ihsmarkit.com/products/lignosulfonates-chemical-economics-handbook.html (accessed on 20 June 2020).
14. Pratt P. F., (1972). *Quality Criteria for Trace Elements in Irrigation Waters* (p. 46). California Agricultural Experiment Station Bulletin; Davis, CA: University of California.

Part IV:
Case Studies

CORRELATION ANALYSIS OF MUNICIPAL SEWAGE DISCHARGE IN RIVER GANGES (VARANASI, INDIA)

GARIMA JHARIYA, DEVENDRA MOHAN, and RAM MANDIR SINGH

ABSTRACT

Quality of water is a major issue so that water can be used for its designated purpose. The extant of pollutants in treated wastewater (WW) relies on the class of water supply, nature of squanders enhanced through use, and the level of treatment of WW. Contaminants in terms of BOD, COD, TDS, suspended solids are of greater significance for the Central Pollution Control Board-India, therefore the STPs are designed to monitor and report the water quality data of treated effluent. The actual composition of WW may differ from community to community and in many cases, the water quality standards for reclaimed WW are like drinking water standards but cannot be utilized for drinking purpose and it is only useful for agricultural and some domestic tasks (flushing, washing of the car, etc.). This research study is based on logical technique so that all analyzed samples for physicochemical properties of municipal WW utilized for irrigation and discharge into river Ganges are within desirable limits that are set by various agencies. The correlation technique helps to know the extent of inter-relationship among variables. Linear relationships were quantified among parameters and were acceptable if the correlation coefficients (CCs) are significant.

12.1 INTRODUCTION

Being one of the main constituents of the environment, water is a crucial element to sustain a high quality of human life and for social and economic development. However, this essential resource is under threat since the demographic growth, urbanization, extensive agriculture, enhanced industrial activities have led to an unprecedented increase in water demand, not only for domestic use but also for agricultural use. In developed countries like India, most of the water resources (96%) are utilized for agriculture purposes, and 3% for domestic and 1% for industrial use [6].

The industrialization, urbanization, and population growth, which our world has faced during recent years, has caused an increase in ecological contamination thus affecting the water quality. In this chapter, water quality describes synthetic, physical, and natural qualities of water for a specific use. Water quality is affected by anthropogenic activities. Natural processes (hydrological, physical, chemical, and biological) may affect the characteristics and concentration of chemicals and compounds in freshwater. Out of the total water available on this globe, only 0.16% is suitable for individual use and the remaining is contaminated due to various environmental factors [17].

Because freshwater availability is a major concern today, therefore reuse of reclaimed wastewater (WW) can be implemented as a key move toward irrigation agriculture to ensure sustainable growth and to reduce water famine in water-scarce regions. WW salvages possibly will involve further treatment of the waste matter to meet several quality criteria. Increasing demand and untreated WW discharge aggravate stress on water bodies, therefore keeping the desired characteristics of water are most important. To maintain the required quality of water, there is a need to prioritize the contaminants in water. Also, there is a consequence of chemical and biological measurements to safeguard the quality of water to avoid health hazards.

This research study was conducted in Bhagwanpur sewage treatment plant (STP) in Varanasi, to break down water quality parameters of city WW release from different sources. The correlation coefficients (CCs) and relationships were established between the pairs of parameters to select important variables for future study.

12.2 CHARACTERISTICS OF WASTEWATER (WW)

Water may be used in a community for a variety of beneficial uses. Depending upon the use and impurities added during this use, the WW acquires different characteristics that can be divided into two categories-industrial WW and domestic (municipal) WW. Typically, domestic sewage from various sources consists of 99.9% water and 0.1% natural and inorganic compounds in different forms [4, 19, 26]. In most of the research, the water quality standards for reclaimed WW are found similar as given for drinking water standards setup by different organizations [7], but its concentrations in sewage effluents vary with space and time [21].

The domestic WW is therefore unstable, decomposable, and unsafe due to presence of potential pathogenic organisms. The industrial WW varies widely in their characteristics depending on the nature of industry and various operations within the industry. The effluents from industries contain toxic substances especially heavy metals. The presence of heavy metals in the environment is of major concern because of their toxicity, bioaccumulating tendency, a threat to human life, and the environment [24].

12.3 MATERIALS AND METHODS

The current research was accomplished in Bhagwanpur village, which is situated in the southern part of Varanasi. Varanasi is an ancient religious city, famous for its cultural heritage, music, art, craft, and education; and lies on the bank of holy river Ganges, one of the largest rivers of the world. Being an important pilgrimage center, thousands of tourists visit the holy city Varanasi from India and abroad. Varanasi is a major commercial and industrial center of eastern U.P. situated in the north-eastern part of India at 25°0' to 25°16' N latitude and 82°5' to 83°1'E longitude.

The region of Bhagwanpur has a well-established sewage treatment plant (STP) unit to treat 8 MLD (million liters per day) sewage daily using Activated Sludge Process. Municipal sewage after treatment in this region is being utilized for irrigation or discharged into the river system; therefore, there is a need for water quality analysis.

12.3.1 SAMPLING OF TREATED MUNICIPAL WASTEWATER (TWW) FOR PHYSICOCHEMICAL CHARACTERISTICS

Secondary treated municipal sewage (STMS) samples were collected in the morning (9.00 AM to 10.00 AM) in acid-washed plastic bottles from STP. Various physicochemical parameters (like temperature, pH, EC, TDS, BOD, COD, DO, alkalinity, chloride, hardness, phosphate, sulfates, nitrates) in the water samples were determined using standard methods by APHA [3]. The correlation was established among all these parameters. The general procedure was followed for water quality analysis (Figure 12.1).

12.4   RESULTS AND DISCUSSION

The bridge of analysis of data from municipal sewage is being utilized for irrigation and discharge into river Ganges, Varanasi district of Uttar Pradesh (Table 12.1).

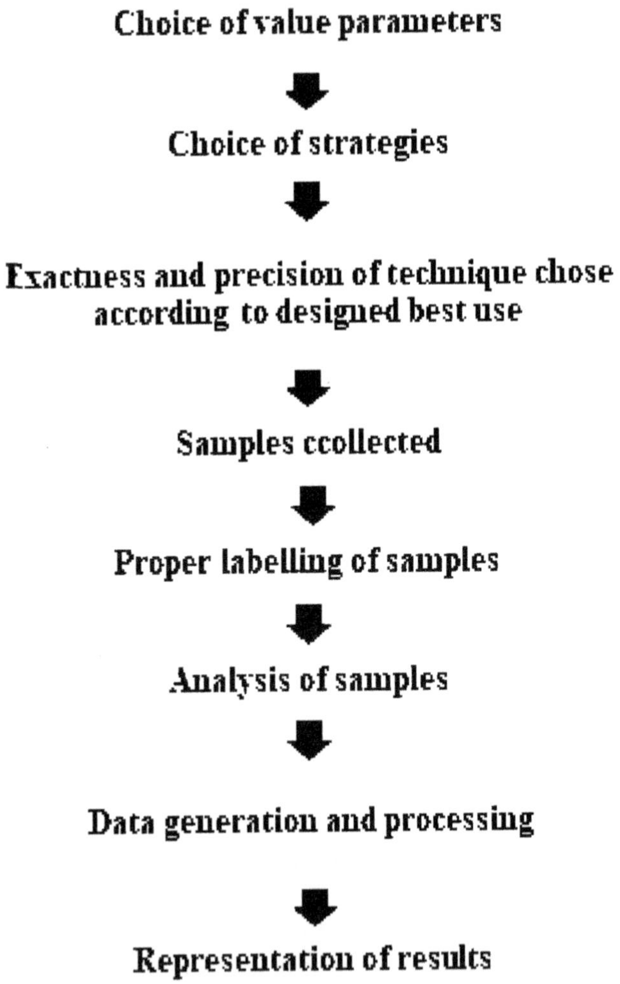

Choice of value parameters

Choice of strategies

Exactness and precision of technique chose according to designed best use

Samples ccollected

Proper labelling of samples

Analysis of samples

Data generation and processing

Representation of results

FIGURE 12.1   General procedure to follow in water quality analysis.

TABLE 12.1 Physicochemical Properties of Secondary Treated Municipal Sewage Samples from Bhawanpur Sewage Treatment Plant

Parameter	Disposal Standards for Quality of Treated Wastewater by CPCB	Mean ± SE	Range
BOD (mg/L)	100[a]	13.5 ± 0.8	8.7–18.6
Cl⁻ (mg/L)	600[a]	42.1 ± 2.6	22.0–55.3
COD (mg/L)	250[a]	67.8 ± 2.0	58.0–79.0
DO (mg/L)	4–6	4.42 ± 0.17	3.8–5.2
EC (dS/m)	2.25	0.713 ± 0.04	0.58–0.95
NO_3 (mg/L)	10[*]	7.70 ± 0.29	5.7–10.1
pH	5.5–9.0[abc]	7.26 ± 0.10	6.31–8.0
PO_4 (mg/L)	5[*]	5.77 ± 0.28	4.30–7.1
SO_4 (mg/L)	1000[abc]	43.8 ± 2.0	34.6–53.6
T. A. (mg/L)	600[d]	213.7 ± 15.2	149.6–285
T.H. (mg/L)	600[d]	139.3 ± 11.5	76–189
TDS (mg/L)	2100.00[abc]	309.83 ± 19.99	210–408
Temp. (°C)	40[b]	29.79 ± 1.69	19.90–38.40

[*] Into inland surface waters;

[a] On land for irrigation Indian Standards: 3307 (1974);

[b] Into inland surface waters Indian Standards: 2490 (1974);

[c] Into public sewers Indian Standards: 3306 (1974).

Permissible limits [*Source:* CPCB–2008, CPCB–20010/IS 10500–2012].

12.4.1 PHYSICAL AND CHEMICAL CHARACTERISTICS OF SEWAGE

12.4.1.1 MEASUREMENT OF TEMPERATURE, PH, AND EC

The rate of chemical activities depends on the physiological behavior and distribution of the organisms, which are controlled by temperature. In this study, water temperature ranged from 19.90°C to 38.40°C. The acidic or basic characteristic of water is determined by its pH of water. The average pH values were 7.26 ± 0.10. All samples for analysis of treated WW had pH values well-within the permissible limits. The electrical conductivity (EC) of water is the capacity to transmit electrical current and can evaluate the purity of water [16]. In the current study, EC values ranged from 0.58 to 0.95 dS/m. High conductivity implies an elevated level of pollution [10, 14, 20, 25] and the tropic levels of the aquatic body [1]. According to the analysis in this study, trends in the variation of temperature, pH, and EC are shown (Figure 12.2) throughout the year.

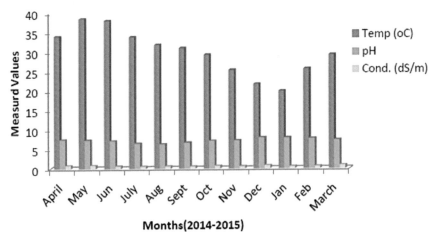

FIGURE 12.2   Trends in temperature (°C), pH, and EC (dS/m) of treated sewage effluents in Bhagwanpur STP.

12.4.1.2   MEASUREMENT OF TDS, ALKALINITY, AND HARDNESS

Total dissolved solids (TDS) were 309.83 ± 19.99 mg/L and it was 426.63 ± 23.15 mg/L for Municipal WWHaridwar district of Uttarakhand [23].

Alkalinity is the acid-neutralizing capacity of water to maintain its pH to a predetermined level. Deprivation of flora and fauna and organic waste add to carbonate and bicarbonate, consequently an increase in alkalinity value [5]. The average value of the alkalinity of WW samples in Bhagwanpur STP was 213.66 ± 15.21 mg/L.

The presence of calcium, magnesium, and chlorides in the domestic WW is the major contributor of hardness [15]. The hardness of WW ranged from 76 to 18,900 mg/L in the study area. This water is classified as hard and therefore is unsuitable for domestic and industrial uses as there is a possibility of scale formation in boilers and pipes. In this study, the variation in TDS, alkalinity, and hardness is shown in Figure 12.3. These mean values of parameters were within the permissible limits.

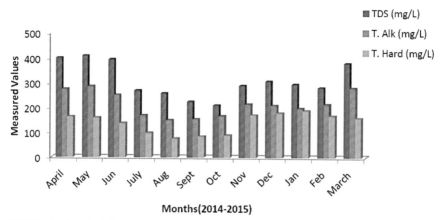

FIGURE 12.3   Variation of TDS, alkalinity, and hardness of treated sewage effluent.

12.4.1.3   MEASUREMENT OF ANIONS, NUTRIENTS PARAMETERS

In the STMS of Bhagwanpur STP, Cl⁻ ions on an average were 42.11 ± 2.57 mg/L. The average concentration of divalent ions SO_4^{2-}, NO_3^{2-}, and PO_4^{2-} were 43.84 ± 1.98, 7.70 ± 0.29, and 5.77 ± 0.28 mg/L, respectively. Among the analyzed parameters, the trend of concentrations of anions was in the order: $Cl^- > SO_4^{2-} > NO_3^{2-} > PO_4^{2-}$ in the effluent discharge from STP (Figure 12.4). Key concentration of Cl⁻ provides further support the basis for NaCl expulsion from domestic wastes [22].

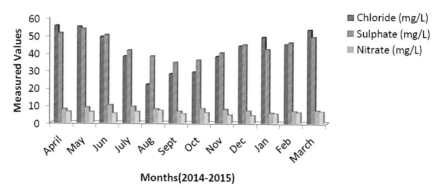

FIGURE 12.4   Variation of major anions and nutrients present in secondary treated municipal sewage.

The increased application of fertilizers, overflow from agricultural land, and addition of detergent wastes from textile industries, greatly attributed to nitrates, sulfate, and phosphorus available in various types of compounds of phosphates in the sewage effluents. The presence of nitrate and the forms of phosphorus in WW could be a good source of plant nutrients.

12.4.1.4   MEASUREMENT OF OXYGEN DEMAND PARAMETERS

Water quality can be assessed by the measurement of dissolved oxygen (DO) and is the main reason for the establishment of organic processes. The higher water temperature increases microbial activity [13] thus affecting DO concentration in water. Whereas testing of BOD indicates the efficiency of the water treatment system thus determining the strength of sewage. The pollution level of the aquatic environment is directly correlated with the increased concentration of BOD [11]. On the other hand, determination COD in water indirectly gives the extent of organic impurity in water and is a dependable parameter for judging the level of pollution in water [2].

In this chapter, average values were 4.42 ± 0.17 mg/L of DO, 8.7 to 18.6 mg/L of BOD, and 67.82 ± 1.97 mg/L of COD in WW. All water samples under study had DO, BOD, and COD concentrations within the prescribed limits (Figure 12.5).

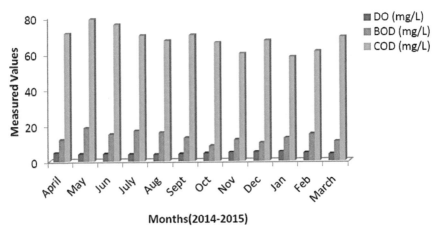

FIGURE 12.5   Recorded monthly variations in DO, BOD, and COD of treated wastewater.

12.4.2   CORRELATION AMONG PARAMETERS OF TREATED WASTEWATER (WW)

Understanding the quality of WW is essential because treated water can be reused for different purposes (domestic and irrigated agriculture) on the basis of examination of water quality. The correlation technique helps to know the extent of inter-relationships among variables with quantification of linear relationships, which can be used to select important variables for further study.

CC (r) among parameters (such as pH, temperature (°C), TDS, BOD, COD, etc., were determined to select the variables that are significant. The parameters selected for treated WW are applied subsequently for irrigation and discharge into waterway Ganges, Varanasi. The significance level of the relationship between any two of the parameters is dependent on the CCs (r), which are shown in Table 12.2 as a 13×13 correlation matrix. A higher value of CC (r) means the existence of a good relationship; and an increase or decrease of a parameter value determines the importance of such a relationship between the parameters [18].

In this study, the CCs (r) among each parameter-pair were computed using average values of the parameters as shown in Figures 12.2–12.5. The WW temperature showed a strong positive correlation between pH and NO_3^{2-} at 5%, pH, and PO_4^{3-} at 1%. Also, there was a strong negative correlation with DO at 5%; and between Cl⁻ and TDS at 1%. The pH of WW showed a strong negative correlation with alkalinity and chloride at 5%, and with DO at 1%.

The TDS showed a positive correlation with DO and a negative one with PO_4^{3-} at 1%. Alkalinity showed a positive correlation with hardness and chloride. Sulfate recorded a positive correlation with the hardness of WW. The COD and PO_4^{3-} were strongly negatively correlated with DO at 5%. The DO also recorded a negative correlation with BOD and positive with Cl⁻ at 1%. The PO_4^{3-} was strongly positive correlated with BOD. The COD of WW was positively correlated with NO_3^{2-} and PO_4^{3-} at 5%; and it was negatively correlated with Cl⁻ at 1%. The NO_3^{2-} was positively correlated with PO_4^{3-} at 5%.

12.5   SUMMARY

WW from industrial or domestic sources plays a vital role in the water (surface or groundwater) quality. In this study, understanding the quality of WW is a major concern. The correlation technique helps us to know the extent of inter-relationships among variables, which can be used to select significant variables for future study. The collected samples were mostly

TABLE 12.2 Pearson Correlation Coefficients for Various Parameters of Treated Wastewater from Bhagwanpur STP

	Temp	pH	EC	TDS	T.Alk	T. Hard	DO	BOD	COD	Cl⁻	SO_4^{2+}	NO_3^{2-}-N	PO_4^{3-}-P
Temp	1												
pH	0.693*	1											
EC	-0.049	-0.373	1										
TDS	-0.647*	-0.511	0.379	1									
T. Alk	-0.518	-0.904**	0.406	0.374	1								
T. Hard	0.168	-0.515	0.48	-0.118	0.693*	1							
DO	-0.822**	-0.599*	-0.141	0.578*	0.43	-0.241	1						
BOD	0.522	0.361	-0.362	-0.414	-0.29	0.079	-0.641*	1					
COD	0.868**	0.554	0.101	-0.461	-0.437	0.281	-0.826**	0.418	1				
Cl⁻	-0.699*	-0.911**	0.308	0.401	0.934**	0.494	0.611*	-0.417	-0.643*	1			
SO_4^{2-}	0.343	-0.365	0.523	-0.193	0.467	0.913**	-0.452	0.302	0.436	0.255	1		
NO_3^{2-}-N	0.706*	0.46	-0.196	-0.427	-0.458	0.04	-0.492	0.464	0.738**	-0.623*	0.272	1	
PO_4^{3-}-P	0.860**	0.612*	-0.156	0	-0.432	0.239	-0.919**	0.729**	0.837**	-0.640*	0.409	0.64*	1

Talk = Total alkalinity; T. Hard = Total hardness.

* Relationship is critical at the 0.05 level (2-tailed).

** Correlation is huge at the 0.01 level (2-tailed).

domestic WW near river Ganges that can be a source of water pollution. The results revealed that the physicochemical characteristics of the treated WW were in the range of permissible limits based on CPCB standards. The parameters like nitrate and phosphate can be a source of plant nutrients if WW is reused for irrigation. The status of the water nature of TSEs may secure the water assets and the people can be attentive about the water pollution status throughout the surrounding environment.

KEYWORDS

- biological oxygen demand
- chemical oxygen demand
- domestic waste
- electrical conductivity
- sewage treatment plant
- treated sewage effluent (TSE)
- wastewater

REFERENCES

1. Ahluwalia, A. A., (1999). *Limnological Study of Wetlands Under Sardar Sarovar Command Area* (p. 236). PhD Thesis; Gujarat University, Ahmedabad.
2. Amirkolaie, A. K., (2008). Environmental impact of nutrient discharged by aquaculture wastewater on the Haraz River. *J. Fish Aquat. Sci., 3*, 275–279.
3. APHA (American Public Health Association), (2017). *Standard Methods for the Examination of Water and Wastewater* (23rd edn., p. 1796). Washington, D.C.: American Public Health Association.
4. Brar, M. S., Mahli, S. S., & Singh, A. P., (2000). Irrigation effects on some potentially toxic trace elements in soil and potato plants in northwestern India. *Can. J. Soil Sci., 80*, 465–471.
5. Chaurasia, M., & Pandey, G. C., (2007). Study of physicochemical characteristic of some water ponds of Ayodhya-Faizabad. *Ind. J. Environ. Protect, 27*(11), 1019–1023.
6. Chavan, A., (1990). *Fundamentals of Environmental Science* (p. 56). Anmol Prakashan, New Delhi, India.
7. Cirelli, G. L., Consoli, S., & Di Grande, V., (2008). Long-term storage of reclaimed water: The case studies in Sicily (Italy). *Desalination, 218*(3), 62–73.
8. CPCB (Central Pollution Control Board), (2010). *Water Quality Trend, Monitoring of Indian Aquatic Resources* (p. 127). Series: MINARS/31/2009–2010; New Delhi: CPCB (Central Pollution Control Board) Ministry of Environment and Forest Ganga. https://

www.cpcb.nic.in/openpdffile.php?id=UmVwb3J0RmlsZXMvTmV3SXRlbV8xMjlfTl
dNUC0yMDA3LnBkZg= (accessed on 20 June 2020).

9. CPCB, (2008). *General Guidelines for Water Quality Monitoring* (p. 127). New Delhi: CPCB (*Central Pollution Control Board*) Ministry of Environment and Forest Ganga.

10. Das, P., & Tamminga, K. R., (2012). The Ganges and the gap: An assessment of efforts to clean a sacred river. *Sustainability, 4*, 1647–1668.

11. Gurumahum, S. D., Daimari, P., & Goswami, B. S., (2000). Physicochemical qualities of water and plankton of selected rivers in Meghalaya. *J. Inland Fisheries Society of India, 34*, 36–42.

12. IS (Bureau of Indian Standards), (2012). *Drinking Water Specifications* (p. 16). Drinking Water Sectional Committee, FAD 25; New Delhi: Bureau of Indian Standards.

13. Kataria, H. C., Singh, A., & Pandey, S. C., (2006). Studies on water quality of Dahod Dam, India. *Poll. Res., 25*(3), 553–556.

14. Mona, R., & Shuchi, M., (2012). Analysis of various physicochemical parameters for the water quality assessment of central region. *Asian J. Eng. Manage., 1*(1), 4–8.

15. Murali, M., & Satyanarayana, T., (2001). Study on source of water pollution at Machilipatnam, AP. *Poll. Res., 20*(2), 471–473.

16. Murugesan, A., Ramu, A., & Kannan, N., (2006). Water quality assessment from Uthamapalayam municipality in Theni District, Tamil Nadu, India. *Poll. Res., 25*(1), 163–166.

17. Mustaqeem, M. S., & Usmani, G. A., (2010). Evaluation of trends in chemical and physical properties of ground water and its modeling: A correlation regression study. *Rasayan J. Chem., 3*(2), 236–239.

18. Patil, V. T., & Patil, P. R., (2010). Physicochemical analysis of selected groundwater samples of Amalner town in Jalgaon district Maharashtra, India. *Electronic Journal of Chemistry, 7*(1), 111–116.

19. Pescod, M., (1992). *Wastewater Treatment and Use in Agriculture* (p. 167). FAO Irrigation and Drainage Paper 47; Rome: FAO.

20. Pradeep, V., Deepika, C., Gupta, U., & Hitesh, S., (2012). Water quality analysis of an organically polluted lake by investigating different physical and chemical parameters. *International J. Research in Chemistry and Environment, 2*(1), 105–111.

21. Rattan, R. K., Datta, S. P., Chandra, S., & Sahran, N., (2002). Heavy metals and environmental quality: Indian scenario. *Fertile News, 47*(11), 21–40.

22. Sharma, P., Meher, P. K., Kumar, A., Gautam, Y. P., & Mishra, K. P., (2014). Changes in water quality index of Ganges River at different locations in Allahabad. *Sustainability of Water Quality and Ecology, 3*(4), 67–76.

23. Shirin, S., & Yadav, A. K., (2014). Physicochemical analysis of municipal wastewater discharge in Ganga River, Haridwar District of Uttarakhand, India. *Current World Environment, 9*(2), 536–543.

24. Singh, A., Sharma, R. K., Agrawal, M., & Marshall, F. M., (2010). Risk assessment of heavy metal toxicity through contaminated 124 vegetables from wastewater irrigated area of Varanasi, India. *International Society for Tropical Ecology, 51*, 375–387.

25. World Health Organization (WHO), (2011). *Guidelines for Drinking Water Quality* (4th edn.).

26. Yadav, R. K., Goyal, B., Sharma, R. K., Dubey, S. K., & Minhas, P. S., (2002). Post-irrigation impact of domestic sewage effluent on composition of soils, crops, and groundwater: A case study. *Environment International, 28*, 481–486.

CHAPTER 13

MAPPING OF NORMALIZED DIFFERENCE DISPERSAL INDEX FOR GROUNDWATER QUALITY STUDY ON PARAMETER-BASED INDEX FOR IRRIGATION: KANCHIPURAM DISTRICT, INDIA

KISHAN S. RAWAT, VINOD KUMAR TRIPATHI, SUDHIR K. SINGH, and SUSHIL K. SHUKLA

ABSTRACT

The present study focuses on the suitability of irrigation water using indices based on hydro-geochemical parameters of groundwater from 44 fixed bore-wells in Kanchipuram district, Tamil Nadu, during pre-monsoon (May 2011) and post-monsoon (January 2012) seasons. The normalized difference dispersal index (NDDI) method was used for mapping of pre- and post-monsoon groundwater quality at this location. NDDI maps of total dissolved solids (TDS), sodium absorption ratio (SAR), percent sodium (%Na)/sodium hazard, residual sodium carbonate (RSC)/residual alkalinity (RA) ratio and magnesium hazard (MH)/magnesium absorption ratio (MAR)) were generated by integrated geochemical analyses with software *Simplex numerica*. Maximum NDDI value enrichment is shown by RSC (1.65), which shows major accretion/major attrition; while $NDDI_{TDS}$, $NDDI_{SAR}$, and $NDDI_{MgHR}$ showed accretion of limits of 0.5. Whereas for RSC (−1.16), there is major dilution while $NDDI_{Na\%}$ showed minor dilution (−0.22) during monsoon (2011–2012).

13.1   INTRODUCTION

The elevated concentration of chemical constituents in the groundwater has detrimental effects on: environment, humans, and aquatic systems, and imparts eco-toxicological impacts and intricacies [5, 13, 17, 22, 33]. The higher residence time of these polluting chemicals has major impacts on aquatic ecosystems. The spatial-temporal variations of elements are based on the geo-environmental backgrounds [3, 4]. Most of these studies focus largely on spatial-temporal statistics and groundwater quality; and a quantitative comparative appraisal of site-specific dispersion and attenuation of chemical elements sparse [10, 18, 19]. Spatial-temporal monitoring of groundwater quality and sediments using multivariate statistics [9, 11, 25, 26, 29–32] is conducted using water quality index of urban lake water [1, 28], mangrove forest water [12], and river water [32].

Groundwater quality appraisal for drinking and irrigation purpose must be conducted [8, 9, 14–16, 20, 21]. Gajbhiye et al. [6] have assessed the impact of change land use/land cover on water quality using multi-temporal Landsat satellite data. Sharma et al. [24] used satellite data for appraisal of surface water quality of river Ganges. Rawat et al. [16] have used statistical index for evaluation of groundwater with special reference to nitrate. Rawat et al. [15] performed a hydro-chemical survey and quantified spatial variations in groundwater quality in coastal region of Chennai, India. With the use of GIS and numerical index, the water quality can be easily evaluated for different purposes [7, 17, 20, 27]. The statistical index like water quality index and fractal dimension is commonly used for assessment of different water resource for their suitability for different purposes [16].

Gautam et al. [9] performed a study of environmental monitoring of water resources with the use of PoS index of Subarnarekha River basin, India. The hydro-chemical research of groundwater quality for drinking and irrigation purposes of two case studies of Koprivnica-Križevci County (Croatia) and district Allahabad (India) was performed by Nemčić-Jurec et al. [14]. The disposal of untreated wastewater (WW) affects the groundwater quality [7] and this untreated WW is used for irrigation of vegetable crops in urban areas [2]. These vegetables have negative health effects.

The factors such as SAR, SSP (or %Na), RSC (or RA), MH, and TDS have caused problems in agriculture.

The objective of the work was to evaluate groundwater quality using NDDI index for irrigation purpose. Therefore, in this study, a binary approach was followed using geochemical data and statistical software (*Simplex numerica*)-based output images.

13.2 MATERIALS AND METHODS

The Chennai Water Metro Board/Central Ground Water Board (CGWB) has continuously monitored water quality parameters of during the pre- and post-monsoon seasons. The authors of this chapter have acquired the data from the Chennai Water Metro Board and CGWB of 44 sampling stations during May 2011 (pre-monsoon) and January 2012 (post-monsoon). The 44 wells (Table 13.1) co-ordinates were recorded using a hand-held GPS (global positioning system) device (eTrex Legend® HCx, having ± 15-meter accuracy) (Figure 13.1).

TABLE 13.1   Well Serial Numbers Along with Name of Location

Wells Serial Number (Z)	Name of Location	Wells Serial Number (Z)	Name of Location
1	Ottivakkam	23	Gudalur
2	Pallikaranai	24	Sembakkam
3	Akkarai	25	Maduramangalam
4	Walajabad	26	Mugaiyur
5	Damal	27	Pondur
6	P. V. Kalathur	28	Thandalam
7	Tambaram	29	Orathi
8	Thiruneermalai	30	Karunilam
9	Panangattucherry	31	Vedal
10	Muttukkadu	32	Malayankulam
11	Vengivasal	33	Govindavadi
12	Pallavaram	34	Chittamur
13	Ponmar	35	Thennampattu
14	Alathur	36	Salaiyur. 193
15	Purisai	37	Sengattur. 171
16	Kunrathur	38	Kayappakkam
17	Pattaravakkam	39	Endathur
18	Puthirankottai	40	Velamur [ramapuram]
19	Injambakkam	41	Magaral
20	Oragadam	42	Muttukkadu
21	Kilambi	43	Madampakkam
22	Thenneri	44	Kottamedu

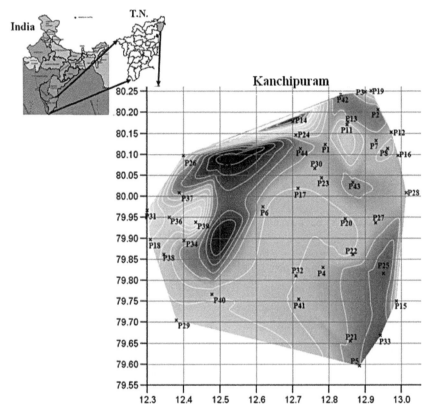

FIGURE 13.1   Location map of sampling stations in the study area (Kanchipuram, Tamil Nadu, India).

13.2.1   STUDY AREA

District Kanchipuram in Tamil Nadu (India) lies between 11°00' to 12°00' North latitudes and 77°28' to 78°50' East longitudes and on the bank of the Vegavathi River, a tributary of the Palar River. The average elevation is 83.2 m above mean sea level (MSL). The study area (Figure 13.1) is flat and slopes towards the south and east and is bounded by Bay of Bengal on the East.

The main occupation of habitants is agriculture with 47% of people engaged. The paddy is a major crop. Other crops such as groundnuts, sugarcane, cereals, and millets and pulses are also cultivated (Table 13.2). The majority of soil is clay, with some loam and sand (Table 13.2). The total forest area is 23,586 ha, which is spread in interior regions and 66.675 ha is a reserved land.

TABLE 13.2   General Information About Study Area

Total area	4393.37 km²
Net sown area	1364.89 km²
Net irrigation area	1236.28 km²
Forest area	426.57 km²
Poromboke area	1553.47 km²
Town area	82.57 km²
Number of villages	1214
Administrative divisions	8 Taluks, 13 blocks, 648 Panchayats, and 1214 villages.
Basin and sub-basin	Kancheepuram district is part of the composite east-flowing river basin and spread over parts of Palar and Cheyyar sub-basins.
Drainage	Palar and Cheyyar are important rivers. The drainage pattern in general is sub-dendritic and radial. All the rivers are seasonal and carry substantial flow during the monsoon period.
Total Canals (13 blocks)	550
Tanks	60,732
Tube and Ordinary wells	12,166 and 47,252
Summer temperature	21.1 to 36.6°C
Winter temperature	19.8 to 28.7°C
Rainfall	1133.0 mm (Actual), 1213.3 mm (Normal)
Major crops	Rice (1,45,966 ha), sugarcane (7586 ha), groundnut (28,766 ha), gingelly (912 ha), pulses (2966 ha), cotton (53 ha), millets, and cereals (1217 ha).
Soil types	Red loam (Kanchipuram, Uthiramerur blocks), Lateritic soil (Pleatus in the district), Blacksoil (Spread in all blocks), Sandy coastal alluvium (Thiruporur and St. Thomas Mount) and Red sandy soil (Kancheepuram and urban blocks),

The study area receives majority of rainfall (54%) during the north-east monsoon and it also receives slightly less rainfall (36%) during the south-west monsoon season. The coastal area receives more rainfall compared to interior parts of the study area. The distribution of pre-monsoon rainfall is uniform. The average rainfall is 1213.3 mm (Table 13.1) during the monsoon season. The climate of the area is hot and humid.

13.2.2   IRRIGATION WATER QUALITY INDEX (IWQI)

Irrigation indices like TDS, SAR, RSC/RA, Na%, and MH were calculated to evaluate the groundwater quality for irrigation use. IWQI was computed using the following chemical elements (Na^+, Mg^{2+}, CO_3^-, Cl^-, SO_4^{2-}). This helps to understand the complex water quality parameters into understandable and useable information to the public. Therefore, water quality parameter-based indices (TDS, SAR, RSC/RA, Na%, and MH) were used as indicators for suitability assessment of groundwater for irrigation purposes.

13.2.2.1   TOTAL DISSOLVED SOLIDS (TDS)

In natural water, dissolved solids are mixtures of CO_3^{2-}, HCO_3^-, Cl^-, SO_4^{2-}, and PO_4^{3-}. Salts are present in irrigation water; and these are released into the water from weathering or dissolution of soil and rocks and from the anthropogenic activities. These salts are carried by water during its use for irrigation. Usually, TDS indicates the general nature of the salinity of any kind of water.

13.2.2.2   SODIUM ABSORPTION RATIO (SAR)

The SAR is a relative ratio of Na^+ ions to Ca^{2+} and magnesium (Mg^{2+}) ions present in the water sample. The SAR is used to estimate the potential for Na^+ to accumulate in the soil primarily (water movement) at the expense of Ca^{2+}, Mg^{2+}, and K^+ because of regular use of sodic water [30, 31]. The SAR is given in Eqn. (13.1):

$$SAR = \frac{Na^+}{\sqrt{\frac{\left(Ca^{2+} + Mg^{2+}\right)}{2}}} \tag{13.1}$$

13.2.2.3   RESIDUAL SODIUM CARBONATE (RSC)/RESIDUAL ALKALINITY (RA)

The residual sodium carbonate (RSC) is the amount of sodium carbonate ($NaCO_3$) and sodium bicarbonate ($NaHCO_3$) present in the irrigation water [30, 31]. RSC is expressed Eqn. (13.2).

$$RSC = \left(HCO_3^- + CO_3^{2-}\right) + \left(Ca^{2+} + Mg^{2+}\right) \tag{13.2}$$

13.2.2.4   PERCENT SODIUM (%NA) OR SODIUM HAZARD

The percent of Na is also used in the classification of sources of water for irrigation use. The conversion of soil into saline is due to the use of water having a high concentration of salts or over-irrigation of by continuous use of groundwater in semi-arid and arid regions [30, 31]. The %Na is expressed in Eqn. (13.3).

$$\%Na = \frac{Na^+}{Ca^{2+} + Mg^{2+} + Na^+ + K^+} \times 100 \tag{13.3}$$

13.2.2.5   MAGNESIUM HAZARD (MH)/MAGNESIUM ABSORPTION RATIO (MAR)

Usually, alkaline earths (Ca^{2+} and Mg^{2+}) are in an equilibrium state in groundwater. The MH/MAR ratio is shown in Eqn. (13.4).

$$MH = \frac{Mg^{2+}}{Ca^{2+} + Mg^{2+}} \times 100 \tag{13.4}$$

13.2.3   COMPUTATION OF NORMALIZED DIFFERENCE DISPERSAL INDEX (NDDI)

The well information is given in Table 13.1. The chemical component, net-difference maps (post-monsoon and pre-monsoon) revealed high degree of variability in both degree and space on temporal scale [2]. The spot specific apportion of chemical component was normalized and quantified with the help of normalized difference dispersal index (NDDI). The NDDI is expressed in Eqn. (13.5).

$$NDDI = \frac{Concentration\ of\ element(post-monsoon-pre-monsoon)}{Concentration\ of\ element(post-monsoon+pre-monsoon)} \tag{13.5}$$

The NDDI ranged from–1 (absolute dilution) to +1 (absolute accretion). The areas with high inconsistencies on NDDI maps could be identified, classified, and linked to spatial data. The pointwise NDDI information does not show spatial variability; hence, software *Simplex numerica* was used for interpolation of NDDI point data for meaningful information.

13.2.4   STATISTICAL ANALYSIS

Descriptive statistics of NDDI was computed using *Microsoft Office Excel*. The statistical parameters like mean (T_1), standard error (T_2), median (T_3), mode (T_4), standard deviation (T_5), range (T_6), minimum (T_7), and maximum (T_8) were calculated. The correlation coefficient (CC) is a basic statistical tool to measure the relationship between two variables. CC matrix of NDDI was also developed using *Microsoft Office Excel*.

13.3   RESULT AND DISCUSSION

13.3.1   DESCRIPTIVE STATISTICS

IWQI at 44 locations and their descriptive statistics of pre- and post-monsoon are tabulated in Table 13.3. Table 13.4 revealed that variation of TDS (221 to 2481), SAR (3.77 to 55.81), RSC (–279 to 558), Na% (27.93 to 86.46) and Mg HR (5.51 to 79.66) during pre-monsoon of year 2011 compared to variations in TDS (258 to 2562), SAR (6.39 to 54.05), RSC (–336 to 374), Na% (27.93 to 86.46) and Mg HR (5.51 to 79.66) during post-monsoon of year 2012. Table 13.5 reveals that there was no co-relationship between groundwater qualities based indexes except SAR and TDS during pre- and post-monsoon, because TDS is a combination of salts (mainly Na, Ca, and Mg) and these salts play a major role in the generation of SAR (Eqn. (13.1)).

TABLE 13.3   Irrigation Groundwater Quality Indexed at Different Well Locations for: (a) Year 2011 and (b) Year 2012

Z	(a) 2011					(b) 2012				
	A	**B**	**C**	**D**	**E**	**A'**	**B'**	**C'**	**D'**	**E'**
1	321	4.11	114	27.93	27.50	280	7.19	105	48.54	35.85
2	956	13.30	241	86.46	68.42	697	25.94	140	69.38	16.46
3	288	5.62	77	44.66	26.32	331	6.44	38	39.45	42.42
4	1247	55.81	451	81.14	39.53	1115	54.05	266	81.62	52.00
5	1092	23.40	437	64.44	19.46	735	30.71	219	81.65	67.35

TABLE 13.3 *(Continued)*

Z	(a) 2011					(b) 2012				
	A	B	C	D	E	A'	B'	C'	D'	E'
6	2077	14.90	−279	33.71	33.33	1932	15.59	−336	35.99	45.23
7	759	36.82	252	78.07	79.66	680	21.23	117	62.01	40.23
8	868	12.73	227	63.22	33.88	906	19.58	166	57.81	37.01
9	1584	42.72	219	69.52	32.58	500	26.87	146	73.08	46.94
10	378	3.77	101	33.33	25.00	258	12.02	31	59.55	33.33
11	1017	28.93	74	65.43	50.41	1136	26.87	122	60.67	47.71
12	1461	52.68	558	80.62	11.50	1408	42.14	374	72.83	65.22
13	695	11.64	107	50.00	34.40	658	11.57	−2	48.29	37.19
14	326	6.50	178	41.94	16.67	1755	31.62	−13	58.89	47.76
15	1145	26.92	4	61.73	40.65	1241	21.95	114	61.75	32.53
16	601	18.30	219	61.26	20.93	556	23.79	126	71.50	47.37
17	364	10.79	140	49.23	45.45	363	12.66	151	54.84	71.43
18	260	4.05	100	41.58	32.20	395	14.12	73	56.12	18.03
19	327	8.20	72	56.14	52.00	475	14.02	130	54.44	35.06
20	581	18.86	230	62.67	33.33	599	17.33	169	69.23	36.11
21	757	25.94	360	68.73	16.48	604	20.93	278	66.67	34.21
22	502	13.35	257	54.30	27.06	492	19.60	223	65.57	52.38
23	661	11.68	163	46.55	32.26	614	13.65	137	50.97	50.50
24	608	11.80	117	43.30	5.51	602	8.56	−2	35.00	35.38
25	549	14.86	335	64.29	22.50	321	7.55	170	41.38	35.29
26	361	8.40	122	57.45	26.67	518	8.62	183	50.00	29.90
27	2481	42.51	−59	64.22	24.58	2229	36.25	−148	60.87	30.56
28	1370	19.90	160	58.47	32.04	1289	21.43	79	60.90	38.80
29	315	4.00	36	51.75	20.00	437	7.06	73	45.51	15.29
30	1214	18.98	83	52.83	45.83	1132	21.39	230	52.96	26.70
31	221	10.09	64	58.23	39.39	521	10.12	105	44.31	52.69
32	765	39.20	321	83.16	64.00	770	44.70	318	83.45	62.50
33	1227	29.66	162	62.13	23.35	662	10.09	5	39.44	47.29
34	459	16.94	69	62.66	49.15	428	11.49	158	54.67	52.94
35	716	22.74	321	67.05	40.23	511	12.67	154	50.00	44.44
36	1734	41.94	45	69.65	37.82	2562	31.28	−180	52.42	49.51
37	639	8.69	88	47.11	41.18	497	10.65	50	50.59	26.19
38	322	18.65	87	70.59	37.14	371	9.95	84	47.69	23.53
39	344	4.70	45	37.72	54.93	690	12.99	93	46.46	58.68

TABLE 13.3   *(Continued)*

Z	(a) 2011					(b) 2012				
	A	B	C	D	E	A'	B'	C'	D'	E'
40	844	36.63	147	78.73	64.18	941	37.72	123	74.72	24.44
41	910	20.44	141	62.88	50.41	872	24.51	260	62.20	20.97
42	378	3.77	101	33.33	25.00	258	12.02	31	59.55	33.33
43	1040	12.66	211	40.37	47.92	829	25.46	128	70.28	71.76
44	367	5.72	145	48.28	22.67	275	6.39	50	38.14	30.00

Legend: Z = Wells serial number; A = TDS; B = SAR; C = RSC; D = Na%; E = MgHr for 2011 pre-monsoon.

A' = TDS; B' = SAR; C' = RSC; D' = Na%; E' = MgHr for 2012 post-monsoon.

TABLE 13.4   Descriptive Statistics Index for Study Area for: (a) year 2011; and (b) Year 2012

T	(a) 2011					(b) 2012				
	A	B	C	D	E	A'	B'	C'	D'	E'
T_1	798.4	19.17	160.07	57.66	35.76	782.84	19.56	107.68	57.30	40.97
T_2	77.06	2.08	21.43	2.20	2.36	78.97	1.70	19.00	1.89	2.16
T_3	678.0	14.88	140.50	59.86	33.33	609.00	16.46	122.50	56.96	37.99
T_4	378	3.77	219.00	33.33	33.33	258.00	12.02	105.00	59.55	33.33
T_5	511.2	13.81	142.14	14.61	15.63	523.82	11.29	126.03	12.51	14.33
T_6	2260	52.05	837	58.53	74.15	2304	47.66	710	48.45	56.47
T_7	221	3.77	−279	27.93	5.51	258	6.39	−336	35.00	15.29
T_8	2481	55.81	558	86.46	79.66	2562	54.05	374	83.45	71.76

Legend: T = Statistical test, T_1 = mean, T_2 = Standard error, T_3 = Median, T_4 = Mode, T_5 = Standard deviation, T_6 = Range, T_7 = Minimum, T_8 = Maximum, T_{12} = Sum;

A = TDS, B = SAR, C = RSC, D = Na%, E = MgHr for 2011 pre-monsoon;

A' = TDS, B' = SAR, C' = RSC, D' = Na%, E' = MgHr for 2012 post-monsoon.

Table 13.6 represents IWQI and NDDI for each well while Tables 13.7 and 13.8 represent descriptive statistics and correlation matrix for different water quality parameter-based index's NDDI in the study area. Table 13.6 indicates variations in NDDI of TDS (or NDDI$_A$), SAR (or NDDI$_B$), RSC (or NDDI$_C$), Na% (or NDDI$_D$) and MhHR (or NDDI$_E$) as −0.52 to 0.69, −0.49 to 0.66, −1.16 to 1.67, 0.22 to 0.28, and −0.61 to 0.73, respectively. Table 13.4c reveals that there does not exist any correlation between any two NDDI, because after applying the NDDI concept (or Eqn. (13.5)) all types of internal relationship vanishes [17, 22]. Figure 13.2 gives an idea of major, minor dilution and accretion phenomena at the study area for different NDDI. According to Figure 13.2, average $NDDI_{RSC}$ (or $NDDI_C$) shows minor dilution in the study area during the monsoon year (2011–2012). Based on Figure 13.2, it can be concluded that the groundwater of the study area is fit for irrigation purposes.

TABLE 13.5   Correlation Matrix for Groundwater Quality Index for Irrigation for: (a) 2011 and (b) 2012

	(a) 2011					(b) 2012				
	A	B	C	D	E	A'	B'	C'	D'	E'
A	1					**A'**	1				
B	0.68	1				**B'**	0.58	1			
C	−0.08	0.41	1			**C'**	−0.42	0.28	1		
D	0.33	0.76	0.53	1		**D'**	0.14	0.80	0.55	1	
E	0.00	0.17	−0.09	0.37	1	**E'**	0.14	0.31	0.18	0.27	1

Legend: A = TDS, B = SAR, C = RSC, D = Na%, E = MgHr for 2011 pre-monsoon; A' = TDS, B' = SAR, C' = RSC, D' = Na%, E' = MgHr for 2012 post-monsoon.

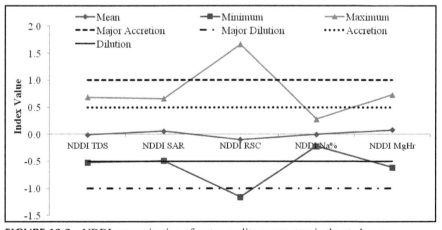

FIGURE 13.2   NDDI categorization of water quality parameters in the study area.

TABLE 13.6   Groundwater Quality Based on NDDI Values

Well Serial Number	NDDI$_A$	NDDI$_B$	NDD$_C$	NDDI$_D$	NDDI$_E$
1	−0.07	0.27	−0.04	0.27	0.13
2	−0.16	0.32	−0.27	−0.11	−0.61
3	0.07	0.07	−0.34	−0.06	0.23
4	−0.06	−0.02	−0.26	0.00	0.14
5	−0.20	0.14	−0.33	0.12	0.55
6	−0.04	0.02	0.09	0.03	0.15
7	−0.05	−0.27	−0.37	-0.11	−0.33
8	0.02	0.21	−0.16	−0.04	0.04
9	−0.52	−0.23	−0.20	0.02	0.18
10	−0.19	0.52	−0.53	0.28	0.14
11	0.06	−0.04	0.24	−0.04	−0.03
12	−0.02	−0.11	−0.20	−0.05	0.70
13	−0.03	0.00	−1.04	−0.02	0.04
14	0.69	0.66	−1.16	0.17	0.48
15	0.04	−0.10	0.93	0.00	−0.11
16	−0.04	0.13	−0.27	0.08	0.39
17	0.00	0.08	0.04	0.05	0.22
18	0.21	0.55	−0.16	0.15	−0.28
19	0.18	0.26	0.29	−0.02	−0.19
20	0.02	−0.04	−0.15	0.05	0.04
21	−0.11	−0.11	−0.13	−0.02	0.35
22	−0.01	0.19	−0.07	0.09	0.32
23	−0.04	0.08	−0.09	0.05	0.22
24	0.00	−0.16	−1.03	−0.11	0.73
25	−0.26	−0.33	−0.33	−0.22	0.22
26	0.18	0.01	0.20	−0.07	0.06
27	−0.05	−0.08	0.43	−0.03	0.11
28	−0.03	0.04	−0.34	0.02	0.10
29	0.16	0.28	0.34	−0.06	−0.13
30	−0.03	0.06	0.47	0.00	−0.26
31	0.40	0.00	0.24	−0.14	0.14
32	0.00	0.07	0.00	0.00	−0.01
33	−0.30	−0.49	−0.94	−0.22	0.34
34	−0.03	−0.19	0.39	−0.07	0.04

TABLE 13.6   *(Continued)*

Well Serial Number	NDDI$_A$	NDDI$_B$	NDD$_C$	NDDI$_D$	NDDI$_E$
35	−0.17	−0.28	−0.35	−0.15	0.05
36	0.19	−0.15	1.67	−0.14	0.13
37	−0.13	0.10	−0.28	0.04	−0.22
38	0.07	−0.30	−0.02	−0.19	−0.22
39	0.33	0.47	0.35	0.10	0.03
40	0.05	0.01	−0.09	−0.03	−0.45
41	−0.02	0.09	0.30	−0.01	−0.41
42	−0.19	0.52	−0.53	0.28	0.14
43	−0.11	0.34	−0.24	0.27	0.20
44	−0.14	0.06	−0.49	−0.12	0.14

TABLE 13.7   Descriptive Statistics of NDDI in the Study Area

T	NDDI$_A$	NDDI$_B$	NDDI$_C$	NDDI$_D$	NDDI$_E$
T$_1$	−0.01	0.06	−0.10	0.00	0.08
T$_2$	0.03	0.04	0.08	0.02	0.04
T$_3$	−0.03	0.05	−0.15	−0.01	0.12
T$_4$	−0.19	0.52	−0.53	0.28	0.14
T$_5$	0.19	0.25	0.50	0.13	0.28
T$_6$	1.21	1.15	2.82	0.51	1.34
T$_7$	−0.52	−0.49	−1.16	−0.22	−0.61
T$_8$	0.69	0.66	1.67	0.28	0.73

Legend: T = Statistical Test; T$_1$ = mean, T$_2$ = Standard Error, T$_3$ = Median, T$_4$ = Mode, T$_5$ = Standard Deviation, T$_6$ = Range, T$_7$ = Minimum, T$_8$ = Maximum, TDS = NDDI$_A$, SAR = NDDI$_B$, RSC = NDDI$_C$, %Na = NDDI$_D$, MgHr = NDDI$_E$.

TABLE 13.8   Correlation Matrix for NDDI of Different Irrigation Groundwater Quality Index

	NDDI$_A$	NDDI$_B$	NDDI$_C$	NDDI$_D$	NDDI$_E$
NDDI$_A$	1				
NDDI$_B$	0.41	1			
NDDI$_C$	0.21	−0.11	1		
NDDI$_D$	0.04	0.77	−0.14	1	
NDDI$_E$	−0.04	−0.09	−0.34	0.13	1

Legend: TDS = NDDI$_A$, SAR = NDDI$_B$, RSC = NDDI$_C$, %Na = NDDI$_D$, MgHr = NDDI$_E$.

13.3.2   CORRELATION COEFFICIENT (CC)

The CC (r) of each irrigation water quality index (IWQI) is given Table 13.5 for both seasons (year 2011 and 2012). The analogous trend was observed among the IWQI. The positive correlation (0.76) was observed for SAR and %Na; and TDS vs. SAR showed (r = 0.68), whereas the negative correlation (−0.08) was observed between total dissolved solids (TDS) vs. RSC(Table 13.5a). A highly significant positive correlation (0.80) was observed between %Na and SAR; and TDS vs. SAR shows (r = 0.58) whereas the highly negative correlation (−0.42) was observed between TDS vs. RSC (Table 13.5b). The strong to good correlation among the various IWQI and NDDIs has been observed. Table 13.8 shows that NDDI CC was highly significant with positive correlation (0.77) between $NDDI_{SAR}$(or $NDDI_B$) vs. $NDDI_{\%Na}$(or $NDDI_D$) and there was highly negative correlation (−0.14) between $NDDI_{SAR}$ vs. $NDDI_{RSC}$.

13.3.3   3-D MAPPING

The contour maps based on NDDI helps in quantification, demarcating, and categorization of spot specific enrichment of any chemical elements, parameters, or any index. The NDDI-based average value of TDS (or $NDDI_A$) is 0.04 and ranges from 0.52 and 0.69 (Table 13.7) due to dilution of groundwater during year 2012 (Figure 13.3a). Moreover, two hot spots appear along P14 and P31 (Figure 13.3a) with the highest index value of 0.69 (P14) and 0.40 (P31) thus showing almost minor accretion; therefore, groundwater in the study area was suitable from TDS point of view during 2011–2012.

NDDI values of SAR (or $NDDI_B$) varied from 0.49 to 0.66 (Table 13.7) with a median of 0.6.3-DNDDI image (Figure 13.3b) of SAR illustrates good amount of spatial variation in accretion plume (reddish patch) with almost equal distributed area by dilution (bluish patch) in diminutive nature. NDDI contour image of SAR (Figure 13.3b) illustrates an abundant diffusion of higher and lower values of monsoon during 2011–2012.

On the contrary, 28 out of 44 well's index values of RSC (or $NDDI_C$) are negative and ranged from 1.67 (P36) to 0.09 (P40) (Tables 13.6 and 13.7) with a median value of- 0.15 (P20) (Tables 13.6 and 13.7). This illustrates consistent and absolute accrual after wet spells (Figure 13.3c) because of only one deeper spot (bluish spot) appearing rest as bluish in color. RSC depends on HCO_3^-, CO_3^{2-}, Ca^{2+}, and Mg^{2+}, and all components are in accrual mode in the study area; therefore, only one dark blue patch existed (Figure 13.3c). From the NDDI image of RSC, there is a clear hot spot location is around the P36 (Figure 13.3c).

$NDDI_{Na\%}$ (or $NDDI_D$) ranges from- 0.22 to 0.28 (Table 13.7) and it shows high dilution (Figure 13.3d, more spatial distributed bluish color). Figure 13.3d clearly reveals that a high dilution contours appear with hot spot positions at P14, P1, P43, and P42.

$NDDI_{MgHR}$ (or $NDDI_E$) varied from- 0.61 to 0.73 (Table 13.7) with a median of 12. It suggests a widespread and almost absolute accrual of average value. Figure 13.3e shows almost more yellow with light reddish patch while a big blue (dilution) area is also reported in Figure 13.3e. Figure 13.3e showed a clear spatial variation of $NDDI_{MgHR}$ in the study area during monsoon year (2011–2012).

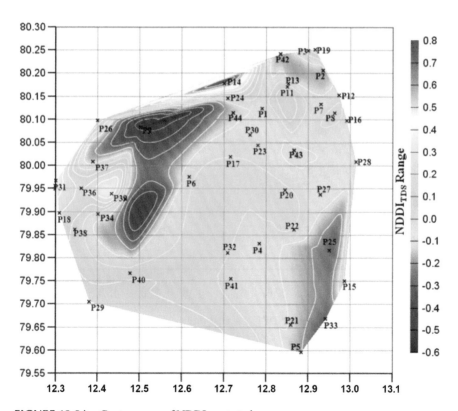

FIGURE 13.3A   Contour map of $NDDI_{TDS}$ at study area.

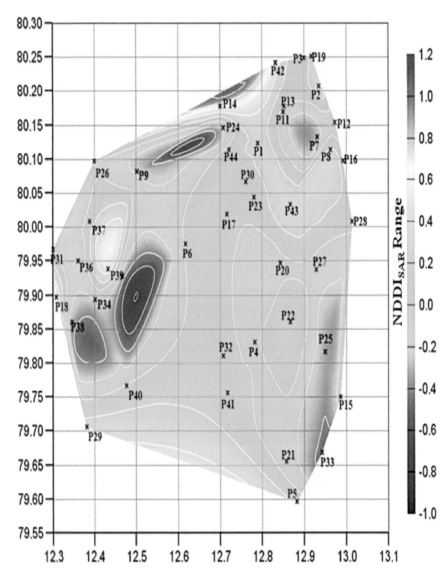

FIGURE 13.3B   Contour map of NDDI$_{SAR}$ at study area.

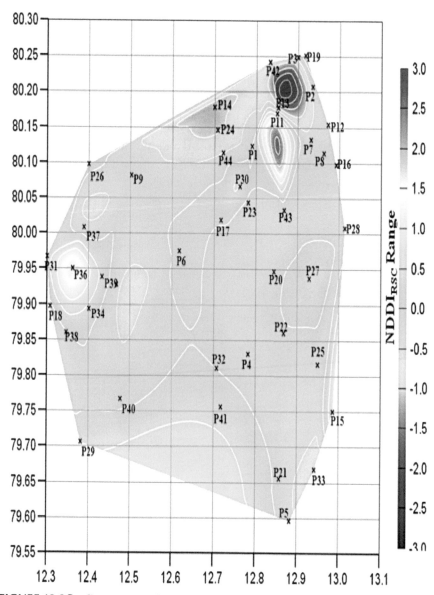

FIGURE 13.3C   Contour map of NDDI$_{RSC}$ at study area.

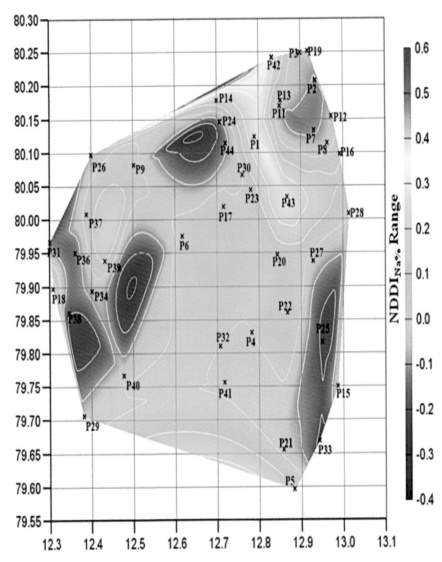

FIGURE 13.3D   Contour map of NDDI$_{\%Na}$ at study area.

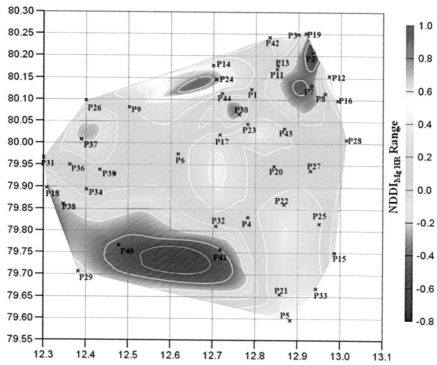

FIGURE 13.3E   Contour map of NDDI$_{Mg\ HR}$ at study area.

13.4   SUMMARY

The groundwater is a major source of irrigation and drinking water. The qualitative information of groundwater is needed for sustainable growth agriculture. Hence, the evaluation of groundwater for irrigation purposes was performed using the NDDI. The NDDI approach helps in the identification of the dilution and enrichment of elements in the water. The NDDI results showed the spot specific enrichment (accretion or attrition and dilution or minor accretion) during the year 2011–2012. The groundwater quality of the area is controlled by season and anthropogenic inputs. The 3-D maps based on NDDI were prepared to identify the hot spots of the pollution.

ACKNOWLEDGMENT

The authors express their sincere thanks to the Ex-Joint Director (Shri K. Santanam) of Water Resource Division, P.W.D, Chennai, Tamil Nadu, and

India for providing the data and insightful discussion during the preparation of the manuscript.

KEYWORDS

- irrigation water quality index
- magnesium absorption ratio
- normalized difference dispersal index
- residual alkalinity
- residual sodium carbonate
- sodium hazard

REFERENCES

1. Amin, A., Fazal, S., Mujtaba, A., & Singh, S. K., (2014). Effects of land transformation on water quality of Dal Lake, Srinagar, India. *J. Indian Soc. Remote Sens., 42,* 119–128.
2. Bharose, R., Singh, S. K., & Srivastava, P. K., (2013). Heavy metals pollution in soil-water-vegetation continuum irrigated with groundwater and untreated sewage. *Bull. Environ. Sci. Res., 2,* 1–8.
3. Chamyal, L. S., Maurya, D. M., Bhandari, S., & Rachna, R., (2002). Late quaternary geomorphic evolution of the lower Narmada valley, Western India; Implications for neo-tectonic activity along the Narmada-Son fault. *Geomorphology, 46,* 177–202.
4. Coynel, A., Schafer, J., Dabrin, A., Girardot, N., & Blanc, G., (2007). Groundwater contributions to metal transport in a small river affected by mining and smelting waste. *Journal of Water Research, 41,* 3420–3428.
5. Dutta, P. S., (2003). *Water Quality Problems of Delhi Region* (p. 142). Annual report of DST Project; New Delhi: Central Pollution Control Board (CBCP).
6. Gajbhiye, S., Singh, S. K., & Sharma, S. K., (2015). Assessing the effects of different land use on water quality using multi-temporal Landsat data. In: Siddiqui, A. R., & Singh, P. K., (eds.), *Resource Management and Development Strategies: A Geographical Perspective* (pp. 81–88). Pravalika Publication, Allahabad, U.P., India.
7. Gautam, S. K., Sharma, D., Tripathi, J. K., Singh, S. K., & Ahirwar, S., (2013). Study of the effectiveness of sewage treatment plants in Delhi region. *Applied Water Science, 3,* 57–65.
8. Gautam, S. K., Singh, A. K., Tripathi, J. K., Singh, S. K., Srivastava, P. K., Narsimlu, B., & Singh, P., (2016). Appraisal of surface and groundwater of the Subarnarekha River Basin, Jharkhand, India: Using remote sensing, irrigation indices, and statistical techniques. *Geospatial Technol. Water Resour. Application,* 144–169.
9. Gautam, S. K., Tziritis, E., Singh, S. K., Tripathi, J. K., & Singh, A. K., (2018). Environmental monitoring of water resources with the use of PoS index: A case study

from Subarnarekha River basin, India. *Environmental Earth Sciences, 77,* 70–77. https://doi.org/10.1007/s12665–018–7245–5 (accessed on 20 June 2020).

10. Goovaerts, P., AvRuskin, G., Meiliker, J., Slotnick, M., Jacquez, G., & Nriagu, J., (2005). Geostatistical modeling of the spatial variability of arsenic in groundwater of southeast Michigan. *Journal of Water Research, 41,* 1–19.

11. Jacintha, T. G. A., Rawat, K. S., Mishra, A., & Singh, S. K., (2016). Hydrogeochemical characterization of groundwater of peninsular Indian region using multivariate statistical techniques. *Applied Water Science, 7*(6), 3001–3013.

12. Kumar, R. P., Ranjan, R. K., Ramanathan, A. L., Singh, S. K., & Srivastava, P. K., (2015). Geochemical modeling to evaluate the mangrove forest water. *Arab J. Geosci., 8,* 4687–4702.

13. Kumaresan, M., & Riyazuddin, P., (2005). Major ion chemistry of environmental samples around the sub-urban of Chennai city. *Current Science, 91,* 1668–1677.

14. Nemcic´-Jurec, J., Singh, S. K., Jazbec, A., Gautam, S. K., & Kovac, I., (2017). Hydro chemical investigations of groundwater quality for drinking and irrigational purposes: Two case studies of Koprivnica-Križevci County (Croatia) and district Allahabad (India). *Sustain. Water Resour. Manag., 5*(2), 467–490. https://doi.org/10.1007/s40899–017–0200-x (accessed on 20 June 2020).

15. Rawat, K. S., Jacintha, T. G. A., & Singh, S. K., (2018). Hydrochemical survey and quantifying spatial variations in groundwater quality in coastal region of Chennai, Tamil Nadu, India-case study. *Indonesian Journal of Geography, 50*(1), 57–69.

16. Rawat, K. S., Jeyakumar, L., Singh, S. K., & Tripathi, V. K., (2018). Appraisal of groundwater with special reference to nitrate using statistical index approach. *Groundwater for Sustainable Development.* E-article. https://doi.org/10.1016/j.gsd.2018.07.006 (accessed on 20 June 2020).

17. Rawat, K. S., Mishra, A. K., & Singh, S. K., (2017). Mapping of groundwater quality using normalized difference dispersal index of Dwarka Sub-city at Delhi National Capital of India. *ISH Journal of Hydraulic Engineering, 50*(10), 1–12. doi: 10.1080/09715010.2016.1277795.

18. Rawat, K. S., Mishra, A. K., Sehgal, V. K., & Tripathi, V. K., (2013). Identification of geospatial variability of Fluoride contamination in ground water of Mathura District, Uttar Pradesh, India. *Journal of Applied and Natural Science, 4*(1), 117–122.

19. Rawat, K. S., Mishra, A. K., Sehgal, V. K., & Tripathi, V. K., (2012). Spatial variability of ground water quality in Mathura District (Uttar Pradesh, India) with Geostatistical Method. *International Journal of Remote Sensing Application, 2*(1), 1–9.

20. Rawat, K. S., & Singh, S. K., (2018). Water quality indices and GIS-based evaluation of a decadal groundwater quality. *Geology, Ecology, and Landscapes,* 1–12. https://doi.org/10.1080/24749508.2018.1452462 (accessed on 20 June 2020).

21. Rawat, K. S., Singh, S. K., Jacintha, T. G. A., Nemcic´-Jurec, J., & Tripathi, V. K., (2018). Appraisal of long-term groundwater quality of peninsular India using water quality index and fractal dimension. *J. Earth Syst. Sci.* E-article. https://doi.org/10.1007/s12040–017–0895-y (accessed on 20 June 2020).

22. Rawat, K. S., Singh, S. K., & Tripathi, V. K., (2017). Groundwater quality evaluation using numerical Indices: A case study (Delhi, India). *Sustainable Water Resources Management,* E-article. doi: 10.1007/s40899–017–0181–9.

23. Rawat, K. S., & Tripathi, V. K., (2015). Hydro-chemical survey and quantifying spatial variations of groundwater quality in Dwarka-Delhi, India. *The Institution of Engineers (India).* E-article. doi: 10.1007/s40030–015–0116–0.

24. Sharma, B., Kumar, M., Denis, D. M., & Singh, S. K., (2018). Appraisal of river water quality using open-access earth observation data set: A study of river Ganga at Allahabad (India). *Sustainable Water Resources Management* (pp. 1–11). https://doi.org/10.1007/s40899-018-0251-7 (accessed on 20 June 2020).

25. Singh, S., Singh, C., Kumar, K., Gupta, R., & Mukherjee, S., (2009). Spatial-temporal monitoring of groundwater using multivariate statistical techniques in Bareilly district of Uttar Pradesh, India. *J. Hydromechanics, 57*, 45–54.

26. Singh, S. K., Singh, C. K., & Mukherjee, S., (2010). Impact of land-use and land-cover change on groundwater quality in the Lower Shiwalik hills: A remote sensing and GIS-based approach. *Cent. Eur. J. Geosci., 2*, 124–131.

27. Singh, H., Singh, D., Singh, S. K., & Shukla, D. N., (2017). Assessment of river water quality and ecological diversity through multivariate statistical techniques, and earth observation dataset of rivers Ghaghara and Gandak, India. *Int. J. River Basin Manag., 15*(3), 347–360. https://doi.org/10.1080/15715124.2017.1300159 (accessed on 20 June 2020).

28. Singh, S. K., Singh, P., & Gautam, S. K., (2016). Appraisal of urban lake water quality through numerical index, multivariate statistics, and earth observation data sets. *International Journal of Environmental Science and Technology, 13*, 445–456.

29. Singh, S. K., Srivastava, P. K., & Pandey, A. C., (2013). Fluoride contamination mapping of groundwater in Northern India integrated with geochemical indicators and GIS. *Water Sci. Technol. Water Supply, 13*, 1513–1523.

30. Singh, S. K., Srivastava, P. K., Pandey, A. C., & Gautam, S. K., (2013). Integrated assessment of groundwater influenced by a confluence river system: Concurrence with remote sensing and geochemical modeling. *Water Resources Management, 27*(12), 4291–4313. https://doi.org/10.1007/s11269-013-0408-y (accessed on 20 June 2020).

31. Singh, S. K., Srivastava, P. K., Singh, D., Han, D., Gautam, S. K., & Pandey, A. C., (2015). Modeling groundwater quality over a humid subtropical region using numerical indices, earth observation datasets, and X-ray diffraction technique: A case study of Allahabad district, India. *Environ. Geochem. Health, 37*(1), 157–180. https://doi.org/10.1007/s10653-014-9638-z (accessed on 20 June 2020).

32. Srivastava, P. K., Mukherjee, S., Gupta, M., & Singh, S. K., (2011). Characterizing monsoonal variation on water quality index of river Mahi in India using Geographical Information System. *Water Qual. Expo. Heal., 2*, 193–203.

33. Tripathi, V. K., Shukla, S. K., & Rawat, K. S., (2016). Calcium and chloride variability assessment in the Semi-arid region: A case study. *International Journal of Science, Environment, and Technology, 5*(5), 2877–2884.

CHAPTER 14

RUNOFF AND SEDIMENT ESTIMATION USING ANN AND ANFIS: CASE STUDY OF GODAVARI BASIN, INDIA

ASHISH KUMAR, PRAVENDRA KUMAR, and VINOD KUMAR TRIPATHI

ABSTRACT

Knowledge of stage-discharge and runoff-sediment relationships are extremely important for the planning and management of water resources. This study focuses on the issues of water resources management through assessment of the runoff and sediment vulnerability in the Godavari basin. To estimate the runoff and suspended sediment daily data of stage, runoff, and sediment for monsoon period (from 1st June to 30th September) of 1995 to 2010 were explored using artificial neural network (ANN) and adaptive neuro-fuzzy inference system (ANFIS) techniques for Pathagudem and Polavaram sites located in Chhattisgarh and Andhra Pradesh, India. Gamma test (GT) was used to select the best combination of input variables for both runoff and sediment prediction. Selected input combinations were then used as input vectors of ANN and ANFIS models for both runoff and sediment estimation. In the case of ANN models, back-propagation algorithm, and log sigmoid activation while in ANFIS models, triangular, trapezoidal, generalized bell, and Gaussian membership functions (MF) were used to train and test the models. The correlation coefficient (CC), root mean square error (RMSE), coefficient of efficiency (CE), and pooled average relative error (PARE) were utilized to validate the developed models. ANFIS model (Gauss, 3) with CC, RMSE, CE, and PARE values of 0.987, 299.75 m^3/ sec, 0.98 and –0.005 for runoff prediction and ANFIS model (Triangular, 3) with CC, RMSE, CE, and PARE values of 0.836, 0.159 g/l, 0.918 and –0.0074 for sediment prediction were selected as the best performing models at the Pathagudem site. While for the Polavaram, ANFIS model (Triangular,

3) with CC, RMSE, CE, and PARE values of 0.994, 859.93 m³/sec, 0.995 and –0.0011 and ANFIS model (Gauss, 3) with CC, RMSE, CE, and PARE values of 0.939, 0.113 g/l, 0.966, –0.0036 were better than other ANN and ANFIS-based models for runoff and sediment prediction, respectively. The sensitivity analysis indicated that the current day stage is the most sensitive parameter for runoff prediction and current-day runoff is the most sensitive parameter for sediment prediction at both sites.

14.1 INTRODUCTION

Due to water scarcity, increasing rate (IR) of degraded land, and IR of population, it has become necessary to make judicious use of available land and water resources for proper planning and efficient management for future scenarios. Runoff is the amount of excess water, which flows after the fulfillment of initial losses, i.e., infiltration, interception, depression storage, etc. After the saturation of the soil and initial losses, water starts to flow along the slope and detaches the soil from its parent material along with transport the detached soil particle. Runoff produces a driving force to transport the soil particles and accelerate the erosion, and the eroded soil is known as sediment. Therefore, accurate information about the rate of runoff in rivers plays a crucial role for a variety of hydrologic applications, such as resources planning and operation, water, and sediment budget analyses, hydraulic, and hydrologic modeling, and design of storage and conveyance structures. Therefore, it has been a common practice to convert water stages into discharges by using pre-established stage-discharge relationships.

Runoff and sedimentation are important factors to accelerate these problems. Forecasting of runoff and sediment is desired for better planning and utilization of land and water resources in various fields such as water supply, flood control, soil, and water conservation, irrigation, drainage, water quality, etc. Sedimentation, as an aggravated process, may lead to irreparable damages in the construction plans like accumulation of sediments behind the barriers, occurrence of reduction in their valuable amount, destruction of constructions, damage to ports and coasts, reduction of capacity and increases in the maintenance expenditure related to the irrigation canals. On one hand, sediment transport influences the quality of agriculture and drinking water, therefore the estimation of the sedimentation in soil conservation projects, design, and execution of the watery constructions, watershed, and the utilization of the water sources is required and considered a very important factor

[1]. On the other hand, runoff estimation also plays a crucial role to transport sediment particles from one place to another. There are many formulae and models, which are used to estimate runoff rate. The most discharge records are derived from converting the measured water levels (stages) to discharges by functional relationships that are expressed as rating curves. A calibrated stage-discharge rating offers an easy, cheap, and fast technique to estimate discharge.

Water resource systems need systematic study with reliable data to arrive at optimal planning and management decision. Any decision regarding the planning or operation of water development system requires the prediction of the characteristic and quantity of water available. The experts are always attempting to estimate the runoff rate and suspended sediment load in the stream of the rivers. For this purpose, first of all, they should care about the runoff rate and the mechanism of moving sediments, and then the amount of the transported sediments should be measured carefully to design and implement the water plans with high confidence coefficient. The history of scientific investigation related to the suspended sediments is more than 100 years old [35]. Hydrological models have been developed since 1930's for describing the processes of rainfall-runoff, runoff-sediment yield, and rainfall-runoff-sediment in a watershed fluvial system, and these are also useful in forecasting.

Watershed models are vital tools for the planning and management of watersheds or river basins. These models can integrate information over a large scale and a long period to simulate various watershed processes such as runoff, soil erosion, and transport of nutrients and pesticides. Some of the commonly used physically based, continuous time-scale watershed models suitable for the simulation of runoff and sediment transport from large watersheds with the capabilities of simulation of the best management practices are: conceptual models, physical models, and data-driven models:

1. **Conceptual Model:** These are models, which are formed after a conceptualization or generalization process. Conceptual models are often abstractions of things in the real-world whether physical or social. Semantics studies are relevant to various stages of concept formation. Conceptual models are generally based on Saint-Venant equations comprising of partial differential equations of continuity and momentum. In the past few decades, many conceptual based models have been proposed for forecasting hydrological phenomena [5, 19, 20]. While these models can well explain, the internal

mechanisms of the hydrological processes but require large amount of calibration data, sophisticated mathematical tools, and expertise with the model.

2. **Physical Models:** These are based on understanding of the physical processes; often require physical data sets and may not be ideal for real-time forecasting due to tremendous data requirements and associated long computation time for model calibration. Applications of physical-based models have been described in the past [9, 30].

3. **Data-Driven Models:** This extract information from the input-output data sets without considering the complex physical process by which they are related and establish a statistical correspondence between input(s) and output(s).

The first fundamental concepts related to neural computing were developed by McCulloch and Pitts, [24] and much of the ANN activities have been centered on back-propagation and its extensions [27]. Data-driven models such as adaptive neuro-fuzzy inference system (ANFIS), artificial neural network (ANN), and fuzzy logic (FL) have been useful in modeling time-series hydrologic problems. The main advantage of these models is that they do not require specifying functional relationships appropriately; self-organize their structure and adapt it in an interactive manner learning the underlying relationships. Data-driven models are preferable for flood forecasting problems, where the main concern is to make accurate and timely forecasts essential for flood damage mitigation. ANN approach provided better results than other methods for the estimation of sediment concentration [33].

ANFIS is the integration of neural networks and FL and can capture the benefits of both these fields in a single framework. ANFIS utilizes linguistic information from the FL as well as the learning capability of an ANN. ANFIS is a fuzzy mapping algorithm that is based on Takagi-Sugeno-Kang (TSK) fuzzy inference system (FIS) [23]. ANFIS has been successfully used for mapping the input-output relationship based on the available data [12]. The system acquires its adaptability by utilizing a hybrid learning method that combines back-propagation and least mean square optimization algorithms and possesses properties such as the capability of learning, constructing, expensing, and classifying.

In the hydrological forecasting context, application of data-driven models can be found in the works of Dawson and Wilby [8], Tokar and

Johnson [34], Liong et al. [21],Bazartseren et al. [2] and Nayak et al. [26] who carried out a study on river flow forecasting based on rainfall and runoff data using ANN, FL, ANFIS, and they reported better performances with the ANFIS model. The ANFIS model works on a set of linguistic rules and can handle imprecision and uncertainty present in the model and the data structure.

Jacquin and Shamseldin [15] explored the application of Takagi-Sugeno FISs to rainfall-runoff modeling. They showed that FISs are a suitable alternative to the traditional methods for modeling the non-linear relationship between rainfall and runoff. Habib et al. [14] developed stage-discharge relations for low-gradient tidal streams using data-driven models. Karl and Lohani [18] developed a flood forecasting system using statistical and ANN techniques for the downstream catchment of Mahanadi basin, India, and found that ANN methods perform better than the statistical methods. Bisht and Jangid [4] developed ANFIS models for the river discharge using the past river stage and discharge as inputs for specified lag time. Chidthong et al. [6] developed a hybrid multi-model to forecast the flood level at Chiang Mai and Koriyama floods in Japan, and they found the satisfactory application of the applied model.

Stage-discharge and sediment concentration relations were developed using the ANN approach for two different sites on the Mississippi River and it was depicted that the ANN outputs were close to the observed values than the traditional techniques [17]. ANN and linear-based regression models were developed by Raghuwanshi et al. [29] for runoff and sediment yield estimation in Nagwan watershed in India. It was observed that ANN models performed better than the linear regression models, for predicting both runoff and sediment yield on a daily and weekly simulation. Bisht et al. [3] developed stage-discharge models using ANN and MLR methods, and they proved that the developed ANN models showed better performance than the MLR models.

The capacity of ANN and SRC methods were studied by Shabani et al. [31] for estimation of daily suspended sediment in Kharestan watershed, Fars province in Iran. The results showed that the estimation of ANN was more accurate than SRC. The performance of ANN model increases with an increase in input parameters and changing combination inputs of parameters. ANN model performed better than the linear regression model for the prediction of sediment yield for Kal River in Maharashtra [13]. Nayak et al. [25] developed an adaptive Neuro-Fuzzy model to forecast the river flow of the Baitarani River in Orissa, India. It was found that ANFIS-based results were

more promising than ANN and other traditional time series-based models
in terms of computational speed, efficiency, forecast errors, and peak flow
estimation. Shafie et al. [32] proposed the ANFIS model as the best model
to estimate the inflow rate for the Nile River at Aswan High Dam (AHD) on
a monthly basis.

ANN, MLR, and ANFIS-based models were developed by Firat et al.
[10] to forecast sediment outflow for Menders basin and it was found that
ANFIS showed better performance than the other models. Stage discharge
models, for Godavari River at Rajahmundry-Dhawalaishwaram Barrage
site in Andhra Pradesh, using ANFIS and MLR methods were developed. A
comparison of both models showed that ANFIS models given better results
than traditional models, such as MLR [4].

Folorunsho et al. [11] developed ANFIS models with 70% of data for
training and 30% for validation and it was concluded that ANFIS models
were good estimators of runoff. Flood at Dharoi Dam on the Sabarmati
River near village Dharoi in Kheralu Taluka of Mehsana District in Gujarat-
India was estimated using ANFIS and it was suggested that ANFIS model
could accurately and reliably be used to forecast flood [28]. Combination
of a genetic input selection algorithm and wavelet transforms using ANFIS
models was explored for streamflow prediction for the Lighvan and Ajichai
basins in Iran and it was found that the accuracy of results was considerably
improved, when same combination was used [7].

The main purpose of this study was to analyze the response of stage-
discharge and runoff-sediment models by comparing the performance of
ANN and ANFIS models at Polavaram and Pathagudem gauging stations
comprising of 3,07,800 km^2 and 40,625 km^2 drainage areas, respectively
in Godavari river basin. The root mean square error (RMSE), correlation
coefficient (CC) (r), coefficient of determination (R^2), and coefficient of efficiency (CE), pooled average relative error (PARE) were used for evaluation
of the performance of models. This chapter focuses on:

- Selection of most effective input parameters, based on gamma test
  (GT) to develop stage-discharge and discharge-sediment models.
- Development of stage-discharge models using ANN and ANFIS
  techniques for Pathagudem and Polavaram sites in the Godavari
  basin.
- Development of discharge-sediment models using ANN and ANFIS
  techniques for the study area.
- Performance evaluation of the developed models based on qualitative and quantitative comparisons.

- Sensitivity analysis of selected models to determine the most sensitive parameters for runoff and sediment prediction.

14.2 MATERIALS AND METHODS

14.2.1 GENERAL DESCRIPTION OF THE STUDY AREA

The Godavari River is the longest river in India after the river Ganga. Godavari River originates in the Sahyadris near Triambakeswar, about 80 km from the shore of the Arabian Sea, at an elevation of 1067 m in the Nasik district of Maharashtra state. After flowing in a general southeast direction through Maharashtra and Andhra Pradesh for about 1465 km, it ends into the Bay of Bengal. The catchment area of the river is 3,12,812 km², which is approximately 10% of the total geographical area of the country. The basin lies in the Deccan Plateau and is situated between latitude 16°16' N to 22°36' N and longitude 73°26' E to 83°07' E.

Except for the hills forming the watershed around the Godavari basin, the entire drainage area comprises rolling and undulating, a series of ridges and valleys interspersed with low hill ranges. The upper reaches of the basin are occupied by the ocean traps containing hypersthene, minerals, augite, diopside, enstatite, magnetite, epidote, biotite, zircon, rutile, apatite, and chlorite. The middle part of the basin is Archean granites and Dharwars composed of phyllites, quartzites, granites, and amphiboles. The downstream of the middle basin is occupied mainly by the Cuddapah and Vindhyan metasediments and rocks of the Gondwana group (Figure 14.1).

The soils of the Godavari basin are grouped as black cotton soils, red earth with loamy sub-soil, and forest soils. Black cotton soils are moderately deep from 15 to 240 cm. These soils are poor in humus, nitrogen, and phosphate content. The soils in the basin are generally fertile and good for growing *Kharif* crops like jowar, bajra, pulses, and groundnut. The climate of the basin is tropical characterized by hot and cold weather, southwest monsoon, and post-monsoon. The southwest monsoon starts from mid-June to the end of mid-October. The average annual surface temperature in the Western Ghats area is approximately 24°C and it increases gradually when move towards in the basin and shows a maximum of 29.4°C on the east coast during January. The mean daily temperature increases from west to east from 15°C on the Western Ghats to about 18°C east coast. The mean daily temperature attains an average maximum value generally exceeding 30°C in the western part of the basin.

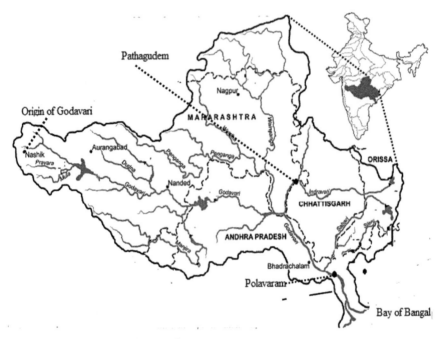

FIGURE 14.1   Location of the study area.

The Godavari basin attains its maximum rainfall during the southwest monsoon. The months of January and February are almost dry with the rainfall being less than 15 mm and during the next three months from March to end of the May, it varies from 20 to 50 mm in most parts of the basin. All parts of the Godavari basin receive maximum rainfall (84%) during the southwest monsoon from June to September.

Pathagudem site is situated on the Indravati River, which is the second-largest tributary of the Godavari river basin (Figure 14.1) with a total catchment area of 40,625 km². The Indravati river lies between 80°16'19" E to 83°07'10" E longitudes and 18°43'25" N to 19°26'46" N latitudes. Whereas Polavaram is in west Godavari district of Andhra Pradesh at 81°38'53" E longitude and 17°15'07" N latitude. It follows the southern path and falls in the Bay of Bengal.

14.2.2   DATA ACQUISITION

The daily stage, runoff, and sediment data for the monsoon season (1st June to 30th September) during 1996 to 2010 at Pathagudem and Polavaram sites were obtained from Krishna and Godavari Basin Organization, Divisional

Office of Central Water Commission, Hyderabad (Andhra Pradesh). Sediment load is the suspended sediment concentration (SSC) measured by taking a one-liter water sample and was reported in g/l.

> **Pre-analysis of data:** The collected daily stage, runoff, and sediment concentration data (June 1st to September 30th) for 1996–2010 were used for model development (training) and validation (testing). Therefore, the total number of data for each period was 122.

14.2.3   METHODOLOGY

In this study, ANNs and ANFIS predicting models were developed to predict runoff and SSC on a daily basis. This section deals with the basic concept of ANNs and ANFIS.

14.2.3.1   ARTIFICIAL NEURAL NETWORKS (ANNS)

McCulloch and Pitts [24] recognized the first computational model for neural networks based on mathematics and algorithm called threshold logic. An ANN is a computational model based on the structure and functions of biological neural networks. ANNs are the system of presenting interconnected "neurons," which send messages to each other. The connections have numeric weights that can be adjusted based on experience, making neural nets adaptive to inputs and capable of learning. ANN has been proven to provide better results for estimation of runoff and sediment.

14.2.3.1.1   Basic Concept of Artificial Neural Networks (ANNs)

An ANN is an information-processing system composed of many nonlinear and densely inter-connected processing elements or neurons (which can possess a local memory and can carry out localized information processing operations). The main function of the ANN paradigms is to map a set of inputs to a set of outputs. A single processing unit or neuron is shown in Figure 14.2. The incoming signals are multiplied by respective weights through which they are propagated toward the neurons or node, where they are aggregated (summed up) and the net input is passed through the activation function to produce the output.

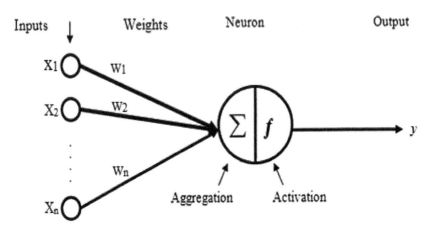

FIGURE 14.2   A single artificial neuron (perceptron).

Let $x_i (i = 1, 2,..., n)$ are inputs and w_i $(i = 1, 2,..., n)$ are respective weights. The net input to the node can be computed as:

$$net = \sum_{i=1}^{n} x_i w_i$$

$$(14.1)$$

The net input is then passed through an activation function (transfer function) f and the output y of the node is determined as below:

$$y = f\ (net) \qquad\qquad (14.2)$$

One neuron cannot perform well for a complex problem; therefore, more than one neuron is used for solving complex problems and is called multilayer perceptron (MLP). In such a network, all neuron outputs of the previous layer relate to neuron inputs of the next layer. There are two fundamental classes of network architecture, namely: recurrent neural network and feed-forward MLP. A recurrent network distinguishes itself from feed-forward network in which there is at least one feedback loop. A signal is transmitted in one direction from the input layer to the output layer and therefore this architecture is called feed-forward. The basic structure of a feed-forward multilayer network usually consists of three or more layers, i.e., the input layer, where the input data is introduced to the network; the hidden layer or layers, where data is processed; and the output layer, where the result of the given input is produced. In these networks, the signal passes in a forward direction from the input layer to the output layer through hidden layer(s). Feed-forward network was used in this chapter and is described as follows:

14.2.3.1.2 Feed-Forward Calculation

In the feed-forward calculation, the nodes in the input layer receive the input signals, which are passed to the hidden layer and then to the output layer. The signals are multiplied by the current values of weights and then the weighted inputs are added to yield net input to each neuron of the next layer. The net input of a neuron is passed through an activation or transfer function to produce the output of the neuron. The procedure for feed-forward calculations in different layers is as follows:

The net input to j^{th} node of the hidden layer is given as:

$$neth_j = \sum_{i=1}^{ni} wh_{ji}\, x_i$$

$$(14.3)$$

where, ni is the number of neurons in the input i^{th} layer; and wh_{ji} is the connection weight between i^{th} node of the input layer and j^{th} node of the hidden layer.

The output of jth node of the hidden layer h_j is:

$$h_j = f(neth_j)$$

$$(14.4)$$

where, $f(.)$ is the activation function, e.g., a sigmoid activation function. Therefore, we get:

$$h_j = \frac{1}{1 + \exp\left(-neth_j\right)}$$

$$(14.5)$$

Similarly, the net input to k^{th} node of the output layer is written as:

$$nety_k = \sum_{j=1}^{nh} wo_{kj} h_j$$

$$(14.6)$$

where, nh is the number of neurons in the hidden layer; and wo_{kj} is the connection weight between j^{th} node of the hidden layer and k^{th} node of the output layer

The output of kth node of the output layer is:

$$y_k = f(nety_k) \qquad (14.7)$$

After the calculation of these outputs, the error between desired and calculated output is computed, which is propagated in the backward direction. The error calculated at the output layer is propagated back to the hidden layers and then to the input layer to determine the updates for the weights. The sum square error ϵ for a single input-output pair data set is given as:

$$E = \frac{1}{2}\sum_{k=1}^{no}\left(y_k - t_k\right)^2 \qquad (14.8)$$

where, t_k is the desired output or target at the kth node; and y_k is the calculated output at the same node.

To minimize the error function (Eqn. (14.8)), weights are updated and changed from their old values. The learning process starts with a random set of weights. During the training process, weights are updated through error back-propagation to reach efficient speed of the learning process. It gives rise to many types of back propagations.

14.2.3.2 ADAPTIVE NEURO-FUZZY INFERENCE SYSTEM (ANFIS)

An Adaptive network-based neuro-fuzzy inference system (ANFIS) is a fuzzy mapping algorithm based on Tagaki-Sugeno-Kang (TSK) FIS [16, 23]. ANFIS combines the advantage of both neural networks (i.e., learning capability, optimization capabilities, and connectionist structures) and FL (i.e., IF-THEN rule-based ease of incorporating expert knowledge) and have the capability to capture the benefits of both these networks in a single framework. ANFIS utilizes linguistic information from the FL and learning capability of an ANN for automatic fuzzy if-then rule base generation and parameter optimization.

An intangible ANFIS consists of five components: input database; output database; a fuzzy system generator; a FIS; and Adaptive Neural Network. The Sugeno-type FIS is the combination of an adaptive neural network. FIS (Figure 14.3) was used in this study for stage-discharge and runoff-sediment modeling. Hybrid learning algorithms is used for optimization.

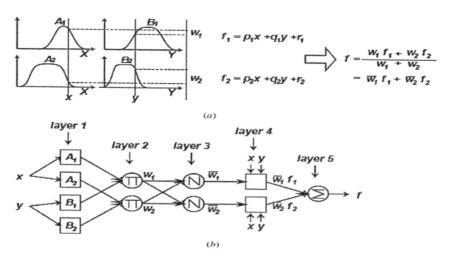

(a)

(b)

FIGURE 14.3 ANFIS architecture.

14.2.3.2.1 *ANFIS Architecture*

The ANFIS is a fuzzy Sugeno model put in the framework of adaptive systems to facilitate learning and adaptation [16]. A first-order Sugeno model, a common-rule set with two fuzzy if-then rules is as follows:

- **Rule 1:** If x_1 is A_1 and x_2 is B_1, then $f_1 = a_1x_1 + b_1x_2 + c_1$
- **Rule 2:** If x_1 is A_2 and x_2 is B_2, then $f_2 = a_2x_1 + b_2x_2 + c_2$

where, x_1 and x_2 are the crisp inputs to the node; and A_1, B_1, A_2, B_2 are fuzzy sets; a_i, b_i, and $c_i (i = 1, 2)$ are the coefficients of first-order polynomial linear function (It is possible to assign different weight to each rule base on the structure of the system); weights w_1 and w_2 are assigned to rules 1 and 2, respectively.

Weighted average is calculated as:

$$f = \text{weighted average} \tag{14.9}$$

The ANFIS consists of the following five layers [16]:

➢ **Layer 1:** Each node output in this layer is fuzzified by membership grades of a fuzzy set corresponding to each input. Fuzzification means using the membership to compute each term's degree of validity at a specific point of the process. The membership function

(MF) for this fuzzy set can be triangular, trapezoidal, generalized bell, and Gaussian MFs. $O_{j,i}$ is the output of the ith node in layer j.

$$O_{1,i} = \mu_{Ai}(x_1) \qquad i = 1, 2 \text{ or} \qquad\qquad (14.10)$$

$$O_{1,i} = \mu_{Bi-2}(x_2) \qquad i = 3, 4 \qquad\qquad (14.11)$$

where, x_1 and x_2 is the input to node i (i = 1, 2 for x_1 and i = 3, 4 for x_2); and A_i(or B_{i-2}) is a fuzzy label.

The MFs for A and B can be any MFs parameterized appropriately, for example:

$$\mu_A(x_1) = \frac{1}{1 + \left[\left(\dfrac{x_1 - c_i}{a_i}\right)^{b_i}\right]} \qquad\qquad (14.12)$$

where; {ai, bi, ci} are the parameters on which bell-shaped function depends, thus exhibiting various forms of MFs on linguistic label A_i.

Parameters in this layer are referred to as foundation parameters. The outputs of this layer are the membership values of the premise part. In this chapter, triangular-shaped, generalized bell-shaped and gauss type, MFs were used.

> **Layer 2:** In this layer, the AND/OR operator is applied to get one output that represents the firing strength of a rule, which performs fuzzy, AND operation. Each node in this layer, labeled Π, is a stable node, which multiplies incoming signals and sends the product out.

$$O_{2,I} = W_i = \mu_{Ai}(x_1)\, \mu_{Bi}(x_2)\ i = 1, 2 \qquad\qquad (14.13)$$

> **Layer 3:** Each node in this layer is a fixed node labeled as N. The ith node calculates the ratio of the ith rule's firing strength to the sum of all rules' firing strength.

$$O_{3,i} = \bar{W}_i = \frac{W_i}{\sum_{k=1}^{n} W_k} i = 1,2 \qquad\qquad (14.14)$$

> **Layer 4:** Each node output in this layer is the normalized value of each fuzzy rule. The nodes in this layer are adaptive. Here is the

output of layer 3, and $\{,\,\}$ is the parameter set. Parameters of this layer are referred to as consequence or output parameters and can be expressed as:

$$O_{4i} = \overline{W}_i\, f_i = \overline{W}_i\left(a_i\, x_i + b_i\, x_2 + c_i\right) \quad i=1,2$$

$$(14.15)$$

➢ **Layer 5:** The single node in this layer is the overall output of the system, which is the summation of all coming signals.

$$Y = \pounds_1^2\, W_i\, f_i = \frac{\pounds_1^2\, W_i\, f_i}{\pounds_1^2\, W_i}$$

$$(14.16)$$

In this way, the input vector is fed through the network layer by layer.

The two major phases for applying the ANFIS for applications are the structure identification phase and the parameter identification phase. The structure identification phase involves finding an appropriate number of fuzzy rules and fuzzy sets and a proper partition feature space. The parameter identification phase involves the adjustment of the suitable and consequence parameters of the system.

14.2.4   DEVELOPMENT OF ANN AND ANFIS MODELS

The development of models consists of system identification and parameter estimation. In this chapter, neural network and neuro-fuzzy inference systems have been used to predict the runoff and sediment concentration on a daily basis for the Godavari river basin at Pathagudem and Polavaram sites in the states of Chhattisgarh and Andhra Pradesh, India. The year-wise hydrological data is shown in Table 14.1.

14.2.4.1   DEVELOPMENT OF ANN AND ANFIS-BASED MODELS FOR RUNOFF PREDICTION

As the daily runoff, prediction is a complex and dynamic process and it needs a proper time lag for its modeling. Therefore, for development of runoff prediction models, different combinations of stage and runoff (i.e., stage of current day, previous one-day, previous two days, previous three days and runoff of previous one-day, previous two days, previous three days) as inputs and runoff of current day as output were used in runoff prediction.

TABLE 14.1 Year-Wise Hydrological Data

Year	Days	Stage	Runoff	Sediment Conc.
1	1	H_{11}	Q_{11}	S_{11}
	2	H_{12}	Q_{12}	S_{12}
	3	H_{13}	Q_{13}	S_{13}
	...	...	...	...
	N	H_{1n}	Q_{1n}	S_{1n}
2	1	H_{21}	Q_{21}	S_{21}
	2	H_{22}	Q_{22}	S_{22}
	3	H_{23}	Q_{23}	S_{23}
	...	...	...	...
	N	H_{2n}	Q_{2n}	S_{2n}
3	1	H_{31}	Q_{31}	S_{31}
	2	H_{32}	Q_{32}	S_{32}
	3	H_{33}	Q_{33}	S_{33}
	...	...	...	...
	N	H_{3n}	Q_{3n}	S_{3n}
...	...	...	...	...
I	1	H_{i1}	Q_{i1}	S_{i1}
	2	H_{i2}	Q_{i2}	S_{i2}
	3	H_{i3}	Q_{i3}	S_{i3}
	...	...	...	...
	N	H_{in}	Q_{in}	S_{in}
...	...	...	...	...
M	1	H_{M1}	Q_{M1}	S_{M1}
	2	H_{M2}	Q_{M2}	S_{M2}
	3	H_{M3}	Q_{M3}	S_{M3}
	...	...	...	...
	N	H_{Mn}	Q_{Mn}	S_{Mn}

$$Q_{i,j} = f(H_{i,j}, H_{i,j-1}, H_{i,j-2,...}H_{i,j-m}, Q_{i,j-1}, Q_{i,j-2}, ...Q_{i,j-m}) \qquad (14.17)$$

where, $H_{i,j}$ is the stage for j_{th} day of the i_{th} year; Q_{ij} is the runoff for j_{th} day of the i_{th} year; and m standards for time lag (which was taken as three in this chapter). The form of year-wise data of stage and runoff data with time lag is shown in Table 14.2.

TABLE 14.2 Input-Output Pairs in Training and Testing for Stage-Discharge Models

Year (i)	Days (j)	Input for j_{th} Day							Output
		X_1	X_2	X_3	X_4	X_5	X_6	X_7	Y
		$Q_{i,j-3}$	$Q_{i,j-2}$	$Q_{i,j-1}$	$H_{i,j-3}$	$H_{i,j-2}$	$H_{i,j-1}$	$H_{i,j}$	$Q_{i,j}$
1	4	$Q_{1,1}$	$Q_{1,2}$	$Q_{1,3}$	$H_{1,1}$	$H_{1,2}$	$H_{1,3}$	$H_{1,4}$	$Q_{1,4}$
	5	$Q_{1,2}$	$Q_{1,3}$	$Q_{1,4}$	$H_{1,2}$	$H_{1,3}$	$H_{1,4}$	$H_{1,5}$	$Q_{1,5}$
	6	$Q_{1,3}$	$Q_{1,4}$	$Q_{1,5}$	$H_{1,3}$	$H_{1,4}$	$H_{1,5}$	$H_{1,6}$	$Q_{1,6}$
	...	...	...	...	...	...	...	...	...
	122	$Q_{1,119}$	$Q_{1,120}$	$Q_{1,121}$	$H_{1,119}$	$H_{1,120}$	$H_{1,121}$	$H_{1,122}$	$Q_{1,122}$
2	4	$Q_{2,1}$	$Q_{2,2}$	$Q_{2,3}$	$H_{2,1}$	$H_{2,2}$	$H_{2,3}$	$H_{2,4}$	$Q_{2,4}$
	5	$Q_{2,2}$	$Q_{2,3}$	$Q_{2,4}$	$H_{2,2}$	$H_{2,3}$	$H_{2,4}$	$H_{2,5}$	$Q_{2,5}$
	6	$Q_{2,3}$	$Q_{2,4}$	$Q_{2,5}$	$H_{2,3}$	$H_{2,4}$	$H_{2,5}$	$H_{2,6}$	$Q_{2,6}$
	...	...	...	...	...	...	...	...	...
	122	$Q_{2,119}$	$Q_{2,120}$	$Q_{2,121}$	$H_{2,119}$	$H_{2,120}$	$H_{2,121}$	$H_{2,122}$	$Q_{2,122}$
3	4	$Q_{3,1}$	$Q_{3,2}$	$Q_{3,3}$	$H_{3,1}$	$H_{3,2}$	$H_{3,3}$	$H_{3,4}$	$Q_{3,4}$
	5	$Q_{3,2}$	$Q_{3,3}$	$Q_{3,4}$	$H_{3,2}$	$H_{3,3}$	$H_{3,4}$	$H_{3,5}$	$Q_{3,5}$
	6	$Q_{3,3}$	$Q_{3,4}$	$Q_{3,5}$	$H_{3,3}$	$H_{3,4}$	$H_{3,5}$	$H_{3,6}$	$Q_{3,6}$
	...	...	...	...	...	...	...	...	...
	122	$Q_{3,119}$	$Q_{3,120}$	$Q_{3,121}$	$H_{3,119}$	$H_{3,120}$	$H_{3,121}$	$H_{3,122}$	$Q_{3,122}$
...	...	...	...	...	...	...	...	...	...
15	4	$Q_{15,1}$	$Q_{15,2}$	$Q_{15,3}$	$H_{15,1}$	$H_{15,2}$	$H_{15,3}$	$H_{15,4}$	$Q_{15,4}$
	5	$Q_{15,2}$	$Q_{15,3}$	$Q_{15,4}$	$H_{15,2}$	$H_{15,3}$	$H_{15,4}$	$H_{15,5}$	$Q_{15,5}$
	6	$Q_{15,3}$	$Q_{15,4}$	$Q_{15,5}$	$H_{15,3}$	$H_{15,4}$	$H_{15,5}$	$H_{15,6}$	$Q_{15,6}$
	...	...	...	...	...	...	...	...	...
	122	$Q_{15,119}$	$Q_{15,120}$	$Q_{15,121}$	$H_{15,119}$	$H_{15,120}$	$H_{15,121}$	$H_{15,122}$	$Q_{15,122}$

14.2.4.1.1 Normalization of Input Data

The values of input data range from a minimum value to a maximum value and differ significantly from one variable to other. To avoid the model from giving more importance to some of the variables as compared with others, input data were normalized within a common range. Both stage and runoff data were normalized within range 0 to 1 as follows:

$$X_N = \frac{X - X_{min}}{X_{max} - X_{min}}$$

(14.18)

where, N is the value of the input or output variable (X) in the input-output data; is the normalized value of the input or output variable $(_x)$ in N^{th} data; X_{max} is maximum value of X_N; and X_{min} is minimum value of X_N.

14.2.4.1.2 *Formulation of Training and Testing Data*

Stage and runoff are represented by H_{ij} and Q_{ij} of ith year and j^{th} day, respectively. For training and testing of the ANN and ANFIS, the required daily stage time series H_{ij} for i = 1 to M year index and for j = 1 to N day index were available, where, M is the total number of years and N is the total number of days in the monsoon season in the data set of ith year.

Similarly, the required daily runoff time series Q_{ij}, i = 1 to M and j = 1 to N, were also available. It was observed that N = 122 days (i.e., 1st June to 30th September) in a year and M = 15 years (1996–2010) for the Godavari river basin at Pathagudem and Polavaram site. The data were divided into two sets: a set of training data for model development, and a set of testing data for validation (testing) of developed model.

The daily stage and runoff data of monsoon season (1st June to 30th September) during 1996–2007 for both sites were used for training of ANN and ANFIS models; and the data for 2008 to 2010 were used for validation (testing) of both models.

14.2.4.2 *DEVELOPMENT OF ANN AND ANFIS-BASED SEDIMENT CONCENTRATION PREDICTION MODELS*

The prediction of daily SSC is a complex procedure, which depends on various parameters, namely: rainfall behavior, runoff, soil characteristics and vegetative cover, etc., and it needs a proper time lag for its modeling. Therefore, for development of sediment concentration prediction models, different combinations of runoff and sediment concentration (i.e., runoff of current day, previous one-day, previous two days, previous three days and sediment concentration of previous one-day, previous two days, previous three days) as inputs and sediment concentration of current day as output were used in sediment concentration prediction. The year-wise runoff and sediment concentration data with time lag is shown in Table 14.3.

$$Q_{i,j} = f(Q_{i,j}, Q_{i,j-1}, Q_{i,j-2,...}, Q_{i,j-m}, S_{i,j-1}, S_{i,j-2}, \ldots, S_{i,j-m}) \qquad (14.19)$$

where, $Q_{i,j}$ is the runoff for j_{th} day of the i_{th} year; S_{ij} is the sediment concentration for j_{th} day of the i_{th} year; and m standards for time lag, which was taken as three in this chapter.

TABLE 14.3 Input-Output Pairs in Training and Testing for Runoff-Sediment Models

Year (i)	Days (j)	Input for j_{th} Day							Output
		X_1	X_2	X_3	X_4	X_5	X_6	X_7	Y
		$S_{i,j-3}$	$S_{i,j-2}$	$S_{i,j-1}$	$Q_{i,j-3}$	$Q_{i,j-2}$	$Q_{i,j-1}$	$Q_{i,j}$	$S_{i,j}$
1	4	$S_{1,1}$	$S_{1,2}$	$S_{1,3}$	$Q_{1,1}$	$Q_{1,2}$	$Q_{1,3}$	$Q_{1,4}$	$S_{1,4}$
	5	$S_{1,2}$	$S_{1,3}$	$S_{1,4}$	$Q_{1,2}$	$Q_{1,3}$	$Q_{1,4}$	$Q_{1,5}$	$S_{1,5}$
	6	$S_{1,3}$	$S_{1,4}$	$S_{1,5}$	$Q_{1,3}$	$Q_{1,4}$	$Q_{1,5}$	$Q_{1,6}$	$S_{1,6}$
	...	...	...	...	...	...	...	...	...
	122	$S_{1,119}$	$S_{1,120}$	$S_{1,121}$	$Q_{1,119}$	$Q_{1,120}$	$Q_{1,121}$	$Q_{1,122}$	$S_{1,122}$
2	4	$S_{2,1}$	$S_{2,2}$	$S_{2,3}$	$Q_{2,1}$	$Q_{2,2}$	$Q_{2,3}$	$Q_{2,4}$	$S_{2,4}$
	5	$S_{2,2}$	$S_{2,3}$	$S_{2,4}$	$Q_{2,2}$	$Q_{2,3}$	$Q_{2,4}$	$Q_{2,5}$	$S_{2,5}$
	6	$S_{2,3}$	$S_{2,4}$	$S_{2,5}$	$Q_{2,3}$	$Q_{2,4}$	$Q_{2,5}$	$Q_{2,6}$	$S_{2,6}$
	...	...	...	...	...	...	...	...	...
	122	$S_{2,119}$	$S_{2,120}$	$S_{2,121}$	$Q_{2,119}$	$Q_{2,120}$	$Q_{2,121}$	$Q_{2,122}$	$S_{2,122}$
3	4	$S_{3,1}$	$S_{3,2}$	$S_{3,3}$	$Q_{3,1}$	$Q_{3,2}$	$Q_{3,3}$	$Q_{3,4}$	$S_{3,4}$
	5	$S_{3,2}$	$S_{3,3}$	$S_{3,4}$	$Q_{3,2}$	$Q_{3,3}$	$Q_{3,4}$	$Q_{3,5}$	$S_{3,5}$
	6	$S_{3,3}$	$S_{3,4}$	$S_{3,5}$	$Q_{3,3}$	$Q_{3,4}$	$Q_{3,5}$	$Q_{3,6}$	$S_{3,6}$
	...	...	...	...	...	...	...	...	...
	122	$S_{3,119}$	$S_{3,120}$	$S_{3,121}$	$Q_{3,119}$	$Q_{3,120}$	$Q_{3,121}$	$Q_{3,122}$	$S_{3,122}$
...	...	...	...	...	...	...	...	...	...
15	4	$S_{15,1}$	$S_{15,2}$	$S_{15,3}$	$Q_{15,1}$	$Q_{15,2}$	$Q_{15,3}$	$Q_{15,4}$	$S_{15,4}$
	5	$S_{15,2}$	$S_{15,3}$	$S_{15,4}$	$Q_{15,2}$	$Q_{15,3}$	$Q_{15,4}$	$Q_{15,5}$	$S_{15,5}$
	6	$S_{15,3}$	$S_{15,4}$	$S_{15,5}$	$Q_{15,3}$	$Q_{15,4}$	$Q_{15,5}$	$Q_{15,6}$	$S_{15,6}$
	...	...	...	...	...	...	...	...	...
	122	$S_{15,119}$	$S_{15,120}$	$S_{15,121}$	$Q_{15,119}$	$Q_{15,120}$	$Q_{15,121}$	$Q_{15,122}$	$S_{15,122}$

14.2.4.2.1 Formulation of Training and Testing Data

The required daily runoff and SSC time series data for both sites are shown in Table 14.3. It was observed that N = 122 days (i.e., 1st June to 30th September) in a year; and M = 15 years (1996–2010) for the Godavari river basin at Pathagudem and Polavaram sites. The data were divided into two sets: a training set for model development and a testing set for validation.

The daily runoff and SSC data of monsoon season (1st June to 30th September) for 1996 to 2007 were utilized for calibration and the data for 2008 to 2010 were used for validation of ANN and ANFIS models for both sites.

14.2.5   SENSITIVITY ANALYSIS

While training a network, the effect of each network inputs on the network output is studied. This provides feedback as to which input parameters are the most significant. Based on this feedback, one may decide to prune the input space by removing the significant parameters. This also reduces the size of the network which in turn reduces the network complexity and the training time. The sensitivity analysis is carried out by removing the each of the parameters in turn from the input parameters used in ANN and ANFIS models and then comparing the performance statistics. The greater the effect observed in the output, the greater is the sensitivity of that particular input parameter.

14.2.6   PERFORMANCE EVALUATION OF DEVELOPED MODELS

14.2.6.1   QUALITATIVE EVALUATION

The qualitative assessment of stage-discharge and runoff-sediment models is based on visual observation and is one of the simplest methods for the evaluation of models. The goodness of fit between observed and predicted runoff and SSC are tentatively checked by qualitative comparison. The visual observation consists of a graphical comparison of various features of observed and predicted runoff and sediment.

14.2.6.2   QUANTITATIVE EVALUATION

The predictive effectiveness of ANN and ANFIS-based models were assessed on the basis of performance indicators. To judge the predictive capability of the developed models, the following performance evaluating indices were used:

1. **Correlation Coefficient (CC):** It is determined using the following equation:

$$CC = \frac{\sum_{j=1}^{n}\left\{\left(Y_j - \bar{Y}_j\right)\left(Y_{ej} - \bar{Y}_{ej}\right)\right\}}{\sum_{j=1}^{n}\left(Y_j - \bar{Y}_j\right)^2 \sum_{j=1}^{n}\left(Y_{ej} - \bar{Y}_{ej}\right)^2} \times 100$$

(14.20)

where, Y_j is the predicted values; $\bar{Y}_j$ is the mean of predicted values; Y_{ej} is the observed values; n is the number of observations; and $\bar{Y}_{ej}$ is the mean of observed values.

The CC measures the statistical correlation between the observed and predicted values. The value of CC closer to one means better model.

2. **Root Mean Square Error (RMSE):** It is the most commonly used for assessment of numeric prediction. The RMSE was calculated with the following equation:

$$RMSE = \sqrt{(1/n)\left(\sum_{i=1}^{n}\left(Y_{ej} - Y_j\right)^2\right)}$$

(14.21)

The value of RMSE closer to zero indicates better fit and increased values indicate higher disagreement between predicted and observed values.

3. **Coefficient of Efficiency (CE):** It is computed using the following equation:

$$CE = \left[1 - \frac{\sum_{i=1}^{n}\left(Y_{ej} - Y_j\right)^2}{\sum_{i=1}^{n}\left(Y_{ej} - \bar{Y}_{ej}\right)^2}\right] \times 100\%$$

(14.22)

4. **Pooled Average Relative Error (PARE):** The accuracy of prediction (under and over-prediction) was judged on the basis of PARE [22]. The positive value of PARE indicates that the model is over predicted, and a negative value indicates that the model is under predicted. The PARE is calculated as below:

$$PARE\,(\%) = \frac{1}{N}\left[\frac{\sum_{i=1}^{n}\left(Y_j - Y_{ej}\right)}{\sum_{i=1}^{n}Y_{ej}}\right] * 100$$

(14.23)

14.3　RESULTS AND DISCUSSION

This section presents a methodology for the development and application of GT, ANNs, and ANFIS models for daily runoff and SSC prediction and the outcomes after evaluating the models qualitatively by visual observation and quantitatively by various statistical and hydrological indices.

14.3.1　SELECTION OF BEST INPUT COMBINATION USING GAMMA TEST (GT)

The GT helps to take decisions on the selection of input data or inputs, which are affecting the results of developed models. Number of inputs was selected based on gamma value (Γ), standard error and V-ratio. The minimum values of GT, standard error, and V-ratio were considered as the best model. A total (2^n-1) combinations of inputs for each stage-runoff and runoff-sediment of each site were explored, from which the best one model was selected on the basis of values of gamma, standard error, and V-ratio given in Tables 14.4–14.7.

At the time of selecting the best runoff prediction model, total seven inputs were considered, namely: stage of current day, stage of previous one-day, stage of previous two days, stage of previous three days, runoff of previous one-day, runoff of two days and runoff of three days, which are represented as "$H_t, H_{t-1}, H_{t-2}, H_{t-3}, Q_{t-1}, Q_{t-2}, Q_{t-3}$," respectively, and 127 combinations were explored. Finally, stage of current day, previous one-day, previous three days, runoff of previous one-day, two days and three days were selected as the best input variables for Pathagudem site (Table 14.4); and stage of current day, previous one-day, previous two days, previous three days, runoff of previous one-day and two days as the best inputs for Polavaram site (Table 14.5).

For selection of best SSC prediction model, total seven inputs were considered, namely: runoff of current day, runoff of previous one-days, runoff of previous two days, runoff of previous three days, sediment of previous one-day, sediment of two days and sediment of three days. The seven inputs were represented as "$Q_t, Q_{t-1}, Q_{t-2}, Q_{t-3}, S_{t-1}, S_{t-2}, S_{t-3}$," respectively, and 127 combinations were explored. Finally, runoff of current day, runoff of previous one-day, runoff of previous two days, runoff of previous three days, sediment of previous one-day and sediment of three days were selected as the best inputs variables for Pathagudem site (Table 14.6); and runoff of current day, runoff of previous one-day, runoff of previous two days, sediment of previous one-day, sediment of two days and sediment of three days were the best inputs for Polavaram site (Table 14.7).

TABLE 14.4 Comparison of Various Models to Select the Best Model for Runoff Prediction at the Pathagudem site

Input	Mask	Gamma (Γ)	Standard Error	V-Ratio
$H_{t-3},H_{t-2},H_{t-1},H_t,Q_{t-3},Q_{t-2},Q_{t-1}$	1111111	0.0049	0.0027	0.020
$H_{t-3}, H_{t-2},H_{t-1}, H_t, Q_{t-3}, Q_{t-2}$	0111111	0.0070	0.0021	0.028
$H_{t-3},H_{t-2},H_{t-1}, H_t,Q_{t-3},Q_{t-1}$	1011111	0.0069	0.0022	0.028
$H_{t-3},H_{t-2},H_{t-1}, H_t, Q_{t-2}, Q_{t-1}$	1101111	0.0057	0.0038	0.022
$H_{t-3},H_{t-2},H_{t-1}, Q_{t-3},Q_{t-2}, Q_{t-1}$	1110111	0.0710	0.0048	0.283
$H_{t-3},H_{t-2},H_t,Q_{t-3},Q_{t-2},Q_{t-1}$	1111011	0.0060	0.0015	0.024
$H_{t-3},H_{t-1},H_t,Q_{t-3},Q_{t-2},Q_{t-1}$	1111101	0.0026	0.0012	0.010
$H_{t-2},H_{t-1},H_t,Q_{t-3},Q_{t-2},Q_{t-1}$	1111110	0.0070	0.0025	0.028

TABLE 14.5 Comparison of Various Models to Select the Best Model for Runoff Prediction at the Polavaram site

Input	Mask	Gamma (Γ)	Standard Error	V-Ratio
$H_{t-3},H_{t-2},H_{t-1},H_t,Q_{t-3},Q_{t-2},Q_{t-1}$	1111111	0.0063	0.0012	0.025
$H_{t-3},H_{t-2},H_{t-1}, H_t,Q_{t-3}, Q_{t-2}$	0111111	0.0056	0.0013	0.022
$H_{t-3},H_{t-2},H_{t-1}, H_t,Q_{t-3},Q_{t-1}$	1011111	0.0069	0.0011	0.028
$H_{t-3},H_{t-2},H_{t-1}, H_t, Q_{t-2}, Q_{t-1}$	1101111	0.0047	0.0009	0.019
$H_{t-3},H_{t-2},H_{t-1}, Q_{t-3},Q_{t-2}, Q_{t-1}$	1110111	0.0270	0.0015	0.110
$H_{t-3},H_{t-2},H_t,Q_{t-3},Q_{t-2},Q_{t-1}$	1111011	0.0059	0.0012	0.023
$H_{t-3},H_{t-1},H_t,Q_{t-3},Q_{t-2},Q_{t-1}$	1111101	0.0062	0.0010	0.025
$H_{t-2},H_{t-1},H_t,Q_{t-3},Q_{t-2},Q_{t-1}$	1111110	0.0061	0.0010	0.024

TABLE 14.6 Comparison of Various Models to Select the Best Model for Suspended Sediment Concentration Prediction at the Pathagudem site

Input	Mask	Gamma (Γ)	Standard Error	V-Ratio
$Q_{t-3},Q_{t-2},Q_{t-1},Q_t,S_{t-3},S_{t-2},S_{t-1}$	1111111	0.064	0.0052	0.256
$Q_{t-3},Q_{t-2}, Q_{t-1}, Q_t, S_{t-3}, S_{t-2}$	0111111	0.095	0.0075	0.380
$Q_{t-3}, Q_{t-2}, Q_{t-1}, Q_t, S_{t-3}, S_{t-1}$	1011111	0.058	0.0021	0.233
$Q_{t-3}, Q_{t-2}, Q_{t-1}, Q_t, S_{t-2}, S_{t-1}$	1101111	0.067	0.0045	0.269
$Q_{t-3}, Q_{t-2}, Q_{t-1}, S_{t-3}, S_{t-2}, S_{t-1}$	1110111	0.087	0.0039	0.350
$Q_{t-3}, Q_{t-2}, Q_t, S_{t-3}, S_{t-2}, S_{t-1}$	1111011	0.063	0.0026	0.250
$Q_{t-3}, Q_{t-1}, Q_t, S_{t-3}, S_{t-2}, S_{t-1}$	1111101	0.061	0.0033	0.245
$Q_{t-2}, Q_{t-1}, Q_t, S_{t-3}, S_{t-2}, S_{t-1}$	1111110	0.071	0.0042	0.285

TABLE 14.7 Comparison of Various Models to Select the Best Model for suspended Sediment Concentration Prediction at the Polavaram site

Input	Mask	Gamma (Γ)	Standard Error	V-Ratio
$Q_{t-3}, Q_{t-2}, Q_{t-1}, Q_t, S_{t-3}, S_{t-2}, S_{t-1}$	1111111	0.025	0.0028	0.101
$Q_{t-3}, Q_{t-2}, Q_{t-1}, Q_t, S_{t-3}, S_{t-2}$	0111111	0.040	0.0031	0.162
$Q_{t-3}, Q_{t-2}, Q_{t-1}, Q_t, S_{t-3}, S_{t-1}$	1011111	0.029	0.0036	0.120
$Q_{t-3}, Q_{t-2}, Q_{t-1}, Q_t, S_{t-2}, S_{t-1}$	1101111	0.305	0.0034	0.121
$Q_{t-3}, Q_{t-2}, Q_{t-1}, S_{t-3}, S_{t-2}, S_{t-1}$	1110111	0.049	0.0026	0.202
$Q_{t-3}, Q_{t-2}, Q_t, S_{t-3}, S_{t-2}, S_{t-1}$	1111011	0.031	0.0043	0.133
$Q_{t-3}, Q_{t-1}, Q_t, S_{t-3}, S_{t-2}, S_{t-1}$	1111101	0.028	0.0024	0.111
$Q_{t-2}, Q_{t-1}, Q_t, S_{t-3}, S_{t-2}, S_{t-1}$	1111110	0.025	0.0026	0.100

"Feed-forward backpropagation algorithm" was used to train the network for runoff and SSC predictions. Various networks of the single hidden layers and double hidden layers were trained for maximum epochs of 1000, on the basis of minimum values of RMSE, PARE, and maximum values of CC and CE for all cases. Soft computing software MATLAB (R2015a) was utilized for the construction of ANN models.

14.3.2 ARTIFICIAL NEURAL NETWORKS (ANNS) MODELS FOR RUNOFF PREDICTION

The daily runoff prediction models for the study areas (i.e., Pathagudem, and Polavaram sites) were developed with the stage of current day, stage of previous one-day, stage of previous three days, runoff of previous one-day, runoff of two days and runoff of three days as input variables and current day runoff as output variable for Pathagudem site; and the stage of current day, stage of previous one-day, stage of previous three days, runoff of previous one-day, runoff of two days and runoff of three days as input variables and current day runoff as output variable for Polavaram site. Using the Log-sigmoid activation function for each hidden layer and Pureline activation function in the output layer and Levenberg-Marquardt learning algorithm, a few of the single hidden layer networks and double hidden layer networks were selected (Table 14.8) having higher values of CC, CE and lower values of RMSE, PARE.

It is clear that the performance of the network (6–6–1) is better than the other networks in single hidden layers and double hidden layers for Pathagudem site; and the network (6–9–1) is better than the other models in single hidden layers and double hidden layers for Polavaram site. In the model architecture (6–6–1), six shows input variables in input node; three indicates the number of neurons in the hidden layer and one shows the number of variables in the output node.

14.3.3   ARTIFICIAL NEURAL NETWORKS (ANNS) MODELS FOR SSC PREDICTION

In case of suspended sediment prediction models, the runoff of current day, runoff of previous one-day, runoff of previous two days, runoff of previous three days, SSC of previous one-day and SSC of three days were input variables and current day runoff as output variable for Pathagudem site; and the runoff of current day, runoff of previous one-day, runoff of previous two days, runoff of previous three days, SSC of previous one-day and SSC of three days were input variables and current day runoff as output variable for Polavaram site.

Using Log-sigmoid activation function (transfer function) for each hidden layer, Pureline activation function in output layer and Levenberg-Marquardt learning algorithm, a few of the single hidden layer and double hidden layer networks were selected based on statistical and hydrological indices (Table 14.9). It is apparent that the performance of the network (6–4–4–1) is better than the other networks in single hidden layers and double hidden layers for the Pathagudem site and the network (6–5–1) is better than the other networks in single hidden layers and double hidden layers for Polavaram site.

14.3.4   ANFIS-BASED RUNOFF AND SUSPENDED SEDIMENT CONCENTRATION (SSC) PREDICTION MODELS

Runoff prediction models were developed using stage of current day and previous day as well as previous day runoff as inputs and current day runoff as the output for two sites, i.e., Pathagudem, and Polavaram sites. SSC prediction models were developed using runoff of current day and previous day as well as previous day SSC as inputs and current day SSC as the output for both the sites.

TABLE 14.8 Comparison of Various ANN Models for Runoff Prediction at Pathagudem and Polavaram Sites

Pathagudem Site				Polavaram Site			
Network	r	RMSE (m³/sec)	CE	Network	r	RMSE (m³/sec)	CE
6–1–1	0.989	386.91	0.9788	6–1–1	0.991	928.39	0.9832
6–2–1	0.991	358.88	0.9817	6–2–1	0.992	903.06	0.9843
6–3–1	0.992	330.23	0.9845	6–3–1	0.992	904.66	0.9840
6–4–1	0.990	366.58	0.9810	6–4–1	0.994	784.09	0.9880
6–5–1	0.992	325.98	0.9849	6–5–1	0.991	929.36	0.9830
6–6–1	0.993	310.70	0.9860	6–6–1	0.991	974.92	0.9810
6–7–1	0.990	350.25	0.9826	6–7–1	0.992	883.18	0.9848
6–8–1	0.988	410.93	0.9760	6–8–1	0.992	915.53	0.9836
6–9–1	0.986	444.38	0.9720	6–9–1	0.994	739.03	0.9893
6–10–1	0.986	451.94	0.9710	6–10–1	0.992	850.29	0.9859
6–1–1–1	0.820	1588.91	0.6410	6–1–1–1	0.990	934.46	0.9829
6–2–2–1	0.989	386.04	0.9788	6–2–2–1	0.990	919.60	0.9835
6–3–3–1	0.992	323.10	0.9850	6–3–3–1	0.990	922.57	0.9834
6–4–4–1	0.989	365.09	0.9810	6–4–4–1	0.994	791.91	0.9877
6–5–5–1	0.993	321.75	0.9850	6–5–5–1	0.990	825.06	0.9867
6–6–6–1	0.977	562.42	0.9550	6–6–6–1	0.990	897.22	0.9840
6–7–7–1	0.992	333.47	0.9840	6–7–7–1	0.990	824.61	0.9867
6–8–8–1	0.990	321.02	0.9850	6–8–8–1	0.990	811.96	0.9870
6–9–9–1	0.992	338.17	0.9838	6–9–9–1	0.993	864.72	0.9854
6–10–10–1	0.970	649.24	0.9403	6–10–10–1	0.993	820.83	0.9868

TABLE 14.9 Comparison of Various ANN Models for SSC Prediction at Pathagudem and Polvaram Sites

Pathagudem				Polavaram			
Network	r	RMSE (g/l)	CE	Networks	r	RMSE (g/l)	CE
6–1–1	0.850	0.1781	0.7222	6–1–1	0.915	0.1943	0.8350
6–2–1	0.851	0.1785	0.7206	6–2–1	0.915	0.1924	0.8385
6–3–1	0.853	0.1770	0.7254	6–3–1	0.932	0.1740	0.8680
6–4–1	0.855	0.1793	0.7180	6–4–1	0.930	0.1763	0.8645
6–5–1	0.870	0.1694	0.7485	6–5–1	0.938	0.1650	0.8808
6–6–1	0.850	0.1790	0.7193	6–6–1	0.930	0.1755	0.8657

TABLE 14.9 *(Continued)*

Pathagudem				Polavaram			
Network	r	RMSE (g/l)	CE	Networks	r	RMSE (g/l)	CE
6–7–1	0.840	0.1860	0.6970	6–7–1	0.937	0.1670	0.8784
6–8–1	0.872	0.1665	0.7570	6–8–1	0.927	0.1799	0.8589
6–9–1	0.875	0.1644	0.7630	6–9–1	0.935	0.1693	0.8750
6–10–1	0.883	0.1596	0.7760	6–10–1	0.925	0.1812	0.8568
6–1–1–1	0.850	0.1777	0.7233	6–1–1–1	0.915	0.1930	0.8370
6–2–2–1	0.840	0.1821	0.7096	6–2–2–1	0.930	0.1792	0.8601
6–3–3–1	0.870	0.1673	0.7547	6–3–3–1	0.930	0.1846	0.8514
6–4–4–1	0.890	0.1561	0.7864	6–4–4–1	0.930	0.1777	0.8620
6–5–5–1	0.850	0.1804	0.7150	6–5–5–1	0.940	0.1713	0.8720
6–6–6–1	0.877	0.1644	0.7632	6–6–6–1	0.930	0.1714	0.8719
6–7–7–1	0.876	0.1640	0.7645	6–7–7–1	0.935	0.1700	0.8740
6–8–8–1	0.880	0.1604	0.7745	6–8–8–1	0.930	0.1836	0.8530
6–9–9–1	0.840	0.2012	0.6450	6–9–9–1	0.920	0.1904	0.8420
6–10–10–1	0.884	0.1589	0.7786	6–10–10–1	0.934	0.1708	0.8727

The daily stage, runoff, and sediment concentration data of monsoon period (June to September) of Pathagudem and Polavaram sites were used to describe daily time series and development of models. For runoff prediction at sites, daily stage and runoff data during 1996 to 2007 were used for the training (calibration) of the developed models; whereas daily data for 2008 to 2010 were used for verification (testing) purpose.

Also for sediment predictions at sites, daily runoff and sediment concentration data for 1996 to 2007 were used for the training of models; whereas daily data for 2008 to 2010 were used for validation (testing) of the developed models. The models were trained using the Takagi and Sugeno inference system, and triangular, trapezoidal, Gaussian, and generalized bell MFs. The number of MF was specified by the "MFs per input" (Tables 14.10 and 14.11).

14.3.4.1 ANFIS-BASED MODELS FOR RUNOFF PREDICTION

Different combinations of stage and runoff were considered as inputs of the model and the runoff of the current day as the output, from which the best model was selected on the basis of GT. For Pathagudem site, stage of

TABLE 14.10 Comparison of Various ANFIS Models to Select the Best Model for Runoff Prediction at Pathagudem and Polavaram Sites

Networks/MFs per Input	Pathagudem			Polavaram		
	r	RMSE (m³/sec)	CE	r	RMSE (m³/sec)	CE
Triangular, 2	0.9875	417.52	0.9753	0.9926	866.46	0.9853
Trapezoidal, 2	0.9389	913.90	0.8817	0.9970	490.65	0.9953
Generalized Bell, 2	0.9902	369.91	0.9806	0.9940	772.35	0.9880
Gauss, 2	0.9917	339.94	0.9836	0.9942	741.22	0.9892
Triangular, 3	0.9929	314.62	0.9859	0.9977	485.37	0.9954
Trapezoidal, 3	0.9622	723.44	0.9258	0.9700	1668.07	0.9458
Generalized Bell, 3	0.9931	312.59	0.9860	0.9969	554.03	0.9940
Gauss, 3	0.9941	283.85	0.9885	0.9976	485.53	0.9954

TABLE 14.11 Comparison of Various ANFIS Models to Select the Best Model for SSC Prediction at Pathagudem and Polavaram Sites

Networks/MFs per Input	Pathagudem			Polavaram		
	r	RMSE (g/l)	CE	r	RMSE (g/l)	CE
Triangular, 2	0.8860	0.1565	0.7850	0.9345	0.1652	0.8734
Trapezoidal, 2	0.6544	0.2554	0.4280	0.8101	0.2717	0.6574
Generalized Bell, 2	0.8612	0.1720	0.7408	0.9060	0.1957	0.8223
Gauss, 2	0.8717	0.1655	0.7599	0.9228	0.1787	0.8518
Triangular, 3	0.9001	0.1507	0.8010	0.9491	0.1461	0.9009
Trapezoidal, 3	0.7612	0.2194	0.5783	0.9023	0.2000	0.8143
Generalized Bell, 3	0.8701	0.1658	0.7591	0.9495	0.1456	0.9016
Gauss, 3	0.8902	0.1538	0.7925	0.9540	0.1392	0.9101

current day, stage of previous one-day, stage of previous three days as well as runoff of previous one-day, runoff of two days and runoff of three days as input variables and current day runoff as output variable were used. For Polavaram site, the stage of current day, the stage of previous one-day, the stage of previous two days, the stage of previous three days, runoff of previous one-day and runoff of previous two days as input variables and current day runoff as output variable were used. Input space partitioning has been carried out using grid partition technique in same way as for runoff prediction ANFIS models. Performance of hybrid learning algorithm utilized for train the model was evaluated on the basis of RMSE, CE, PARE, and CC [26].

14.3.4.2   ANFIS-BASED MODELS FOR SSC PREDICTION

Various combinations of runoff and SSC were explored as the inputs of the model and current day SSC as the output, from which the best model was selected on the basis of GT. Runoff of current day, Runoff of previous one-day, Runoff of previous two days, Runoff of previous three days, SSC of previous one-day and SSC of three days as input variables and current day runoff as output variable for Pathagudem were utilized. For Polavaram site, the runoff of current day, runoff of previous one-day, runoff of previous two days, sediment of previous one-day, and sediment of three days as input variables and current day SSC as output variable were utilized. Input space partitioning has been carried out using grid partition technique in same way as for runoff prediction ANFIS models. Performance of hybrid learning algorithm utilized for train the model was evaluated on the basis of RMSE, CE, PARE, and CC [26].

14.3.5   PERFORMANCE EVALUATION OF DEVELOPED MODELS

14.3.5.1   QUALITATIVE EVALUATION

The qualitative evaluation of the model is based on the visual comparison, i.e., overall shape of the observed and predicted graphs. The qualitative assessment of models was made by comparing the predicted daily runoff with observed ones and by comparing the predicted SSC with observed ones at both sites to verify and validate the equivalence between the catchment and models.

Using ANN-based models for Pathagudem site, the plots of observed and predicted values of runoff for the corresponding training (1996–2007) and testing (2008–2010) periods are shown in Figures 14.4–14.7, respectively. Also, the observed and predicted values of runoff for the corresponding training and testing periods of Pathagudem site using ANFIS-based models are shown in Figures 14.8–14.11, respectively.

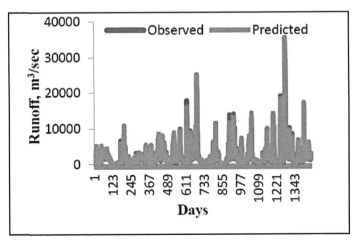

FIGURE 14.4 Observed and predicted runoff using ANN model (6–6–1) Pathagudem site during training period.

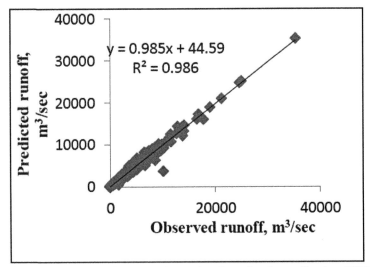

FIGURE 14.5 Scatter plot between observed and predicted runoff using ANN model (6–6–1) at the Pathagudem site during training period.

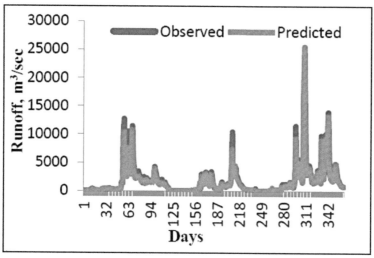

FIGURE 14.6   Observed and predicted runoff using ANN model (6–6–1) at the Pathagudem site during testing period.

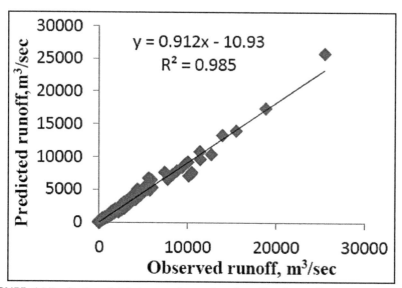

FIGURE 14.7   Scatter plot between observed and predicted runoff using ANN model (6–6–1) at the Pathagudem site during testing period.

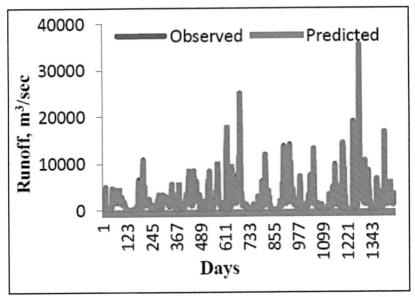

FIGURE 14.8   Observed and predicted runoff using ANFIS model (Gauss, 3) at the Pathagudem site during training period.

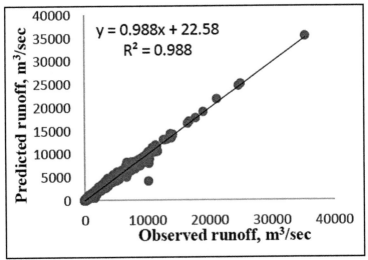

FIGURE 14.9   Scatter plot between observed and predicted runoff using ANFIS model (Gauss, 3) at the Pathagudem site during training period.

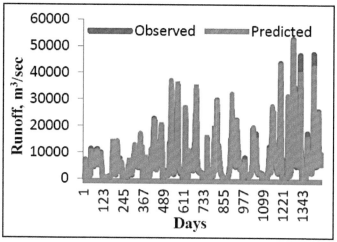

FIGURE 14.10 Observed and predicted runoff using ANFIS model (Gauss, 3) at the Pathagudem site during testing period.

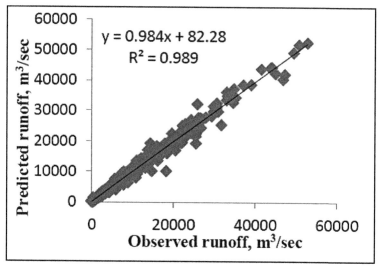

FIGURE 14.11 Scatter plot between observed and predicted runoff using ANFIS model (Gauss, 3) at the Pathagudem site during testing period.

The observed and predicted values of runoff of Polavaram site using ANN models for the corresponding training (1996–2007) and testing (2008–2010) periods are presented in Figures 14.12–14.15, respectively. For ANFIS-based models, the observed and predicted values of runoff during corresponding calibration and validation periods for Polavaram site are depicted in Figures

14.16–14.19, respectively. Based on these figures, there is a good agreement between the predicted and observed runoff, and the overall shape of predicted runoff hydrograph is similar to that of the observed runoff hydrograph. The graphs show that the qualitative performance of the ANN and ANFIS models are satisfactory. ANFIS model performance is better than ANN-based model during runoff prediction at both sites.

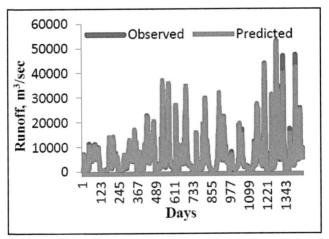

FIGURE 14.12 Observed and predicted runoff using ANN model (6–9–1) at the Polavaram site during training period.

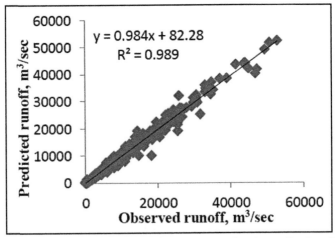

FIGURE 14.13 Scatter plot between observed and predicted runoff using ANN model (6–9–1) at the Polavaram site during training period.

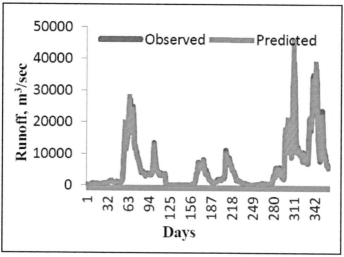

FIGURE 14.14   Observed and predicted runoff using ANN model (6–9–1) at the Polavaram site during testing period.

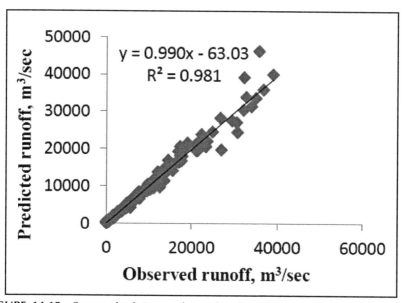

FIGURE 14.15   Scatter plot between observed and predicted runoff using ANN model (6–9–1) at the Polavaram site during testing period.

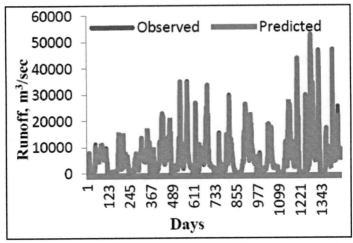

FIGURE 14.16   Observed and predicted runoff using ANFIS model (Triangular, 3) at the Polavaram site during training period.

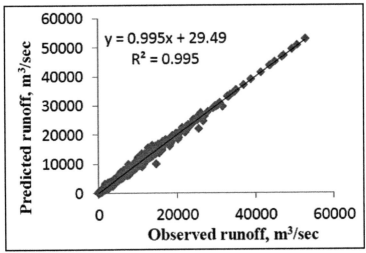

FIGURE 14.17   Scatter plot between observed and predicted runoff using ANFIS model (Triangular, 3) at the Polavaram site during training period.

The plots of observed and predicted values of SSC using ANN during training (1996–2007) and testing (2008–2010) periods for Pathagudem site are depicted in Figures 14.20–14.23, respectively. Based on ANFIS models, the observed and estimated (predicted) values of SSC during training and testing periods are presented in Figures 14.24–14.27, respectively.

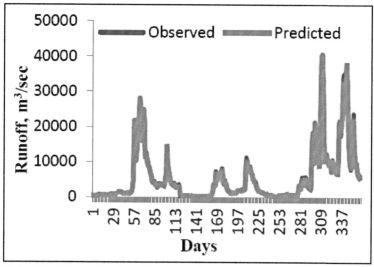

FIGURE 14.18   Observed and predicted runoff using ANFIS model (Triangular, 3) at the Polavaram site during testing period.

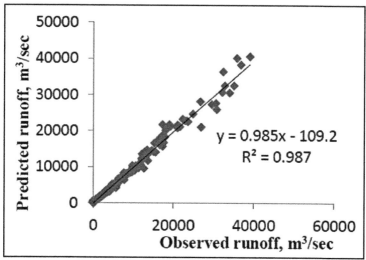

FIGURE 14.19   Scatter plot between observed and predicted runoff using ANFIS model (Triangular, 3) at the Polavaram site during testing period.

For Polavaram site with ANN models, the observed and predicted values of SSC for the corresponding training (1996–2007) and testing (2008–2010) periods are depicted in Figures 14.28–14.31, respectively. For ANFIS-based models, the plots of observed and predicted values of SSC during training and testing periods are represented in Figures 14.32–14.35, respectively. The plots show fair agreement between observed and predicted SSC values.

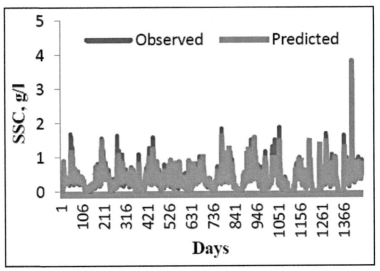

FIGURE 14.20 Observed and predicted SSC using ANN model (6–4–4–1) at the Pathagudem site during training period.

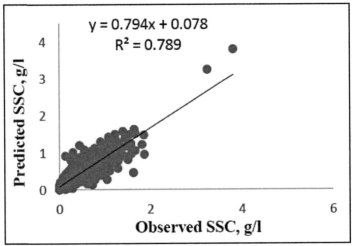

FIGURE 14.21 Scatter plot between observed and predicted SSC using ANN model 6–4–4–1) at the Pathagudem site during training period.

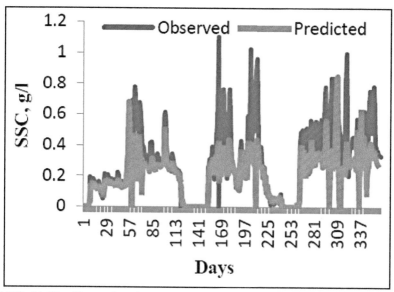

FIGURE 14.22 Observed and predicted SSC using ANN model (6–4–4–1) at the Pathagudem site during testing period.

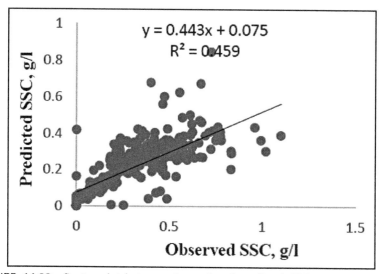

FIGURE 14.23 Scatter plot between observed and predicted SSC using ANN model (6–4–4–1) at the Pathagudem site during testing period.

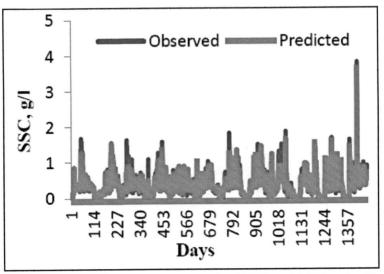

FIGURE 14.24   Observed and predicted SSC using ANFIS model (Triangular, 3) at the Pathagudem site during training period.

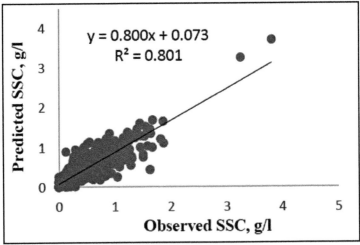

FIGURE 14.25   Scatter plot between observed and predicted SSC using ANFIS model (Triangular, 3) at the Pathagudem site during training period.

It is clear from the visual observation that there is a good agreement between the estimated (predicted) and observed SSCs, and the overall shape of predicted SSC graph is similar to that of the observed sediment concentration graph. It is also apparent from the figures that the qualitative performances of ANN and ANFIS models during training and testing periods are satisfactory. The performance of ANFIS model is better than ANN models in estimation of SSC prediction at Pathagudem and Polavaram sites.

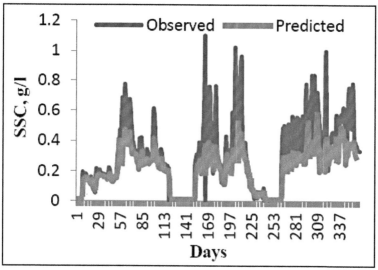

FIGURE 14.26 Observed and predicted sediment concentration using ANFIS model (Triangular, 3) at the Pathagudem site during testing period.

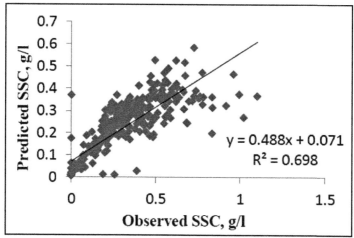

FIGURE 14.27 Scatter plot between observed and predicted SSC using ANFIS model (Triangular, 3) at the Pathagudem site during testing period.

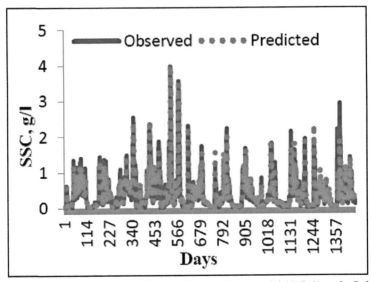

FIGURE 14.28   Observed and predicted SSC using ANN model (6–5–1) at the Polavaram site during training period.

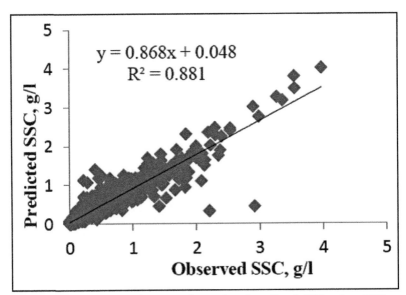

FIGURE 14.29   Scatter plot between observed and predicted SSC using ANN model (6–5–1) at the Polavaram site during training period.

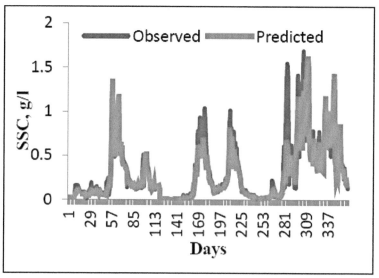

FIGURE 14.30 Observed and predicted SSC ANN model (6–5–1) at the Polavaram site during testing period.

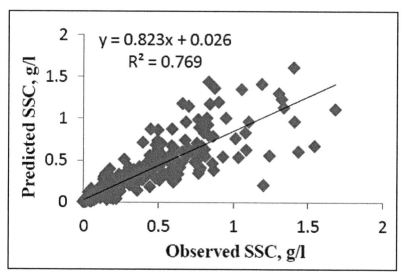

FIGURE 14.31 Scatter plot between observed and predicted SSC using ANN model (6–5–1) at the Polavaram site during testing period.

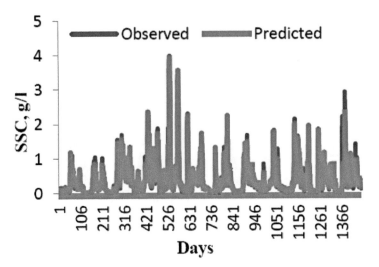

FIGURE 14.32 Observed and predicted SSC using ANFIS model (Gauss, 3) at the Polavaram site during training period.

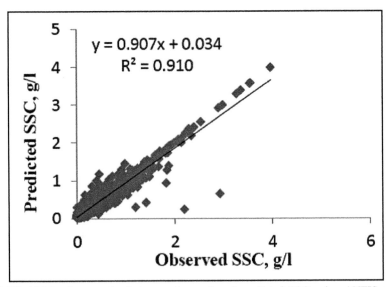

FIGURE 14.33 Scatter plot between observed and predicted SSC using ANFIS model (Gauss, 3) at the Polavaram site during training period.

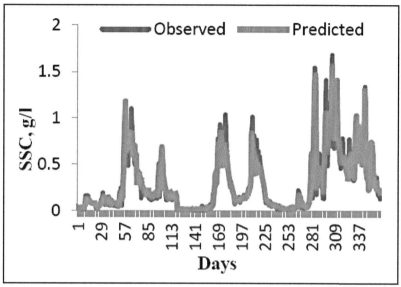

FIGURE 14.34 Observed and predicted SSC using ANFIS model (Gauss, 3) at the Polavaram site during testing period.

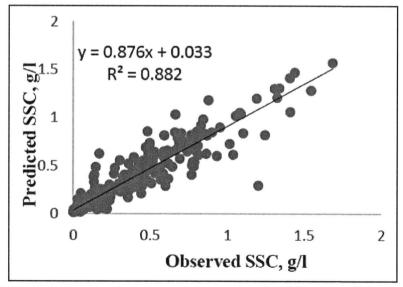

FIGURE 14.35 Scatter plot between observed and predicted SSC using ANFIS model (Gauss, 3) at the Polavaram site during testing period.

14.3.5.2 QUANTITATIVE EVALUATION

The quantitative performance of developed models was also evaluated by applying various statistical and hydrological indices, such as CC, RMSE, CE and PARE. The values of these performance indices are presented in Tables 14.12–14.21.

1. Correlation Coefficient (CC): The values of CC for ANN and ANFIS-based models were computed using Eqn. (14.20) and obtained values are presented in Tables 14.12–14.21. For AN-based runoff prediction model, the values of CC for Pathagudem site during calibration (1996–2007) and validation (2008–2010) periods for the selected network (6–6–1) are 0.993 and 0.992, respectively (Table 14.12); whereas for Polavaram site during calibration (1996–2007) and validation (2008–2010) periods for the selected network (6–9–1) are 0.994 and 0.9905, respectively (Table 14.12).

   The values of CC for Pathagudem site based on ANN SSC prediction model during calibration and validation periods for the selected network (6–4–4–1) are 0.890 and 0.667, respectively (Table 14.13); whereas for Polavaram site during training and testing periods for the selected network (6–5–1) are 0.938 and 0.877, respectively (Table 14.13).

   The values of CC for the selected ANFIS (Gauss, 3) model for Pathagudem site during training and testing periods are 0.9941 and 0.9874, respectively; while for the selected ANFIS (Triangular, 3) model for Polavaram site during training and testing periods are 0.998 and 0.994, respectively in runoff prediction (Table 14.12).

   CC values using ANFIS-based SSC prediction model (Triangular, 3) model for Pathagudem site during training and testing periods are 0.900 and 0.8356, respectively (Table 14.13); while for the selected (Gauss, 3) model of Polavaram site during training and testing periods are 0.954 and 0.939, respectively (Table 14.13). The higher values of CC for training and testing periods show good agreement between observed and predicted values of runoff.

2. Root Mean Square Error (RMSE): RMSE values between observed and predicted values of runoff and SSC using ANN and ANFIS-based models were determined using Eqn. (14.21) and determined values are presented in Tables 14.12 and 14.13 for both sites. The values of RMSE based on ANN model for runoff prediction for Pathagudem

TABLE 14.12 Performance Evaluation of Developed ANN and ANFIS Models for the Best-Chosen Network for Runoff Prediction at Pathagudem and Polavaram Sites

Performance Indices	ANN Network 6-6-1		ANFIS Model Gauss, 3		ANN Network 6-9-1		ANFIS Model Triangular, 3	
	Training Period	Testing Period	Training Period	Testing Period	Training Period	Testing Period	Training Period	Testing Period
	Pathagudem Site				**Polavaram Site**			
CC	0.993	0.992	0.994	0.987	0.994	0.990	0.998	0.994
RMSE (m³/sec)	310.70	427.46	283.85	299.75	739.03	1092.80	485.37	859.93
CE	0.986	0.989	0.988	0.980	0.989	0.992	0.995	0.995
PARE	0.00062	-0.0257	-0.0005	-0.005	-0.0001	-0.0064	-0.0002	-0.0011

TABLE 14.13 Performance Evaluations of Developed ANN and ANFIS Models for the Best-Chosen Network for Suspended Sediment Concentration Prediction at Pathagudem and Polavaram Sites

Performance Indices	ANN Network 6-4-1		ANFIS Model Triangular, 3		ANN Network 6-5-1		ANFIS Model Gauss, 3	
	Training Period	Testing Period	Training Period	Testing Period	Training Period	Testing Period	Training Period	Testing Period
	Pathagudem Site				**POLAVARAM SITE**			
CC	0.890	0.666	0.900	0.836	0.938	0.877	0.954	0.939
RMSE (g/l)	0.156	0.193	0.151	0.159	0.165	0.162	0.139	0.113
CE	0.786	0.881	0.801	0.918	0.881	0.930	0.910	0.966
PARE	0.00049	-0.0823	-0.0004	-0.0074	-0.0007	-0.0242	-0.0003	-0.0036

site during calibration and validation periods for network (6–6–1) are 310.70 and 427.46 m³/sec, respectively while for Polavaram site during calibration and validation periods for network (6–9–1) were 739.03 and 1092.80 m³/sec, respectively (Table 14.12). For ANN-based SSC model, the values of RMSE for Pathagudem site during training and testing periods for network (6–4–4–1) are 0.1561 and 0.1931 g/l; while for Polavaram site during calibration and validation periods for network (6–5–1) are 0.165 and 0.162 g/l (Table 14.13).

For ANFIS-based runoff prediction model, the values of RMSE for the selected network (Gauss, 3) of Pathagudem site during training and testing periods range from 283.85 and 299.75 m³/sec; whereas for the selected network (Triangular, 3) of Polavaram during training and testing periods are 485.37 and 859.93 m³/sec (Table 14.12).

The values of RMSE using ANFIS-based SSC prediction model, for the selected network (Triangular, 3) of Pathagudem site during training and testing periods, are 0.1507 and 0.1590 g/l, respectively; whereas for the selected network (Gauss, 3) of Polavaram during training and testing periods are 0.1392 and 0.1135 g/l, respectively (Table 14.13). The lower values of RMSE for training and testing periods suggest the goodness of fit for the ANFIS-based models.

3. Coefficient of Efficiency (CE): CE values for asserting the applicability of ANN and ANFIS-based models were calculated using Eqn. (14.22) and presented in Tables 14.12 and 14.13. It is clear from Table 14.12 for ANN-based runoff prediction model, the values of CE corresponding to calibration and validation periods for selected network (6–6–1) are 0.9860 and 0.9894, respectively at the Pathagudem site; while for Polavaram site during calibration and validation periods for network (6–9–1) are 0.9893 and 0.992, respectively.

The values of CE for ANN-based SSC model corresponding to calibration and validation periods for selected network (6–4–4–1) are 0.7864 and 0.8811, respectively; whereas for Polavaram site for calibration and validation periods for selected network (6–5–1) are 0.8808 and 0.9302, respectively.

The values of CE for ANFIS model (Gauss, 3) during training and testing periods are 0.9885 and 0.9805, respectively for Pathagudem site; while for ANFIS model (Triangular–3) during training and testing periods are 0.9954 and 0.995, respectively of Polavaram site for runoff prediction (Table 14.12).

In case of ANFIS-based SSC model (Triangular, 3), the values of CE of Pathagudem site during training and testing periods are 0.9001 and 0.8356, respectively; while for the ANFIS model (Gauss, 3) of Polavaram site during calibration and validation periods are 0.9101 and 0.9663, respectively (Table 14.13).

4. Pooled Average Relative Error (PARE): This was computed using Eqn. (14.23) and obtained values are shown in Tables 14.12 and 14.13. Using the ANN-based runoff prediction model, the values of PARE of Pathagudem during calibration and validation periods for the selected network (6–6–1) are 0.00062 and –0.0257, respectively; while for Polavaram site during calibration and validation periods for the selected network (6–9–1) are –0.0001 and –0.0064, respectively (Table 14.12).

In case of SSC model based on ANN, the values of PARE of Pathagudem during calibration and validation periods for the selected network (6–4–4–1) are 0.00049 and –0.0823, respectively; whereas for Polavaram site for calibration and validation periods for the selected network (6–5–1) are –0.0007 and –0.0242, respectively (Table 14.13).

Applying the ANFIS-based technique (Tables 14.12 and 14.13) in case of runoff prediction, the values of PARE for the selected network (Gauss, 3) of Pathagudem site during training and testing periods are –0.0005 and –0.005, respectively; while for the selected network (Triangular, 3) of Polavaram site for training and testing periods are –0.00023 and –0.0011, respectively.

For ANFIS-basedSSC model, the values of PARE for the selected network (Triangular, 3) of Pathagudem site for training and testing periods are –0.0004 and –0.0074, respectively; whereas for selected network (Gauss, 3) of Polavaram site for training and testing periods are –0.0003 and –0.0036, respectively. The positive value of PARE for model indicates that the model slightly overestimated the runoff and SSC; and negative value indicates that runoff and SSC were underestimated.

The performance evaluation of the models revealed that the models were able to predict the runoff and sediment with adequate accuracy. From the results discussed in preceding sections and the figures, the qualitative evaluation of the models based on plots between observed and predicted values of daily runoff indicates that ANFIS models performed better than ANN-based models for runoff prediction. The graph between observed and predicted values of daily SSC indicates that ANFIS models performed better than ANN-based models for SSC prediction. Likewise, the values of the performance indicators for ANN rule and ANFIS-based models indicate much better performance of ANFIS over ANN rule-based models.

14.3.6   SENSITIVITY ANALYSIS

Runoff and SSC are affected by various parameters, which were evaluated using sensitivity analysis using RMSE, CE, CC, and PARE values for this analysis.

14.3.6.1   SENSITIVITY ANALYSIS OF ANN-BASED RUNOFF PREDICTION MODEL FOR PATHAGUDEM SITE

Based on performance evaluation of the ANNs models, the network (6–6–1) has the highest CC, CE, and the lowest error between observed and predicted runoff. Thus, this model was chosen for the sensitivity analysis. The model has six inputs (viz. stage height of present day, previous one-day, previous three days, runoff of previous one-day, previous two days, previous three days) and present-day runoff as the output. While training a network, the network finds the significant relationship between input parameter and output parameter. This provides feedback as to which input parameters are most significant. Based on this feedback, it mainly decides to prune the input space by removing the input parameter one by one.

The CC for ANN-based case (with all input parameters), without current day stage, without previous one-day stage height, without previous three days stage height, without previous one-day runoff, without previous two days runoff and without previous three days runoff were evaluated using Eqn. (14.20) and the calculated values are 0.992, 0.805, 0.963, 0.991, 0.969, 0.991 and 0.99, respectively (Table 14.14). The ANN-based case has the highest value of the CC. The case without current day stage has lowest value compared to other cases. The current day stage is the most sensitive parameter for runoff prediction and followed by previous one-day stage height.

The values of RMSE for ANN base case without current day stage, without previous one-day stage height, without previous three days stage height, without previous one-day runoff, without previous two days runoff and without previous three days runoff were determined using Eqn. (14.21) and are 427.46, 1659.30, 613.22, 537.28, 525.57, 479.17 and 475.00 m³/sec, respectively (Table 14.14). The AN-based case has the lowest RMSE value compared to other cases. The parameter current day stage has the highest RMSE value.

TABLE 14.14 Performance Indices of ANN Model (6–6–1) for Sensitivity Analysis During Testing Period

Perfor-mance Indices	Base case Model (With All Input Parameters) (6–6–1)	Without Current Day Stage Height	Without Previous				
			One-day Stage Height	Three Days Stage Height	One-day Runoff	Two Days Runoff	Three Days Runoff
CC	0.992	0.805	0.963	0.991	0.969	0.991	0.990
RMSE (m³/sec)	427.46	1659.3	613.22	537.28	525.57	479.17	475.00
CE	0.989	0.6466	0.952	0.971	0.963	0.970	0.964
PARE	–0.0257	0.0301	–0.0495	–0.0391	–0.0469	–0.0378	–0.0355

The calculated values of CE for AN-based case, without current day stage, without previous one-day stage height, without previous three days stage height, without previous one-day runoff, without previous two days runoff and without previous three days runoff were calculated using Eqn. (14.22) and were 0.989, 0.6466, 0.952, 0.971, 0.963, 0.970 and 0.964, respectively (Table 14.14). It is observed that the AN-based case has the highest value of CE as compared with the other cases. The parameter current day stage has the highest sensitivity with runoff.

The PARE was computed using Eqn. (14.22). The values of PARE for the AN-based case, without current day stage, without previous one-day stage height, without previous three days stage height, without previous one-day runoff, without previous two days runoff and without previous three days runoff are –0.0257, 0.0301, –0.0495, –0.0391, –0.0469, –0.0378 and –0.0355, respectively (Table 14.14).

According to analysis of ANN model, the runoff is the most sensitive with current day stage followed by previous one-day stage height, previous one-day runoff, previous three days runoff, previous two days runoff and previous three days stage height.

14.3.6.2 SENSITIVITY ANALYSIS OF ANFIS-BASED RUNOFF PREDICTION MODEL FOR PATHAGUDEM SITE

In case of ANFIS model, the network (Gauss, 3) was the best network based on statistical and hydrological indices. The model has six inputs (viz. stage height of present day, previous one-day, previous three days, runoff of previous one-day, previous two days, previous three days) and one parameter in the output (present day runoff) as indicated in Table 14.15.

TABLE 14.15   Performance Indices of ANFIS Model (Gauss, 3) for Sensitivity Analysis during Testing Period for Polavaram site

Perfor-mance Indices	Base Case Model (With All Input Parameters) (Gauss, 3)	Without Current Day Stage Height	Without Previous				
			One-day Stage Height	Three Days Stage Height	One-day runoff	Two Days Runoff	Three Days Runoff
CC	0.987	0.647	0.981	0.986	0.965	0.983	0.979
RMSE (m³/sec)	299.75	1365.67	539.67	329.67	877.56	396.57	439.76
CE	0.981	0.683	0.963	0.980	0.937	0.976	0.974
PARE	−0.005	0.013	−0.004	−0.006	−0.008	−0.006	−0.0064

The CC for ANFIS-based case, without current day stage, without previous one-day stage height, without previous three days stage height, without previous one-day runoff, without previous two days runoff and without previous three days runoff were evaluated using Eqn. (14.20) and are 0.987, 0.647, 0.981, 0.986, 0.965, 0.983 and 0.979, respectively (Table 14.15). The ANFIS-based case has the highest value of the CC. The case without current day stage has lower value compared to other cases.

The values of RMSE were determined using Eqn. (14.21). The values for ANFIS-based case, without current day stage, without previous one-day stage height, without previous three days stage height, without previous one-day runoff, without previous two days runoff and without previous three days runoff are 299.75, 1365.67, 539.67, 329.67, 877.56, 396.57, and 439.76 m³/sec, respectively (Table 14.15). The ANFIS-based case has the lowest RMSE value compared to other cases. The parameters current day stage height and previous three-day stage height have the highest and lowest RMSE values.

The values of CE for ANFIS-based case, without current day stage, without previous one-day stage height, without previous three days stage height, without previous one-day runoff, without previous two days runoff and without previous three days runoff are 0.981, 0.683, 0.963, 0.980, 0.937, 0.976, and 0.974, respectively (Table 14.15). AN-based model performed better than other models when model was selected on the basis of CE values. The parameter current day stage height has the highest sensitivity with runoff.

The PARE was calculated using Eqn. (14.22). The values of PARE for the ANFIS-based case, without current day stage, without previous one-day stage height, without previous three days stage height, without previous one-day runoff, without previous two days runoff and without previous three days runoff are: –0.005, 0.013, –0.004, –0.006, –0.008, –0.006, and –0.0064, respectively (Table 14.15). It is found that the values of PARE are higher and lower in cases without current day stage and without previous one-day runoff.

Based on analysis of ANFIS model, the runoff is the most sensitive with current day stage followed by previous one-day runoff, previous one-day stage height, previous three days runoff, previous two days runoff and previous three days stage height.

14.3.6.3   SENSITIVITY ANALYSIS OF AN-BASED RUNOFF PREDICTION MODEL FOR POLAVARAM SITE

The ANN network (6–9–1) has the highest CC, CE, and the lowest error. Thus, this network has been chosen for the sensitivity analysis. The model has six inputs (viz. stage height of present day, previous one-day, previous two days, previous three days, runoff of previous one-day and previous two days) and present-day runoff as output (Table 14.16).

The CC for AN-based case, without current day stage, without previous one-day stage height, without previous three days stage height, without previous three days stage height, without previous one-day runoff and without previous two days runoff are: 0.991, 0.836, 0.867, 0.913, 0.946, 0.842 and 0.842, respectively (Table 14.16).

RMSE values for AN-based case with all parameters, without current day stage, without previous one-day stage height, without previous three days stage height, without previous three days stage height, without previous one-day runoff and without previous two days runoff are: 1092.80, 2756.70, 2189.5, 1573.60, 1367.50, 2312.61 and 1916.80 m^3/sec, respectively (Table 14.16).

TABLE 14.16 Performance Indices of ANN Model (6–9–1) for Sensitivity Analysis During Testing Period

Perfor-mance Indices	Base case Model –(With All Input Para-meters) (6–9–1)	Without Current Day Stage Height	Without Previous One-day Stage Height	Without Previous Three Days Stage Height	Without Previous One-day Runoff	Without Previous Two Days Runoff	Without Previous Three Days Runoff
CC	0.991	0.836	0.867	0.913	0.946	0.842	0.842
RMSE (m³/sec)	1092.8	2756.7	2189.5	1573.6	1367.5	2312.61	1916.8
CE	0.992	0.693	0.764	0.863	0.935	0.763	0.831
PARE	–0.0064	–0.0021	–0.0039	–0.0053	–0.0061	–0.0036	–0.0049

The calculated values of CE for AN-based case, without current day stage, without previous one-day stage height, without previous three days stage height, without previous three days stage height, without previous one-day runoff and without previous two days runoff are: 0.992, 0.693, 0.764, 0.863, 0.935, 0.763 and 0.831, respectively (Table 14.16).

The PARE was determined using Eqn. (14.23). The values of PARE for the AN-based case, without current day stage, without previous one-day stage height, without previous three days stage height, without previous three days stage height, without previous one-day runoff and without previous two days runoff are: –0.0064, –0.0021, –0.0039, –0.0053, –0.0061, –0.0036, and –0.0049, respectively (Table 14.16).

Based on analysis of ANN model, the runoff is the most sensitive with current day stage height followed by previous one-day runoff, previous one-day stage height, previous two days runoff, previous two days stage height, and previous three days stage height.

14.3.6.4 SENSITIVITY ANALYSIS OF ANFIS-BASED RUNOFF PREDICTION MODEL FOR POLAVARAM SITE

The ANFIS network (Triangular, 3) has been chosen for the sensitivity analysis. The model has six inputs viz. stage height of present day, previous one-day, previous two days, and previous three days, runoff of previous one-day and previous two days and present-day runoff as the output (Table 14.17). The most sensitive parameter was decided by removing the input parameter one by one.

The ANFIS-based case, without current day stage, without previous one-day stage height, without previous two days stage height, without previous three days stage height, without previous one-day runoff and without previous two days runoff have the CC values, such as 0.994, 0.6123, 0.687, 0.815, 0.795, 680, and 0.799, respectively (Table 14.17).

The values of RMSE values for ANFIS-based case, without current day stage, without previous one-day stage height, without previous two days stage height, without previous three days stage height, without previous one-day runoff and without previous two days runoff are: 859.93, 13951.19, 5955.31, 4431.56, 5100.42, 6655.05, and 4606.21, respectively (Table 14.17). The ANFIS-based case has the lowest RMSE value compared to other cases.

TABLE 14.17   Performance Indices of ANFIS Model (Triangular, 3) for Sensitivity Analysis During Testing Period for Pathagudem site

Performance Indices	Base case Model (With All Input Parameters) (Triangular, 3)	Without Current Day Stage Height	Without Previous One-day Stage Height	Without Previous Two Days Stage Height	Without Previous Three Days Stage Height	Without Previous One-day Runoff	Without Previous Two Days Runoff
CC	0.994	0.6123	0.687	0.815	0.795	0.680	0.799
RMSE (m³/sec)	859.93	13951.2	5955.3	4431.56	5100.4	6655.0	4606.2
CE	0.995	0.286	0.766	0.870	0.828	0.707	0.860
PARE	−0.0011	0.0225	−0.0114	−0.0108	0.0192	0.0161	−0.0193

The computed values of CE for ANFIS-based case, without current day stage, without previous one-day stage height, without previous two days stage height, without previous three days stage height, without previous one-day runoff and without previous two days runoff are: 0.995, 0.286, 0.766, 0.870, 0.828, 0.707 and 0.860, respectively (Table 14.17). It is observed that the ANFIS-based case has the highest value of CE compared with the other cases.

The PARE values are −0.0011, 0.0225, −0.01137, −0.0108, 0.0192, 0.0161, and −0.0193 (Table 14.17) for the ANFIS-based case, without current day stage, without previous one-day stage height, without previous two days stage height, without previous three days stage height, without previous one-day runoff and without previous two days runoff, respectively.

Finally, it is observed that runoff is the most sensitive with current day stage followed by previous one-day runoff, previous one-day stage height, previous three days runoff, previous two days runoff and previous three days stage height.

14.3.6.5 SENSITIVITY ANALYSIS OF ANN-BASED SSC PREDICTION MODEL FOR PATHAGUDEM SITE

The AN-based network (6–4–4–1) was selected for the sensitivity analysis. The model has six inputs (viz. runoff of present day, previous one-day, previous two days, and previous three days, SSC of previous one-day, previous three days) and present-day SSC as the output (Table 14.18).

The CC values for AN-based case, without current-runoff, without previous one-day runoff, without previous two days runoff, without previous three days runoff, without previous one-day SSC and without previous three days are: 0.666, 0.613, 0.636, 0.651, 0.663, 0.646, and 0.652, respectively (Table 14.18).

The values of RMSE for AN-based case, without current day runoff, without previous one-day runoff, without previous two days runoff, without previous three days runoff, without previous one-day SSC, and without previous three days, SSC were evaluated using Eqn. (14.21) and these values are: 0.193, 0.243, 0.219, 0.201, 0.195, 0.213 and 0.197 g/l, respectively (Table 14.18).

TABLE 14.18 Performance Indices of ANN Model (6–4–4–1) for Sensitivity Analysis During Testing Period

Perfor-mance Indices	Base Case Model (With All Input Para-meters) (6–4–4–1)	Without Current Day Runoff	Without Previous One-day Runoff	Without Previous Two Days Runoff	Without Previous Three Days Runoff	Without Previous One-day SSC	Without Previous Three Days SSC
CC	0.66`6	0.613	0.636	0.651	0.663	0.646	0.652
RMSE (g/l)	0.193	0.243	0.219	0.201	0.195	0.213	0.197
CE	0.881	0.813	0.837	0.856	0.881	0.847	0.877
PARE	−0.0823	−0.062	−0.066	−0.079	−0.082	−0.073	−0.082

The calculated values of CE for AN-based case, without current day runoff, without previous one-day runoff, without previous two days runoff, without previous three days runoff, without previous one-day SSC, and without previous three days SSC were: 0.881, 0.813, 0.837, 0.856, 0.880, 0.847 and 877 (Table 14.18).

The PARE was calculated using Eqn. (14.23). The values of PARE for the AN-based case, without current day runoff, without previous one-day runoff, without previous two days runoff, without previous three days runoff, without previous one-day SSC, and without previous three days SSC are: −0.0823, −0.062, −0.066, −0.079, −0.0821, −0.073, and −0.082, respectively (Table 14.18).

The AN-based case has the highest value of the CC, CE, RMSE, and PARE. The case without current day runoff has lower values of CC, CE, and higher values of RMSE and PARE compared to other cases. The current day runoff is the most sensitive parameter for SSC prediction and followed by previous one-day runoff

As per analysis of ANN model, the runoff is the most sensitive with current day runoff followed by previous one-day runoff, previous one-day SSC, previous two days runoff, previous three days SSC, and previous two days runoff.

14.3.6.6 SENSITIVITY ANALYSIS OF ANFIS-BASED SSC PREDICTION MODELS FOR PATHAGUDEM SITE

After analyzing the ANFIS models, it was concluded that the network (Triangular, 3) applicable for sensitivity analysis and model has six inputs (viz. runoff of present day, previous one-day, previous two days, previous three days, SSC of previous one-day, previous three days) and present day SSC as the output (Table 14.19).

The values of correlation coefficient (CE) were calculated using Eqn. (14.20) and these are 0.836, 0.731, 0.765, 0.792, 0.796, 0.746 and 0.826, respectively (Table 14.19) for ANFIS-based case, without current-runoff, without previous one-day runoff, without previous two days runoff, without previous three days runoff, without previous one-day SSC and without previous three days SSC, respectively.

The RMSE values for ANFIS-based case, without current day runoff, without previous one-day runoff, without previous two-day runoff, without previous three days runoff, without previous one-day SSC, and without previous three days SSC are: 0.159, 0.325, 0.253, 0.231, 0.196, 0.306, and 0.171 g/l, respectively (Table 14.19).

TABLE 14.19 Performance Indices of ANFIS Model (Triangular, 3) for Sensitivity Analysis During Testing Period for Pathagudem site

Performance Indices	Base Case Model (With All Input Parameters) (Triangular, 3)	Without Current Day Runoff	Without Previous One-day Runoff	Without Previous Two Days Runoff	Without Previous Three Days Runoff	Without Previous One-day SSC	Without Previous Three Days SSC
CC	0.836	0.731	0.765	0.792	0.796	0.746	0.826
RMSE (g/l)	0.159	0.325	0.253	0.231	0.196	0.306	0.171
CE	0.918	0.762	0.813	0.846	0.873	0.793	0.896
PARE	−0.0074	0.013	−0.0036	−0.0029	−0.0023	−0.0052	−0.0017

The values of CE for ANFIS-based case, without current day runoff, without previous one-day runoff, without previous two days runoff, without previous three days runoff, without previous one-day SSC, and without previous three days SSC were determined using Eqn. (14.22) and are: 0.918, 0.762, 0.813, 0.846, 0.873, 0.793, and 0.896, respectively (Table 14.19).

The PARE values for the ANFIS-based case, without current day runoff, without previous one-day runoff, without previous two days runoff, without previous three days runoff, without previous one-day SSC, and without previous three days SSC are: −0.0074, 0.013, −0.0036, −0.0029, −0.0023, −0.0052, and −0.0017, respectively (Table 14.19).

The ANFIS-based case has the highest values of the CC, CE, and the lowest values of RMSE, PAre: the case without current day runoff has lower values of CC, CE, and higher values of RMSE, PARE compared to other cases. The current-runoff is the most sensitive parameter for SSC prediction and followed by previous one-day SSC.

Based on sensitivity analysis of SSC ANFIS model, the SSC is most sensitive with current day runoff followed by, previous one-day SSC, previous one-day runoff, previous two days runoff, previous three days runoff, and previous three days SSC.

14.3.6.7 SENSITIVITY ANALYSIS OF ANN-BASED SSC PREDICTION MODEL FOR POLAVARAM SITE

The AN-based network (6–5–1) has been adopted for the sensitivity analysis. The model has six inputs (viz. runoff of present day, previous one-day, previous two days, SSC of previous one-day, previous two days, previous three days) and present day SSC as the output Based on this feedback, it was mainly decided to prune the input space by removing the input parameter one by one (Table 14.20).

TABLE 14.20 Performance Indices of ANN Model (6–5–1) for Sensitivity Analysis During Testing Period

Perfor-mance Indices	Base Case Model (With All Input Para-meters) (6–5–1)	Without Current Day Runoff	Without Previous One-day Runoff	Without Previous Two Days Runoff	Without Previous One-day SSC	Without Previous Two Days SSC	Without Previous Three Days SSC
CC	0.877	0.643	0.693	0.794	0.735	0.829	0.847
RMSE (g/l)	0.162	0.296	0.273	0.227	0.249	0.193	0.160
CE	0.930	0.731	0.833	0.891	0.867	0.907	0.921
PARE	−0.0242	0.046	−0.0191	−0.0223	−0.0212	−0.0226	−0.0231

The CC values for AN-based case, without current day runoff, without previous one-day runoff, without previous two days runoff, without previous one-day SSC, without previous two days SSC and without previous three days SSC are 0.877, 0.643, 0.693, 0.794, 0.735, 0.829, and 0.847, respectively (Table 14.20). The AN-based case has the highest value of the CC and the case without current day runoff has lower value compared to other cases.

The values of RMSE for AN-based case, without current day runoff, without previous one-day runoff, without previous two days runoff, without previous one-day SSC, without previous two days SSC and without previous three days SSC were calculated using Eqn. (14.21) and these values are: 0.162, 0.296, 0.273, 0.227, 0.249, 0.193 and 0.160 g/l, respectively (Table 14.20). The AN-based case has the lowest RMSE value compared to other cases.

The determined values of CE for AN-based case, without current day runoff, without previous one-day runoff, without previous two days runoff, without previous one-day SSC, without previous two days SSC and without previous three days SSC are: 0.930, 0.731, 0.833, 0.891, 0.867, 0.907, and 0.921, respectively (Table 14.20). AN-based model performed better than other models when model was selected on the basis of CE values. The parameter current-runoff has the highest sensitivity with SSC.

The PARE was calculated using Eqn. (14.23). The values of PARE for the AN-based case, without current day runoff, without previous one-day runoff, without previous two days runoff, without previous one-day SSC, without previous two days SSC and without previous three days SSC are: -0.0242, 0.046, -0.0191, -0.0223, -0.0212, -0.0226, and -0.0231, respectively (Table 14.20). It is found that the values of PARE were higher and lower in cases without current-runoff and without previous three days SSC.

Results in Table 14.20 indicate that the most significant parameter for SSC prediction is current day runoff followed by previous one-day runoff, previous one-day SSC, previous two days runoff, previous two days SSC and previous two three days SSC.

14.3.6.8 SENSITIVITY ANALYSIS OF ANFIS-BASED SSC PREDICTION MODEL FOR POLAVARAM SITE

After selecting the best ANFIS network (Gauss, 3) based on the performance evaluation of models, the network has the highest CC, CE, and the lowest error between calibrated and validated SSC. Thus, this model has been chosen for the sensitivity analysis. Runoff of present day, previous one-day, previous two days, SSC of previous one-day, previous two days, and previous three days as input parameters and present day SSC as output parameter (Table 14.21) have been used for model development. The most significant parameter is decided by removing the input parameter one by one.

The CC for ANFIS-based case, without current day runoff, without previous one-day runoff, without previous two days runoff, without previous one-day SSC, without previous two days SSC and without previous three days SSC were determined using Eqn. (14.20) and calculated values are 0.939, 0.835, 0.920, 0.921, 0.879, 0.924, and 0.921, respectively (Table 14.21). The ANFIS model with all input parameters has the highest value of the CC compared to other cases; the case without current day runoff has lower value as compared to other cases. The current day runoff is the most sensitive parameter for SSC prediction and followed by previous one-day SSC.

The RMSE values for ANFIS-based case, without current day runoff, without previous one-day runoff, without previous two days runoff, without previous one-day SSC, without previous two days SSC and without previous three days SSC are: 0.113, 0.184, 0.126, 0.159, 0.158, 0.126, and 0.128 g/l, respectively (Table 14.21). The ANFIS-based case has the lowest RMSE value compared to other cases. The parameters current day runoff has the highest RMSE value.

TABLE 14.21 Performance Indices of ANFIS Model (Gauss, 3) for Sensitivity Analysis During Testing Period for Polavaram site

Perfor-mance Indices	Base Case Model (With All Input Para-meters) (Gauss, 3)	Without Current Day Runoff	Without Previous One-day Runoff	Without Previous Two Days Runoff	Without Previous One-day SSC	Without Previous Two Days SSC	Without Previous Three Days SSC
CC	0.939	0.835	0.920	0.921	0.879	0.924	0.921
RMSE (g/l)	0.113	0.184	0.126	0.159	0.158	0.126	0.128
CE	0.966	0.912	0.958	0.931	0.934	0.958	0.957
PARE	−0.0036	0.0006	−0.0028	−0.00019	−0.0029	−0.00074	−0.0021

The determined values of CE for ANFIS-based case, without current day runoff, without previous one-day runoff, without previous two days runoff, without previous one-day SSC, without previous two days SSC and without previous three days SSC are: 0.966, 0.912, 0.958, 0.931, 0.934, 0.958, and 0.957, respectively (Table 14.21). It is clear that the ANFIS-based case has the highest value of CE compared with the other cases. The parameter current-runoff has the highest sensitivity with SSC.

The PARE was determined using Eqn. (14.23). The PARE values for the ANFIS-based case, without current day runoff, without previous one-day runoff, without previous two days runoff, without previous one-day SSC, without previous two days SSC and without previous three days SSC are: −0.0036, 0.0006, −0.0028, −0.00019, −0.0029, −0.00074, and −0.00215, respectively (Table 14.21). It is observed that the values of PARE are higher and lower in cases without current-runoff and without previous three days SSC.

Results in Table 14.21 indicate that the runoff is the most sensitive with current day runoff followed by previous one-day SSC, previous one-day runoff, previous three days SSC, previous two days SSC and previous two days runoff.

14.4 SUMMARY

The present study focuses to develop and evaluate ANN and ANFIS models for daily runoff and SSC prediction of monsoon period for the Godavari basin at Pathagudem and Polavaram sites located in Chhattisgarh and Andhra Pradesh, respectively. Total catchment areas of Pathagudem and Polavaram sites are 40,625 and 3,07,800 m^3, respectively. Most of the rainfall occurs during 1st June to 30th September due to south-west monsoon. Daily hydrological data required for the stage-discharge and runoff-sediment modeling was collected for 1st June to 30th September of 1996–2010 from the Krishna and Godavari Basin Organization of Central Water Commission, Hyderabad, Andhra Pradesh. Daily data for 1996–2007 and 2008–2010 were utilized for the training and validation of the developed models.

The performance of developed models was evaluated ANNs and ANFIS qualitatively and quantitatively. The quantitative performance of developed models was assessed using statistical and hydrological indices, such as: CC, RMSE, CE, and PARE.

The sensitively analysis was carried out by removing one by one input parameter from the input parameters used for ANN and ANFIS-based runoff and SSC models and after that comparing the performance indices. The greater the effect observed in the output, the greater is the sensitivity of that particular input parameter.

The MATLAB (2015a) software was used to train a multilayer feed-forward neural network with a back-propagation algorithm. The log sigmoid activation function was used for each hidden layer with a range from 0 to 1 and pure line activation function for the output layer. In the AN-based runoff and SSC prediction models, the six input parameters and one output parameter were considered. ANFIS model was also applied for the same inputs. Different structures of the ANN model were trained and tested for maximum iterations of 1000 for single and double hidden layers network to predict runoff and SSC. Triangular, trapezoidal, Gaussian and generalized bell MFs were used to train and test the ANFIS model for runoff and SSC prediction. Since there is no specific rule available to determine the best structure of the network, a trial-and-error method was used for the selection of the network among various structures of the ANN and ANFIS models.

During runoff prediction, one network was selected among single and double hidden layers of ANN networks based on the performance indices for each site. Also, at the time of SSC prediction, one network was selected among single and double hidden layers of ANN networks for each site. From various networks of ANFIS, one model was chosen for the runoff prediction, and one model was selected for the SSC prediction for each site on the basis of performance indices.

Two networks were selected for runoff prediction for each site and also for SSC prediction; and two networks were selected for each site. Thus, total of eight models were selected for runoff and SSC predictions for Pathagudem and Polavaram sites.

The following conclusions are drawn from the present study:

- A comparative study between the ANN model (6–5–1) and the ANFIS model (Gauss, 3) suggests that the later one shows the best result for SSC prediction at the Polavaram site.
- Among all SSC prediction models, the ANFIS model (Gauss, 3) performs better at the Polavaram site.
- Among single and double hidden layers ANN models, the model (6–5–1) has given better performance for SSC prediction at the Polavaram site.
- ANFIS model (Triangular, 3) gives more precise results than other SSC prediction models at the Pathagudem site.
- ANFIS model (Triangular, 3) performs better than other runoff prediction models at the Polavaram site.
- Based on the goodness of fit, the ANN-based single hidden layer model (6–6–1) performs better than a double hidden layers model for runoff prediction at the Pathagudem site.
- Performance of AN-based SSC model (6–4–4–1) is compared with the ANFIS model (Triangular, 3) for runoff prediction and the ANFIS model has shown better results than ANN model. ANFIS model (Triangular, 3) is selected for SSC prediction at the Pathagudem site.
- The ANFIS model (Gauss, 3) shows better performance among other ANFIS models for runoff prediction at the Pathagudem site.
- The ANFIS model (Gauss, 3) is the best model for runoff prediction at the Pathagudem site.
- The ANN and ANFIS models can be used for the runoff and SSC predictions of the study area.
- The ANN model (6–9–1) is superior to other ANN single and double hidden layer models for runoff prediction at the Polavaram site.
- The model (6–4–4–1) has resulted in better performance among ANN single and double hidden layers models for SSC prediction at the Pathagudem site.

- The relationship between observed and predicted values of runoff and SSC shows a very good correlation both during training and testing periods using ANN and ANFIS models.
- It was observed that (Triangular, 3) model has better performance than the ANN model (6–9–1). Therefore, the ANFIS model (Triangular, 3) is chosen for runoff prediction at the Polavaram site.
- The sensitivity analysis for SSC prediction models suggests that current day runoff is the most sensitive parameter using ANN and ANFIS models for both sites among the other considered parameters (viz. runoff of current day, previous one-day, previous two days, previous three days and SSC of previous one-day, previous two days and previous three days).
- The sensitivity analysis was performed for both ANN and ANFIS-based runoff prediction models considering input parameters (viz. current day stage, previous one-day stage, previous two days stage, previous three days stage, previous one-day runoff, previous two days runoff and previous three days runoff) at Pathagudem and Polavaram sites. The current day stage was the most sensitive parameter for both the sites.

ACKNOWLEDGMENT

This chapter is a modified and edited version of: "Ashish Kumar (2016). Runoff and sediment estimation using ANN and ANFIS-based techniques for Godavari Basin, India. MTech Thesis; Department of Soil and Water Conservation; College of Technology, G.B. Pant University of Agriculture and Technology; Pantnagar – 263145, Uttarakhand, India, p. 110.

KEYWORDS

- adaptive neuro-fuzzy inference system
- artificial neural network
- correlation coefficient
- gamma test
- pooled average relative error
- sediment

REFERENCES

1. Abbasi-Shoshtari, S., & Kashefipoor, M., (2006). Estimation of suspended sediment using artificial neural networks (Case study: Ahwaz station). *7ᵗʰ International River Engineering Conference* (p. 813). Ahwaz, IR IRAN.
2. Bazartseren, B., Hildebrandt, G., & Holz, K. P., (2003). Short-term water level prediction using neural networks and neuro-fuzzy approach. *Neuro Computing,55*, 439–450.
3. Bisht, D. C. S., Raju, M. M., & Joshi, M. C., (2010). AN-based river stage-discharge modeling for Godavari River, India. *Computer Modeling and New Technologies,14*(3), 48–62.
4. Bisht, D. C. S., & Jangid, A., (2011). Discharge modeling using neuro-fuzzy inference system. *International Journal of Advanced Science and Technology, 31,* 99–114.
5. Chau, K. W., & Zhang, X. N., (1995). An expert system for flow routing in a river network. *Advances in Engineering Software, 22*(3), 139–146.
6. Chidthong, Y., Tanaka, H., & Supharatid, S., (2009). Developing a hybrid multi-model for peak flood forecasting. *Hydrological Processes*, *23*(12), 1725–1738.
7. Dariane, A. B., & Azimi, S., (2016). Forecasting stream flow by combination of a genetic input selection algorithm and wavelet transforms using ANFIS models. *J. Hydrologic. Science,61*, 585–600.
8. Dawson, C. W., Harpham, C., Wilby, R. L., & Chen, Y., (2002). Evaluation of artificial neural network techniques for flow forecasting in the river Yangtze, China. *Hydrology and Earth System Sciences,6*(4), 619–626.
9. Feyen, L., Dankers, R., Bodis, K., Salamon, P., & Barredo, J. I., (2010). *Fluvial Flood Risk in Europe in Present and Future Climates* (Vol. 112, p. 48). Switzerland AG: Springer Nature. E-article. doi: 10.1007/s10584–011–0339–7.
10. Firat, M., & Gungor, M., (2007). River flow estimation using adaptive neuro fuzzy inference system. *Mathematics and Computers in Simulation,75*, 87–96.
11. Folorunsho, J. O., Iguisi, E. O., Mu'azu, M. B., & Garba, S., (2012). Application of adaptive neuro-fuzzy inference system (ANFIS) in river Kaduna discharge forecasting. *Research Journal of Applied Science, Engineering and Technology*, *4*(21), 4275–4283.
12. Gallo, D. A., Roediger, H. L., Watson, J. M., & McDermott, K. B., (2001). Factors that determine false recall: A multiple regression analysis. *Psychonomic Bulletin and Revive,8*(3), 385–407.
13. Gharde, K. D., Kothari, M., Mittal, H. K., Singh, P. K., & Dahiphale, P. A., (2015). Sediment yield modeling of Kal River in Maharashtra using artificial neural network model. *Research Journal of Recent Sciences, 4*, 120–130.
14. Habib, E. H., & Meselhe, E. A., (2006). Stage-discharge relations for low-gradient tidal streams using data driven models. *J. Hydraul. Eng.,132*(5), 482–492.
15. Jacquin, A. P., & Shamseldin, A. Y., (2006). Sensitivity analysis of Takagi-Sugeno-Kang rainfall-runoff fuzzy models. *Hydrol. Earth Syst. Sci.,13*, 41–55.
16. Jang, R., (1995). Neuro-fuzzy modeling. *Proceedings of the IEEE,83*(3), 378–388.
17. Jain, S. K., (2001). Development of integrated sediment rating curves using ANNs. *J. Hydraul. Eng., ASCE, 127*(1), 30–37.
18. Karl, A. K., Lohani, A. K., Goel, N. K., & Roy, G. P., (2010). Development of flood forecasting system using statistical and ANN techniques in the downstream catchment of Mahanadi basin, India. *J. Water Resource and Protection,2*, 880–887.

19. Kitanidis, P. K., & Bras, R. L., (1980). Real-time forecasting with a conceptual hydrologic model: Applications and results. *Water Resour. Res.,16*, 1034–1044.

20. Lee, H., Seo, D. J., & Koren, V., (2011). Assimilation of stream flow and in situ soil moisture data into operational distributed hydrologic models: Effects of uncertainties in the data and initial model soil moisture states. *Advances in Water Resources,34*, 1597–1615.

21. Liong, S., Lim, W., & Paudyal, G., (2000). River stage forecasting in Bangladesh: Neural network approach. *Journal of Computing in Civil Engineering, 14*(1), 1–8.

22. Lohani, A. K., (2007). *ANN and Fuzzy Logic in Hydrological Modeling and Flow Forecasting* (p. 218). PhD Thesis; Indian Institute of Technology, Roorkee.

23. Loukas, Y. L., (2001). Adaptive neuro-fuzzy inference system: An instant and architecture free predictor for improved QSAR studies. *J. Med. Chem.,44*(17), 2772–2783.

24. Mcculloch, W. S., & Pitts, W., (1943). A logical calculus of the ideas immanent in nervous activity. *Bulletin of Mathematical Biophysics, 5,* 115–133.

25. Nayak, P. C., Sudheer, K. P., Rangan, D. M., & Ramasastri, K. S., (2004). A neuro-fuzzy computing technique for modeling hydrological time series. *Journal of Hydrology,291*, 52–66.

26. Nayak, P. C., Sudheer, K. P., & Ramasastri, K. S., (2005). Fuzzy computing-based rainfall-runoff model for real time flood forecasting. *Hydrolog. Process, 19*, 955–968.

27. Salas, E., Paris, C. R., & Connon-Bowers, J. A., (2000). Teamwork in multi-person systems: A review and analysis. *Economics,43*(8), 1052–1075.

28. Patel, D., & Parekh, F., (2014). Flood forecasting using adaptive neuro-fuzzy inference system (ANFIS). *International Journal of Engineering Trends and Technology (IJETT),12*(10), 510–514.

29. Raghuwanshi, N., Singh, R., & Reddy, L., (2006). Runoff and sediment yield modeling using artificial neural networks: Upper Siwane River, India. *Journal of Hydrologic Engineering,11*(1), 71–94.

30. Rosenberg, J. S., (2011). KNL1/Spc105 recruits PP1 to silence the spindle assembly checkpoint. *Curr. Biol.,21*(11), 942–947.

31. Shabani, M., & Shabani, N., (2012). Estimation of daily suspended sediment yield using artificial neural network and sediment rating curve in Kharestan Watershed, Iran. *Australian Journal of Basic and Applied Sciences,6*(11), 157–164.

32. Shafie, A. E., Taha, M. R., & Noureldin, A., (2007). Neuro-fuzzy model for Inflow forecasting of the Nile River at Aswan high dam. *Water Resource Management, 21*, 533–556.

33. Sudheer, K. P., Gosain, A. K., & Ramasastri, K. S., (2002). A data-driven algorithm for constructing artificial neural network rainfall-runoff models. *Hydrol. Processes,16*, 1325–1330.

34. Tokar, A. S., & Johnson, P. A., (1999). Rainfall-runoff modeling using artificial neural networks. *J. Hydrol. Eng.,ASCE, 4*(3), 232–239.

35. Walling, D. E., (1977). Assessing the accuracy of suspended sediment rating curves for a small basin. *Wat. Resour. Res., 13*(3), 531–538.

INDEX

nd by CPI Group (UK) Ltd, Croydon, CR0 4YY

23/10/2024

01777701-0010